Applied Probability and Statistics (Continued)

BHAT · Elements of Applied Stochastic Processes
BOX and DRAPER · Evolutionary Operation: A Statistical Method for Process Improvement
BROWNLEE · Statistical Theory and Methodology in Science and Engineering, *Second Edition*
CHAKRAVARTI, LAHA and ROY · Handbook of Methods of Applied Statistics, Vol. II
CHERNOFF and MOSES · Elementary Decision Theory
CHIANG · Introduction to Stochastic Processes in Biostatistics
CLELLAND, deCANI, BROWN, BURSK and MURRAY · Basic Statistics with Business Applications, *Second Edition*
COCHRAN · Sampling Techniques, *Second Edition*
COCHRAN and COX · Experimental Designs, *Second Edition*
COX · Planning of Experiments
COX and MILLER · The Theory of Stochastic Processes
DANIEL and WOOD · Fitting Equations to Data
DAVID · Order Statistics
DEMING · Sample Design in Business Research
DODGE and ROMIG · Sampling Inspection Tables, *Second Edition*
DRAPER and SMITH · Applied Regression Analysis
ELANDT-JOHNSON · Probability Models and Statistical Methods in Genetics
FLEISS · Statistical Methods for Rates and Proportions
GOLDBERGER · Econometric Theory
GUTTMAN, WILKS and HUNTER · Introductory Engineering Statistics, *Second Edition*
HAHN and SHAPIRO · Statistical Models in Engineering
HALD · Statistical Tables and Formulas
HALD · Statistical Theory with Engineering Applications
HOEL · Elementary Statistics, *Third Edition*
HOLLANDER and WOLFE · Nonparametric Statistical Methods
HUANG · Regression and Econometric Methods
JOHNSON and KOTZ · Distributions in Statistics
Discrete Distributions
Continuous Univariate Distributions-1
Continuous Univariate Distributions-2
Continuous Multivariate Distributions
JOHNSON and LEONE · Statistics and Experimental Design: In Engineering and the Physical Sciences, Volumes I and II
LANCASTER · The Chi Squared Distribution
LEWIS · Stochastic Point Processes
MILTON · Rank Order Probabilities: Two-Sample Normal Shift Alternatives
OTNES and ENOCHSON · Digital Time Series Analysis
RAO and MITRA · Generalized Inverse of Matrices and Its Applications
SARD and WEINTRAUB · A Book of Splines
SARHAN and GREENBERG · Contributions to Order Statistics

continued overleaf

Applied Probability and Statistics (Continued)

SEAL · Stochastic Theory of a Risk Business
SEARLE · Linear Models
THOMAS · An Introduction to Applied Probability and Random Processes
WHITTLE · Optimization under Constraints
WILLIAMS · Regression Analysis
WOLD and JUREEN · Demand Analysis
WONNACOTT and WONNACOTT · Econometrics
YOUDEN · Statistical Methods for Chemists
ZELLNER · An Introduction to Bayesian Inference in Econometrics

Tracts on Probability and Statistics

BILLINGSLEY · Ergodic Theory and Information
BILLINGSLEY · Convergence of Probability Measures
CRAMÉR and LEADBETTER · Stationary and Related Stochastic Processes
JARDINE and SIBSON · Mathematical Taxonomy
KINGMAN · Regenerative Phenomena
RIORDAN · Combinatorial Identities
TAKÁCS · Combinatorial Methods in the Theory of Stochastic Processes

Stochastic Geometry

Rollo Davidson, June 1970

Stochastic Geometry

A Tribute to the Memory of Rollo Davidson

Edited by
E. F. HARDING
and
D. G. KENDALL

John Wiley & Sons

London *New York* *Sydney* *Toronto*

Library of Congress catalog card number 72-8603

ISBN 0 471 35141 5

Printed in Great Britain by John Wright & Sons Ltd., at The Stonebridge Press, Bristol, England

IN loving remembrance of ROLLO DAVIDSON, and in gratitude for his vivid imagination, his great clarity and integrity of mind, the gentle courtesy of his spirit, and the affection this engendered in all who knew him

Preface

This volume, and the companion work *Stochastic Analysis*, have been compiled by the friends of Rollo Davidson and by one or two others who had not met him but hoped to become his friends. The two books are very closely linked, and those who wish to explore either one in depth will almost certainly need to consult the other. The degree of unity within each single book is however much greater than that which could have been attained by bringing them together, and we believe that most people will find the two-volume arrangement a great convenience.

We have taken some pains to try to avoid the 'miscellaneous' character often to be found in such cooperative works; both books contain much that is new, even to the specialist, but we believe that each will be found valuable as an introduction to the field named in its title, especially by those interested in the possibility of undertaking research in either area and daunted by the scattered nature of the periodical literature and the absence of any comparable synthesis of it.

Each book is furnished with an introductory chapter surveying the field, and setting the stage for the specialist articles which follow it.

This volume contains all of Davidson's published and the greater part of his unpublished work on stochastic line-processes; indeed we have tried to include as much as is reasonable of all significant work by any author bearing on the theoretical aspects of this exciting new chapter in Probability Theory, although we have not attempted to cover the contributions by Professor Bartlett to the problem of the empirical spectral analysis of observed line-processes. We have supplemented this material by a substantial number of new contributions to Stochastic Geometry, including within the range of this the theory of random sets, and the theory of random measures.

One of Davidson's previously unpublished manuscripts has kindly been edited by Professor F. Papangelou, to whom we wish to express our best thanks. We further wish to thank Miss Mary Brooks, who drew the pictures.

Chapters 2.4, 2.5, 3.1 and 3.4 were first published in the *Izvestiza* of the Academy of Sciences of the Armenian Soviet Socialist Republic; 2.6 by Springer-Verlag; 3.6 by the International Statistical Institute; and 5.1 by the Sailplane and Gliding Association. To all these bodies we are most grateful for their kindness in making possible the complete coverage attempted in this book.

E. F. H
D. G. K.

Contents

7 CONCLUSION

1

INTRODUCTION

1.1

An Introduction to Stochastic Geometry

D. G. KENDALL

(1) I have attempted to survey what is called 'stochastic analysis' in the introductory chapter to the companion volume to this one (*Stochastic Analysis*), and I shall now attempt to deal similarly with 'stochastic geometry', but it will be found that the situation is strikingly different in these two fields. For example, there is an enormous literature concerned with stochastic analysis, only partially surveyed in the opening chapter to *Stochastic Analysis*, whereas almost the whole existing literature concerned with stochastic geometry (in the narrow sense used here) is to be found within the pages of the present book, and some of it indeed is now being presented publicly for the first time. At first one might be tempted to say that stochastic geometry is much the older of the two subjects, for there is a sense in which it takes its origin in Crofton's famous article in the IXth edition of the *Encyclopaedia Britannica*, but it is, I think, more correct to say that it is an entirely novel subject, deriving its inspiration rather from an attempt to eschew Crofton's approach than to imitate it.

What Crofton and his contemporaries did was to perfect a technique for computing probabilities of occurrence of simple geometrical phenomena, and the expectations of simple geometrical enumerations, associated with *uniform* patterns of *independent* random lines in the plane. (Of course they studied higher-dimensional problems of a similar kind, but there was no qualitative difference in the nature of the task when carried out in higher dimensions.) The vital adjectives in this description of the nineteenth-century work, later to flower into the contemporary disciplines of Geometrical Probability and Integral Geometry, are *uniform* and *independent*. The plane (or higher-dimensional Euclidean space) is thought of as carrying the Euclidean group, and the stochastic structures envisaged are all invariant under this group, while the geometrical questions asked are relative to the geometry implied by the group. Further, the stochastic elements, normally lines, are not only 'randomly' and 'uniformly' distributed, but are also in some sense (never made quite precise) 'independent'. It is in fact characteristic of this parent (or anti-parent) of our subject that what is spoken of is never defined, and when careful workers (and there have been many) venture into the field they normally do so under the protection of some such device as the following:

$C(L)$ is a circle centred at a fixed origin and with perimeter $2L$, and N independent 'random' lines cut the circle; E is a geometrical event associated with these lines, which may or may not 'happen'. If N is a Poisson-distributed random variable with expectation λL (where λ is fixed) find $\lim \operatorname{pr}(E)$ *as* $L \to \infty$.

When one encounters an examination question formulated in this way, one knows quite well what additional explanations have to be supplied; thus properly the randomness of N should be mentioned first, and the distribution of these lines is conditional upon the σ-algebra generated by the random variable N. Also, while what is meant by 'independent' is now clear, a convention is concealed by 'uniform'; this means, when applied to exactly one line, that if R is the length of the perpendicular from the origin onto the line, and if Φ is an angle specifying the direction of the line, then R has range $(-L/\pi, L/\pi)$, Φ has range $(0, \pi)$, and R and Φ are random variables independently and uniformly distributed over their respective ranges, so that the corresponding two-dimensional distribution is absolutely continuous and has the density $1/2L$ within the rectangle defined by the ranges. (This supposes that the lines are not 'sensed'; if they are sensed then an obvious adjustment has to be made.)

With this expansion of the original statement it then often becomes a relatively simple matter to compute $\operatorname{pr}(E)$, either for fixed L or as $L \to \infty$. A typical E is the event that the origin is bounded away from the perimeter of the circle by the lines. A typical random variable, for which see Miles [16], is the area of the polygon containing the origin which is defined in this way.

Once all practicable problems about lines have been solved, one proceeds to formulate similar problems about random geodesics on manifolds, and so on. Sometimes the chicken comes before the egg, as when problems about random lines in the plane are solved by looking first at a parallel problem involving random great circles on a sphere, and then letting the radius of the sphere tend to infinity.

Other parallel developments which may still be regarded as being in the tradition of what we here call 'the classical theory' consist in replacing the plane by a homogeneous space and letting the active group play the role formerly assigned to the Euclidean group, so that Haar measure replaces Lebesgue measure, and so on. For details of the classical theory see Kendall and Moran [11], and for its more sophisticated variants just mentioned see Santaló [19].

(2) One principle of stochastic geometry is that the difficulties which are here being evaded should be confronted face to face, and solved 'naturally'. Another is that the geometry (i.e. the group) has previously been allowed to control the stochastic situation too much, and that a richer variety of stochastic situations should be envisaged. Accordingly we adopt an entirely different approach.

Whatever 'random geometric elements' we are considering (random lines in the plane, or in space, random circles in the plane, random orthogonal triads in three dimensions, etc.) any one such element can be represented as a *point* on an appropriate manifold.

For example, if we are considering 'sensed' lines in the plane, a natural (though not of course the only possible natural) representation can be constructed in the following way. Let first the direction of the line be noted; as the line is sensed, this can be represented by a point on a circle (with angular coordinate θ where $0 \leqslant \theta < 2\pi$). It is important to notice that the deficiencies of the coordinatization (via θ) have to do with the representation of the manifold and are not in any way 'the fault of the line' (or of the circle). Thus we could if we preferred use a complex coordinate z, with $|z| = 1$. Next, the direction of the line being fixed, we drop a perpendicular from the origin onto it, say of length R, and we give to this a positive sign if the origin lies to the left of the line, as we 'look along it', and a negative sign if the origin lies to the right of the line (of course $R = 0$ if the origin lies on the line). This provides us with a second coordinate $x = \pm R$, according to the geometrical situation, and now the pair (θ, x) (or (z, x), if preferred) determine the line uniquely, and in fact we have set up a one-to-one correspondence between the set of all lines in the plane, and the Cartesian product of a circle (labelled by θ or z) and a line (labelled by x). In other words, from this point of view there is a one-to-one correspondence between the sensed *lines* in the plane and the *points* on the cylinder.

This being agreed, it will be seen that a *stochastic process of sensed lines in the plane* is equivalent to a stochastic process of points on a cylinder, i.e. to a point-process carried by a cylinder or, to use yet another form of words, to a *random set on a cylinder. This transformation of the original problem into one concerning a random set on an appropriate manifold is the essence of what we here understand by stochastic geometry.*

(3) We have in fact skipped an important point in the above discussion, and the acute reader will have noticed this. A random variable, in the classical real-valued case, is *not* just a mapping from a probability space to a real line R, but rather is such a mapping whose inverse sends Borel subsets of R into measurable sets in the probability space. We know (see *Stochastic Analysis* for this) why it is that the Borel (or rather, more fundamentally, the Baire) subsets of R are the natural ones. (Remember that Baire = Borel on R.) The question therefore arises, since any realization of our stochastic process is a set of points on the representing manifold M, so is a point in the space 2^M, *what is the natural σ-algebra in* 2^M? If we can answer this question, our reformulation of stochastic geometry will work; we then have only to name any probability measure on the said σ-algebra, and we shall have before us a generic example of the mathematical object we wish to study. But until the σ-algebra has been identified, our subject can hardly be said to exist.

(4) At this stage it will be profitable to take a look at the usual approach to point-processes in the simplest case, when the carrier manifold is the line, and immediately we see that our difficulties have not been surmounted even there, for the best known theory ([18]; see, however, Harris [7], Hoffmann-Jørgensen [8] and Meyer [15], and also Dellacherie [6]) is formulated in terms of random variables which enumerate the number of points of the random set which lie in a generic subinterval of the line. Manifestly this theory will not work in a situation where such random variables are degenerate, although it will work quite adequately for simple examples like the Poisson process, the renewal processes and so on. It is quite useless if the random set is, say, the set of zeros of a continuous version of Brownian motion, or the occupation time-set for a Markov process. The latter can to some extent be dealt with by using descriptive random variables which record the Lebesgue measure, rather than the cardinality, of the intersection of the random set (supposed Lebesgue measurable) with a generic interval, but this approach is again useless in the Brownian motion example [20], and loses valuable detail in some Markovian contexts.

(5) Davidson's key idea, through which he is ultimately responsible for the basis of the theory of random sets, as presented here in [9], is in essence very simple. To produce the required σ-algebra we must produce a system of descriptive real-valued random variables whose realized values suffice to identify the realized random set, and then we take the basic σ-algebra to be the smallest one with respect to which all these descriptive real-valued random variables will be measurable.

The descriptive random variables will be associated with testing-sets, like the intervals on the line, and will record some relevant feature of the intersection $E \cap T$, where E is the realized random set and T is the testing-set. We have seen that Card $(E \cap T)$ is not always adequate, and that the Lebesgue measure of $E \cap T$, even when it can be defined, need not be adequate either. So let us be generous and ask for very little; let our descriptive random variables merely record, for each testing-set T, *whether E is disjoint from T, or not.* If we have enough representing random variables (i.e. enough testing-sets, or *traps* as I shall later call them) and if something is known about the type of the random set, e.g. that it is closed, or convex, or topologically or geometrically limited in some other way, then with a sufficiency of traps of the right kind we should be able to reconstruct the realized set from the realized representing random variables (i.e. from the incidence data of the realized situation), and the method will work.

This idea was implicit in Davidson's Smith's Prize Essay, 'Some Geometry and Arithmetic in Probability Theory', of 1967, but was not elaborated any further *as a theory* by him, although he *used* the technique in some important particular cases. I have built what I hope may be adequate as a first shot at a general theory of random sets, by taking what we might call *Davidson's Trapping Principle* as a starting point, and a full account of the first phase of this investigation will be found in [9].

As might be expected from the introductory remarks above, the first stage of the work involves a careful study of what sorts of set can be described by a given trapping system. In the simplest case, when the representing manifold (and by the way, this 'carrier space' does not *have* to be a manifold) is the real line, then it is normally convenient to take the traps to be the open intervals, and what then emerges is a satisfactory general theory of random closed sets which admits an almost immediate and very sweeping generalization to a theory of random closed sets in reasonably well-behaved abstract topological spaces. Much remains to be done in finding an appropriate formulation for random sets of other types, but some progress has been made there also.

(6) Now that we can describe at any rate random closed sets on (for example) nice manifolds, there is no difficulty in giving a satisfactory definition of a Poisson process on such a manifold, and Davidson has indicated how this is to be done. If the manifold is the cylinder and if the Poisson process has a constant density, we then obtain (without the customary subterfuges) the 'random uniform ensemble' of 'independent' lines in the plane treated in the IXth edition of the *Encyclopaedia*

Britannica. Notice that by 'constant density' we mean constant density relative to the Haar measure on the cylinder associated with the group thereon which is induced by the Euclidean group of motions in the plane. This is, in fact, Lebesgue measure on the cylinder (the first theorem in 'integral geometry').

We can, however, do much better than this. Let us call a random closed set on the manifold (on the straight line, in the simplest case) *invariant* when the joint incidence distributions associated with any finite collection of traps (intervals) are unaltered if the traps are 'moved about' (translated) by the active group. This leads one to a consideration of invariant stochastic processes of lines in the plane (or of great circles on the sphere, or of rigid triads in three dimensions, etc.) which is far more general than any of the concepts of the classical theory, because the basic idea of independence has been dropped. It is, however, still 'geometry', because the ruling group has not been dethroned.

In his very first investigation [5] and also in his very last work Davidson was intensely interested in such invariant random processes of lines, and his results will be found in this book. Fortunately he was able to communicate his enthusiasm to R. V. Ambartzumian (who had already been thinking along similar lines), to Klaus Krickeberg and to Fredos Papangelou, and by their efforts, at least, we may hope that this fascinating programme of work may be developed further.

(7) To round off this sketch of stochastic geometry, mention should be made of the work by Lee [13] and Matthes [14] on infinitely divisible point-processes, which suggested the possibility of a generalization to infinitely divisible random sets on the line or on more general carrier spaces [10], of the fundamental work by Miles [16, 17] on the classical problems of geometrical probability viewed from a modern standpoint, of work by Bartlett [1, 2] on the empirical spectral analysis of point- and line-processes (with applications to medicine and to road traffic), of Daley's theoretical approach [4] to the spectral analysis of point-processes, and of a recent data-analytic technique [3] which can be used for point-process diagnostics associated with, say, the random ensembles of lines which are of importance in the paper-making industry, or the ensembles of rigid triads which arise in the statistical analysis of the orbits of comets.

Proper acknowledgement should also be made of the fact that other fundamental approaches to a theory of random sets were being developed by Dellacherie [6], Harris [7], Hoffmann-Jørgensen [8], Krylov and Yuskevic [12] and Meyer [15] at about the same time as the one which is reviewed here and reported on in more detail elsewhere in this book. A careful study of the interrelations of these different attacks on random set-theory has yet to be begun, but would surely be fruitful.

There is little more in the way of general introduction that I can usefully say. The time has now come to let the authors speak for themselves.

REFERENCES

1. M. S. Bartlett, "The spectral analysis of two-dimensional point processes", *Biometrika* **51** (1964), 299–311.
2. ——, "The spectral analysis of line processes", *Proc. 5th Berkeley Symp.* **3** (1967), 135–153.
3. Liliana I. Boneva, D. G. Kendall and I. Stefanov, "Spline transformations: three new diagnostic aids for the statistical data-analyst", *J. R. statist. Soc.* B **33** (1971), 1–71.
4. D. J. Daley, "Weakly stationary point processes and random measures", *J. R. statist. Soc.* B **33** (1971), 406–425.
5. R. Davidson, *Some arithmetic and geometry in probability theory*, Ph.D. thesis, Cambridge (1968).
6. C. Dellacherie, "Ensembles aléatoires, I et II", *Sém. de Prob.* III (1969), 97–136, *Lecture Notes*, Springer.
7. T. E. Harris, "Counting measures, monotone random set functions", *Ztschr. Wahrsch'theorie & verw. Geb.* **10** (1968), 102–119.
8. J. Hoffmann-Jørgensen, "Markov sets", *Math. Scand.* **24** (1969), 145–166.
9. D. G. Kendall, "Foundations of a theory of random sets", 6.2 of this book.
10. ——, "Infinitely divisible random sets", in preparation.
11. M. G. Kendall and P. A. P. Moran, *Geometrical Probability*, Griffin, London (1963).
12. N. V. Krylov and A. A. Yuskevic, "Markov random sets", *Trans. Moscow Math. Soc.* **13** (1965), 127–153.
13. P. M. Lee, "Infinitely divisible stochastic processes", *Ztschr. Wahrsch'theorie & verw. Geb.* **7** (1967), 147–160.
14. K. Matthes, "Unbeschränkt teilbare Verteilungsgesetze stationäre zufälliger Punktfolgen", *Wiss. Z. Hochschule Elektrotechn. Ilmenau* **9** (1963), 235.
15. P. A. Meyer, "Ensembles régénératifs d'après Hoffmann-Jørgensen", *Sém. de Prob.* IV (1970), 133–150, *Lecture Notes*, Springer.
16. R. E. Miles, "Random polygons determined by random lines in a plane, I and II", *Proc. Nat. Acad. Sci. USA* **52** (1964), 901–907 and 1157–1160.
17. ——, "Poisson flats in euclidean spaces, I", *Adv. Appl. Prob.* **1** (1969), 211–237.
18. C. Ryll-Nardzewski, "Remarks on processes of calls", *Proc. 4th Berkeley Symp.* **2** (1961), 455–465.
19. L. A. Santaló, *Introduction to Integral Geometry*, Hermann, Paris (1953).
20. S. J. Taylor, "The α-dimensional measure of the graph and the set of zeros of a Brownian path", *Proc. Camb. Phil. Soc.* **51** (1955), 265–274.

2

GENERAL THEORY

2.1

Stochastic Processes of Flats and Exchangeability

ROLLO DAVIDSON

2.1.1 INTRODUCTION

A 'flat' in n-dimensional space is a subspace of $(n-1)$ dimensions, so that a flat in the plane is a line. Each line in the plane can be viewed as a point on a (right circular, infinite) cylinder, and similarly each flat in n-dimensional space can be viewed as a point on some more complicated manifold. So we can talk of flat- and line-processes, meaning the point-processes that they induce on the appropriate manifolds. A Euclidean motion of n-dimensional space appears as a motion of the manifold: let the set of motions of the manifold that arise from rotations of the space be $\mathscr{M}$. The interest in flat-processes lies mostly in the interaction of their probabilistic properties with the structure of $\mathscr{M}$. I only consider flat-processes certain properties of which are left invariant by all the motions in $\mathscr{M}$: we demand that the process be either strictly or nth order stationary under $\mathscr{M}$.

I start by considering line-processes in the plane, and for these I prove some geometric properties: for example, an entropy property of the mixed Poisson process; and the porism that a strictly stationary line-process has a.s. no or infinitely many pairs of parallel lines. Examples of strictly stationary line-processes are given.

The intersections of the flats of a flat-process in n-dimensional space with an arbitrary line in the space form a point-process on the line. I show that the point-process inherits the stationarity properties of the flat-process that begets it. If an nth-order stationary flat-process in n-space has an nth-order product-moment density, then the point-process it induces on the line has also an nth-order product-moment density which is a constant. So we become interested in point-processes on the line that have constant nth-order densities. I show that if such a process is doubly stochastic

Poisson and $n \geqslant 2$, the process is mixed Poisson. Suppose, however, that we do not know that the process is doubly stochastic Poisson but we do know that it has constant densities of all orders. Then I show that there is a non-negative random variable whose nth moment is the nth-order density for each n. If a certain rate-of-growth condition is put on the sequence of constants, I am able to show that the process is 'exchangeable': that is, the probability that any r non-overlapping intervals on the line of common length l are simultaneously void of points of the process depends on r and l but not on the intervals chosen. It is an interesting outstanding problem to find out to what extent the rate-of-growth condition above can be relaxed: it has to do with the problem of characterizing a distribution by its moments—or so I think at the moment.

The main theorem that I prove about exchangeable point-processes is, that an a.s. locally finite point-process on the line is exchangeable if and only if it is mixed Poisson. The same is true if the carrier space of the process is n-dimensional space; but if the carrier space is the circle (or a product of circles), the point-process is exchangeable if and only if a certain property of conditional independence holds. These results are quite important: mixed Poisson processes arouse continual interest because they have many nice properties and their probabilistic mechanism is exceptionally well understood. Also, my results showed that it is profitable, when considering point-processes, to observe simply whether certain sets contain points of the process or not; the usual approach is, of course, to look at the numbers of points in these sets.

Note. All stochastic point- (line-, flat-) processes with which we deal are supposed geometric: that is, we identify multiple points. Equivalently, we can assume that there are almost surely no multiple points.

2.1.2 STOCHASTIC PROCESSES OF FLATS: PROCESSES OF LINES IN THE PLANE

Let X be the set of all lines in the Euclidean plane. X is a homogeneous space under the group $\mathcal{M}$ of motions induced on it by the Euclidean rotations (which also we shall call $\mathcal{M}$) of the plane. We can represent X as a space of points on a cylinder thus.

Let an origin O be chosen in the plane, and let a fixed axis Ox be given. Let l be an (oriented) line, cutting Ox at the point x_1 and at the angle φ_1 (see Figure 1). Define

$$\theta_1 = \varphi_1 - \tfrac{1}{2}\pi; \quad p_1 = x_1 \cos \theta_1,$$

so that $0 \leqslant \theta_1 < 2\pi$ and $-\infty < p_1 < \infty$. Should we be dealing with unoriented lines, we should still have $0 \leqslant \theta < 2\pi$, but instead of $-\infty < p < \infty$, we should have $0 \leqslant p < \infty$.

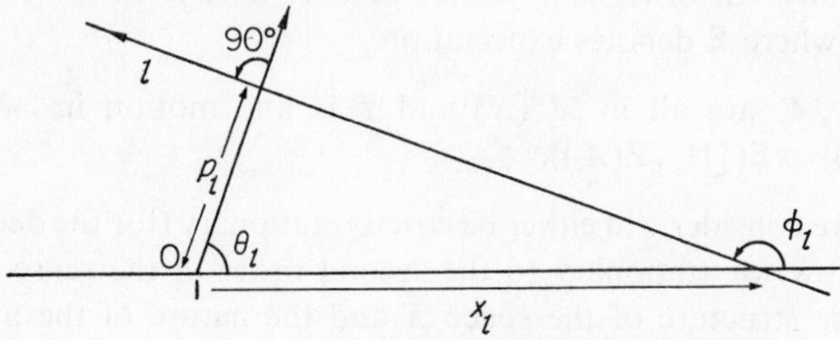

Figure 1.

The representing cylinder will be the set of points $w = (p,\theta)$; we shall refer to the cylinder as X, with the group $\mathscr{M}$ of motions $\mathscr{M}$ on it. It is important to note that $\mathscr{M}$ is *not* the group of Euclidean motions on the cylinder: in fact a set of generators of $\mathscr{M}$ is

(1) $$T(p, \theta) = (p, \theta - \alpha) \quad (\alpha \text{ arbitrary}),$$

(2) $$T(p, \theta) = (p - d \cos \theta, \theta) \quad (d \text{ arbitrary}).$$

From this one sees easily that if $S \subset X$ is a band encircling the cylinder, then $TS \cap S \neq \emptyset$ for any T in $\mathscr{M}$, so $\mathscr{M}$ cannot be the Euclidean group on the cylinder.

Santaló [34] shows that the Haar measure m on X under $\mathscr{M}$ (exists and) is proportional to $dw = dp\, d\theta$; we shall take the normalizing constant to be unity. By the formula of Crofton [6] we see that this choice of normalizing constant yields:

The set of lines S cutting a convex curve of length L in the plane has m-measure $2L$.

(the factor of 2 comes in because of the orientation of the lines). The fact that the Haar measure is $dp\, d\theta$ suggests that the (p, θ) coordinates of a line in the plane are the natural ones.

Let the Borel field of X be $\mathscr{B}(X)$ and let the class of sets in $\mathscr{B}(X)$ that are of finite m-measure be $\mathscr{B}^*(X)$.

Now we are considering processes of lines in the plane, analogous to processes of points on the line. Formally we have a countably additive non-negative-integer-valued random measure Z on $\mathscr{B}(X)$; and by the assumption at the end of Section 2.1.1, the mass that Z gives each of its atoms is a.s. unity. We assume that Z is a.s. locally finite: that is, if S is in $\mathscr{B}^*(X)$ then $Z(S)$ is a.s. finite.

Z is said to be stationary to the nth order if, for all $r \leqslant n$,

(*a*) Z is locally rth order summable: that is, if A is in $\mathscr{B}^*(X)$, then $\mathbf{E}(Z(A))^r < \infty$, where $\mathbf{E}$ denotes expectation.

(*b*) If $A_1, \ldots, A_r$ are all in $\mathscr{B}^*(X)$ and T is any motion in $\mathscr{M}$, then $\mathbf{E}(\prod_{i=1}^r Z(TA_i)) = \mathbf{E}(\prod_{i=1}^r Z(A_i))$.

The Z that we consider will either be strictly stationary (for the definition of this see below) or stationary to the second order in the sense above. In this case the structure of the space X and the nature of the motions $\mathscr{M}$ imply

$$\mathbf{E}(Z(A)) = km(A) \text{ for some constant } k \geqslant 0; \tag{3}$$

for a non-trivial Z we can assume that $k > 0$.

Z is said to be strictly stationary if all its finite-dimensional distributions are invariant under the motions $\mathscr{M}$.

For any Z we define the moment-generating functional b_Z by

$$b_Z(f) = \mathbf{E} \exp\left(-\int_X f(w)\, dZ(w)\right) \tag{4}$$

for all functions f on the cylinder that are uniform limits of non-negative step-functions of compact support. Then b corresponds to Z uniquely; and Z is strictly stationary if and only if for all such f and all motions T in $\mathscr{M}$

$$b_Z(f^T) = b_Z(f),$$

where $f^T(w) \equiv f(Tw)$. We note that if Z is strictly stationary and locally square-summable, then Z is second-order stationary.

For any $w^* = (p^*, \theta^*)$ and any $d > 0$ and α such that $0 < \alpha < \pi$, define $S(\alpha, d, w^*) = (w : w = (p, \theta);\ p^* - d < p < p^* + d,\ \theta^* - \alpha < \theta < \theta^* + \alpha)$. It is clear from equation (3) that if Z is second-order stationary and T is any motion in $\mathscr{M}$ and α, d, w^* are given, then

$$\lim_{t \downarrow 0} \frac{\mathbf{E}(Z(T^{-1} S(t\alpha, td, Tw^*)))}{mS(t\alpha, td, Tw^*)} = k \tag{5}$$

independent of T, α, d.

A second-order stationary Z is said to have a second-order product-moment density (or *g-function*) if for all w and $w' \neq w$ and all T in $\mathscr{M}$ and all $d > 0$ and α such that $0 < \alpha < \pi$,

$$\lim_{t \downarrow 0} \frac{\mathbf{E}(Z(T^{-1} S(t\alpha, td, Tw))\, Z(T^{-1} S(t\alpha, td, Tw')))}{mS(t\alpha, td, Tw)\, mS(t\alpha, td, Tw')} \tag{6}$$

exists (equal to $g(w, w')$, say) independent of α, d and T. In this case from equations (5) and (6) we have the differential form

$$\frac{\mathbf{E}(dZ(w)\,dZ(w'))}{dw\,dw'} = k\delta(w, w') + g(w, w') \tag{7}$$

(see, e.g., Bartlett [1]). If Z is a doubly stochastic Poisson process—that is, a Poisson process whose rate is a realization of a non-negative-valued stochastic process $V(w)$—we have (Bartlett [1])

$$b_Z(f) = \mathbf{E}_V\left\{\exp\left(-\int_X V(w)\,(1-e^{-f(w)})\,dw\right)\right\}; \tag{8}$$

if V is continuous in mean square with a covariance function $h(w, w')$, say, then Z has a g-function g and $g(w, w') \equiv h(w, w') + k^2$.

Proposition 1. *Let Z be second-order stationary and have a g-function g. Then if $\theta \not\equiv \theta'$ (mod π), $g(w, w') = g(\theta - \theta') = g(\theta' - \theta)$.*

Proof. Since $\theta \not\equiv \theta'$ (mod π), the lines l, l' corresponding to w, w' intersect, at the angle $\theta - \theta' \equiv \alpha$, say. Now for any fixed α, $\mathcal{M}$ is transitive on pairs of lines whose θ-values differ by α; so, since the definition of second-order stationarity and equation (6) imply that $g(w, w') = g(Tw, Tw')$ for all T in $\mathcal{M}$, we have the first equality of the proposition.

On the other hand, if we put $T = I$ (the identity of the motions $\mathcal{M}$) in equation (6), the symmetry of the definition ensures that $g(w, w') = g(w', w)$; the second equality of the proposition follows at once.

Corollary. *If Z is second-order stationary under the motions $\mathcal{M}$ and has a g-function g, and if Z has a.s. no parallel lines, then Z is second-order stationary under reflections also.*

Proof. Every reflection R can be represented as $R = T_1 R_0 T_2$, where T_1, T_2 are in $\mathcal{M}$ and R_0 is the reflection of the plane in the line $\theta \equiv 0$. Now R_0 sends the point (p, θ) into $(-p, \pi - \theta)$; so if $\theta \not\equiv \theta'$ (mod π), we have from the proposition $g(w, w') = g(\theta - \theta') = g(\theta' - \theta) = g(R_0 w, R_0 w')$. Hence also $g(w, w') = g(Rw, Rw')$ for all reflections R, provided $\theta \not\equiv \theta'$ (mod π). But in view of Z having a.s. no parallel lines, we can use the formula of Bartlett [1, p.84] to obtain the corollary by integration:

$$\mathbf{E}(Z(RA)\,Z(RB)) = \int_{RA}\int_{RB} g(w, w')\,dw\,dw' + \int_{R(A\cap B)} k\,dw$$

$$= \int_A\int_B g(w, w')\,dw\,dw' + \int_{A\cap B} k\,dw = \mathbf{E}(Z(A)\,Z(B)).$$

***Proposition* 2.** *If Z is strictly stationary, then Z has a.s. either infinitely many or no pairs of parallel lines.*

Proof. Suppose the event E that there be at least one pair of parallel lines has $\mathbf{P}(E)>0$. Take an arbitrary line in the plane; let N be the process of points on this line given by its intersections with lines of Z which have at least one other line of Z parallel to them. Then N is strictly stationary and has at least two points in it. Let I, J be disjoint intervals on the line. Then

$$\mathbf{E}_E\{\exp(-N(I\cup J))\}\leqslant\sqrt{[\mathbf{E}_E\{\exp(-2N(I))\}\,\mathbf{E}_E\{\exp(-2N(J))\}]}$$

(the Cauchy–Schwartz inequality). If I is half-left-open and J is a translate of, and contiguous to, I, we get

$$\mathbf{E}_E\{\exp(-N(2I))\}\leqslant\mathbf{E}_E\{\exp(-2N(I))\}.$$

Repeating this doubling process *ad infinitum* and using bounded convergence we obtain

$$\mathbf{E}_E\{\exp(-N(R))\}\leqslant\mathbf{P}(N(I)=0\,|\,E) \tag{9}$$

for every half-left-open I, where R stands for the whole line. Since $\mathbf{P}(.\,|\,E)$ is a probability measure it is continuous at the null set, so

$$\lim_{I\uparrow R}\mathbf{P}(N(I)=0\,|\,E)=\mathbf{P}(N(R)=0\,|\,E)=0 \tag{10}$$

by the definition of E. So from equation (9) we find $\mathbf{E}_E\{\exp(-N(R))\}=0$; that is, $\mathbf{P}(N(R)=\infty\,|\,E)=1$. Hence a.s. Z has either infinitely many or no pairs of parallel lines.

***Example* 1.** *Take an origin* O *in the plane and a fixed axis* O*x. Take a random direction* O*y making an angle* α *with* O*x, where* α *is distributed uniformly round the circle. Put a point P on* O*y:* OP *is to be uniformly distributed on* $[0,1[$ *and independent of* α*. The process Z is to consist of parallel lines regularly spaced at distance* 1 *from their neighbours, perpendicular to* O*y; one of them is to pass through P. This process is clearly strictly stationary and second-order stationary, but it does not possess a g-function.*

***Example* 2.** *We can generalize Example* 1 *by taking some other distribution of the distance between adjacent lines: e.g. the distances between adjacent lines could be independent negative exponential random variables.*

***Example* 3.** *The Poisson process of lines in the plane is defined by its being a Poisson process of points on the cylinder: strict stationarity is then obvious, and the process is second-order stationary and has a g-function.* (This example and Example 1 are due to Kendall.)

***Example* 4.** *The railway-line process: Take a Poisson process of oriented lines. Drop a perpendicular of fixed length d to the right of each line, and put another line anti-parallel to the first at the foot of the perpendicular. This process has a built-in 'handedness' and it is easy to see that it is not stationary to the second order under reflections of the plane, but it is locally square-summable, and by looking at its moment-generating functional we see that it is strictly stationary under the motions* $\mathcal{M}$.

The question arises: Are there any strictly stationary Z that have g-functions and are not doubly stochastic Poisson processes? We do not know the answer to this, but at least Example 5 below shows that there are Z of the type that we want that are not mixed Poisson processes (in the sense of Nawrotzki [28].

***Example* 5.** *Define a real-valued stochastic process on the cylinder to have the value* $k(1+\cos(\theta-\alpha))$ *at the point* $w=(p,\theta)$, *where* k *and* α *are independent,* $k>0$ *a.s.,* $\mathbf{E}(k^2)<\infty$, *and* α *is distributed uniformly round the circle. Now we put a Poisson process on the cylinder with this random rate: let this doubly stochastic Poisson process be Z. Then from equation* (8) *and the definition of Z we have*

$$(11)\qquad b_Z(f)=\mathbf{E}_k\,\mathbf{E}_\alpha\{\exp(-kA(f)+B(f)\cos\alpha+C(f)\sin\alpha)\},$$

where

$$A(f)=\int_X(1-e^{-f})\,dw,\quad B(f)=\int_X(1-e^{-f})\cos\theta\,dw$$

and

$$C(f)=\int_X(1-e^{-f})\sin\theta\,dw.$$

Let us put $D(f)=\sqrt{\{B^2(f)+C^2(f)\}}$; *then* $A(f)$ *and* $D(f)$ *are invariant under the motions* $\mathcal{M}$: $A(f^T)=A(f)$ *and* $D(f^T)=D(f)$ *for all T in* $\mathcal{M}$. *Reduction of equation* (11) *yields*

$$(12)\qquad b_Z(f)=\mathbf{E}_k\left\{\exp(-kA(f))\oint\exp(-kD(f)\sin(\alpha-\beta))\frac{1}{2\pi}\,d\alpha\right\},$$

where $\beta=\tan^{-1}\{B(f)/C(f)\}$. *But because the* α-*integration is right round the circle,* β *is immaterial and we have*

$$(13)\qquad b_Z(f)=\mathbf{E}_k\{\exp(-kA(f))\,J_0(-ikD(f))\}\quad (i=\sqrt{-1}).$$

The invariance of A and D under the motions $\mathcal{M}$ *then shows that*

$$b_Z(f^T)=b_Z(f)$$

for all f and T, so that Z is strictly stationary. Since $\mathbf{E}(k^2) < \infty$, *Z is second-order stationary also with a g-function g*:

$$g(w, w') = \mathbf{E}(k^2)(1 + \cos(\theta - \theta')).$$

[*Note A.* (This unfinished 'comment' was found among Davidson's papers.) *The question of stationarity of the induced process of parallel lines.*

Let Z be a strictly stationary line-process in the plane, and let Z^* be the process of those lines of Z which have other lines of the process parallel to them. Then Z^* is strictly stationary.

Let A be any bounded Borel set of lines. Then

$$\mathbf{P}(Z^*(A) = 0) = \sum_k \mathbf{P}(Z(A) = k)\, \mathbf{P}(\text{there are no lines of } Z \text{ parallel to the } k \text{ lines of } Z \text{ in } A, \text{ given that there are } k \text{ lines of } Z \text{ in } A).$$

Now

$$\mathbf{P}(.\,|\,\text{there are } k \text{ lines of } Z \text{ in } A) = \mathbf{E}(\mathbf{P}(.\,|\,\text{the positions of the } k \text{ lines})\,|\,k \text{ lines}).$$

So if C is any event and T is a motion in $\mathcal{M}$,

$$\mathbf{P}(C\,|\,\text{there are } k \text{ lines of } Z \text{ in } A) = \mathbf{P}(C^T\,|\,\text{there are } k \text{ lines of } Z \text{ in } A^T).$$

At once, $\mathbf{P}(Z^*(A) = 0) = \mathbf{P}(Z^*(A^T) = 0)$. This is but a long-winded way of saying that the event

There are exactly k_i lines of Z in the set A_i which have other lines of Z parallel to them $(1 \leqslant i \leqslant n)$

is transformed by the motion T in $\mathcal{M}$ into the event

There are exactly k_i lines of Z in the set $T(A_i)$ which have other lines of Z parallel to them $(1 \leqslant i \leqslant n)$.

In view of this transformation and the stationarity of Z, these two events have the same probability. But this is the same as the statement

$$\mathbf{P}(Z^*(A_i) = k_i (1 \leqslant i \leqslant n)) = \mathbf{P}(Z^*(T(A_i)) = k_i (1 \leqslant i \leqslant n)),$$

and this in its turn is the same as the statement that Z^* is strictly stationary, in view of the arbitrariness of n and k_i $(1 \leqslant i \leqslant n)$.]

2.1.3 AN ENTROPY PROPERTY OF THE POISSON LINE-PROCESS

There is a slight practical interest in line-processes: one can consider them as models for the arrangement of fibres in a sheet of paper. It being clear that the strength of a piece of paper depends largely on the number of crossings of its fibres, it becomes of interest to find that process which has the largest number of intersections per unit area relative (in some sense, which we shall consider later) to its density.

We deal here with unoriented lines, so that $0 \leqslant \theta < 2\pi$ and $0 \leqslant p < \infty$; the line-process will be Z. We assume

(*a*) Z is doubly stochastic Poisson with random rate $V(w)$.

(*b*) V is square-summable, continuous in mean square and second-order stationary under the motions $\mathscr{M}$.

For any set A in the plane that is Lebesgue measurable, let $c(A)$ be its measure, and define

$$L(A) = \mathbf{E}\{\text{number of intersections of lines of } Z \text{ lying in } A\}.$$

By (*a*) and (*b*) L is finite if $c(A)$ is; and it is clear that L is a countably additive set function.

Lemma 1. *Under conditions* (*a*) *and* (*b*), *Z is second-order stationary and has a g-function g such that g is continuous and $g(w, w') = g(\theta - \theta')$ for all w, w' and g has the Fourier expansion*

$$g(\theta) = \sum_{n=0}^{\infty} a_n \cos(2n\theta),$$

where $a_n \geqslant 0$ for all n, $a_0 > 0$ and $\sum_{n=0}^{\infty} a_n < \infty$.

Proof. By (*b*), V has a covariance function $f(w, w')$ that is non-negative definite and bounded uniformly in w and w'; the second-order stationarity of V implies that of Z; and we have from (*a*) and (*b*) that Z has a g-function g with $g(w, w') = f(w, w') + k^2$ for some finite positive k. By Proposition 1 we have $g(w, w') = g(\theta - \theta')$ if $\theta \not\equiv \theta' \pmod{\pi}$, so the same holds for f.

The square-summability and continuity in mean square of V and the non-negative definiteness of f imply, by the method of Loève [22, p. 206] that f is everywhere continuous in each of its arguments, so we obtain

$$f(w, w') = f(\theta - \theta') \quad \text{for all } w \text{ and } w'.$$

So $f(\theta)$ is continuous, whence $g(w, w') = f(w, w') + k^2 = g(\theta - \theta')$ for all w and w' and $g(\theta)$ is continuous and non-negative definite as well as f.

Since g is periodic with period 2π, continuous and non-negative definite, it is a positive multiple of the characteristic function of a probability law on the integers. Because we are considering unoriented lines, we have

$$g(\theta) = g(2\pi - \theta) = g(\pi - \theta)$$

for all θ; so we conclude that

$$g(\theta)=\sum_{n=0}^{\infty}a_n\cos(2n\theta),$$

where $a_n\geqslant 0$ for all n,

$$\sum_{n=0}^{\infty}a_n<\infty,$$

and $a_0>0$ because g is got by adding a positive constant to a non-negative definite function.

***Lemma* 2.** *There is a constant k_Z such that $L(A)=k_Z c(A)$ for all measurable sets A in the plane.*

Proof. Santaló [34] shows that the density of pairs of lines in the plane can be written not only as $dw\,dw'$ but also as

$$|\sin(\theta-\theta')|\,dP\,d\theta\,d\theta',$$

where dP is plane measure, corresponding to the point of intersection of the two lines. Hence from formula (7) and the fact proved in the first lemma, that $g(w,w')=g(\theta-\theta')$, we find that

$$L(A)=c(A)\oint\oint g(\theta-\theta')|\sin(\theta-\theta')|\,d\theta\,d\theta'=k_Z c(A). \tag{14}$$

Note. All Lemma 2 says is that the process of line-intersections induced by Z is first-order stationary under the motions $\mathscr{M}$.

Proposition 3. $$k_Z=a_0-\sum_{n=1}^{\infty}\frac{\pi a_n}{2(4n^2-1)}.$$

Proof. By Lemma 2, k_Z is determined if we know $L(A)$ for any one A: let then A be the unit circle, so that $c(A)=\pi$. From equation (14) we obtain

$$k_Z=4\int_{t=0}^{\pi/2}\sin t\,dt\int_{u=0}^{t}\sin u\,du\int_{v=t-u}^{t+u}g(v)\,dv. \tag{15}$$

Putting in the representation of g derived in Lemma 1 and integrating, we get

$$k_Z=a_0-\sum_{n=1}^{\infty}\frac{\pi a_n}{2(4n^2-1)}, \tag{16}$$

as required. We now have the problem of deciding what normalization to use in saying that processes are of the same density with respect to pairs of lines: for it is clear that it is the set of pairs of lines of Z, not the set of these lines taken singly, that is relevant to the study of the process of points of intersection induced by Z.

For any set A in the plane define A^* to be the set of lines that intersect A, so that A^* is a subset of the cylinder X.

Lemma 3.

(i) *If* $A \subset B$ *then* $A^* \subset B^*$.
(ii) $(\bigcup_\alpha A_\alpha)^* = \bigcup_\alpha A_\alpha^*$.
(iii) *If* A *is open, so is* A^*.
(iv) *If* A *is compact, so is* A^*.

Proof. (i) and (ii) are trivial. As to (iii): let A be any open disk of radius $r > 0$. If we take the origin to be the centre of the disk, we find that $A^* = \{w: w = (p, \theta), p < r\}$. So A^* is open. But every motion T in $\mathscr{M}$ is a homeomorphism on X, so the choice of origin is immaterial and (iii) follows from (ii). As to (iv): let A be compact and let (w_n) be a sequence of points in A^* corresponding to lines l_n. Then there is a sequence (a_n) of points in A such that l_n goes through a_n for every n. Since A is compact there is a convergent subsequence $(a_{n'})$ with limit a in A; since the circle is compact there is a sub-subsequence $(a_{n''})$ such that $(a_{n''})$ converges to a and the angles $\theta_{n''}$ of the lines $l_{n''}$ converge to θ, say. Since the line l of angle θ passing through a is such that the corresponding w is in A^* and since $(w_{n''})$ converges to w (this is easy to verify), we see that A^* is sequentially compact. But the topology on the cylinder is the Euclidean metric, so we have that A^* is compact.

Corollary. *If A is compact, or a countable union of compacta, or open, then A^* is measurable and so $Z(A^*)$ is a random variable. If A is compact, then $\mathbf{E}(Z(A^*))^2$ is finite.*

It is, however, not true that the map $A \to A^*$ preserves measurability. If we take A to be a non-measurable subset of a line in the plane, then of course A is plane-measurable with measure zero. But it is easy to see that A^* is as non-measurable on the cylinder as A was on its line. (This example is due to Mr. Swinnerton-Dyer.)

For any compact A, then, define $n(A) \equiv \mathbf{E}\{Z(A^*)(Z(A^*) - 1)\}$; that is, $n(A)$ is the expected number of pairs of lines of Z that cut A. For any A, $n(A)$ is a measure of the average density of pairs of lines of Z (provided that $n(A) > 0$, which will certainly be the case if A has a non-void interior). But n is not an additive set-function, so it is necessary to specify in advance which A we are going to use. Let it be B, and let $n(B) = \tilde{n}$ say. Then we have

(17) $$\tilde{n} = \int_{B^*} \int_{B^*} g(\theta - \theta')\, dw\, dw' = \sum_{r=0}^{\infty} a_r c_r,$$

where for each r

$$c_r = \int_{B^*} \int_{B^*} \cos(2r(\theta - \theta'))\, dw\, dw' = \oint\oint \cos(2r(\theta - \theta')) f(\theta) f(\theta')\, d\theta\, d\theta',$$

where f is a non-negative bounded integrable function. But since $\cos(2n\theta)$ is a non-negative definite function, by taking approximating sums we see that $c_r \geqslant 0$ for all r.

For any non-negative random variable L we define the process Z_L by

$$b_{Z_L}(f) = \mathbf{E}_L\{b_Z(Lf)\},$$

where L and Z are supposed independent. This is in effect a multiplication of the rate of the process Z by a random factor L; we assume always that $0 < L < \infty$. Equivalently, we are operating a random change of scale on the plane in which the lines lie. We demand that the specific intersection-rate of Z be independent of this change of scale, because any realization of Z_L will appear exactly the same as a realization of Z, save that we are looking at it from a different distance.

We assume from now on that $\mathbf{E}(L^2)$ is finite.

Lemma 4. *For all r,* $a_r(Z_L) = \mathbf{E}(L^2)\, a_r(Z)$. This is trivial on considering the rate-process V of Z.

Corollary. $k_{Z_L} = \mathbf{E}(L^2)\, k_Z$ *and* $\tilde{n}(Z_L) = \mathbf{E}(L^2)\, \tilde{n}(Z)$, *for they are both fixed linear combinations of the* a_r. *Hence* $(k_Z/\tilde{n})$ *is invariant under random changes of scale, so we conclude that* $K_Z \equiv (k_Z/\tilde{n})$ *is (not the only, but) a good measure of the specific intersection-rate of the process Z.*

Proposition 4. *Among the line-processes Z that satisfy (a) and (b), the mixed Poisson process has the largest specific intersection-rate (as measured by* K_Z*).*

Proof. The mixed Poisson process is simply the Poisson process under a random change of scale of the sort that we have been considering; and the Poisson process has

(18) $$a_0 > 0,\ a_n = 0$$

for all $n \geqslant 1$. In view of this, equations (16) and (17) and the definition of K_Z show that K_Z is largest when Z is mixed Poisson. But if equation (18) holds for a process Z satisfying (*a*) and (*b*), the fact that the correlation between the values of V at two points w, w' is unity, whatever w, w' are, and the stationarity of V show that there is a random variable L such that $V(w) = L$ a.s. for each fixed w (see the proof of Proposition 6); so Z is mixed Poisson. That is, there is no other process Z satisfying (a) and (b) that has as large a specific intersection-rate as the mixed Poisson process.

2.1.4 HIGHER DIMENSIONS: PROCESSES OF FLATS

By a *flat* in Euclidean n-space we mean a Euclidean $(n-1)$-space. Take an origin in n-space: then a flat is specified by its distance p from the origin and a generalized angle $\boldsymbol{\theta}$, which has $(n-1)$ components. Let the set of all flats in n-space be X_n. We can define processes of flats Z on X_n in the same way that we defined processes of lines in the plane, for we can regard X_n as the product of the real line R with the $(n-1)$-sphere S_{n-1}.

The Z that we consider will be assumed to have the following properties:

(a) Z is locally summable to the nth order;

(b) Z is nth-order stationary under the group $\mathscr{M}_n$ of motions induced on X_n by the Euclidean rotations of n-space;

(c) Z has an nth-order product-moment density g_n;

(d) a.s. there is no set of n flats of Z that fail to meet in a point.

Let L be an arbitrary line (without loss of generality through the origin) in n-space. Because of the stationarity of Z it is almost sure that no flat of Z is parallel to L; this implies that a.s. all the flats of Z cut L. Let the flat F of Z meet L at the point x and at the generalized angle $\boldsymbol{\omega}$; then by (c) we have, if the F_i $(1 \leqslant i \leqslant n)$ are all different,

$$(19) \qquad g(F_1, \ldots, F_n) = \left[\mathbf{E}\left\{\prod_{i=1}^{n} dZ(x_i, \boldsymbol{\omega}_i)\right\} \Big/ \prod_{i=1}^{n} dx_i\, d\boldsymbol{\omega}_i\right] \prod_{i=1}^{n} f(\boldsymbol{\omega}_i),$$

where $f(\boldsymbol{\omega})$ is the Jacobian of the transformation from the $(p, \boldsymbol{\theta})$ coordinates to the $(x, \boldsymbol{\omega})$ coordinates. It is important to note that f depends only on the generalized angle: this is trivial, for we have

$$x = p \sec(\psi(\boldsymbol{\theta})) \quad \text{and} \quad \boldsymbol{\omega} = \boldsymbol{\omega}(\boldsymbol{\theta}), \quad \text{for some function } \psi.$$

***Lemma* 5.** *$g_n(F_1, \ldots, F_n)$ is not a function of the x_i if they are all distinct.*

Proof. Let $F_1, \ldots, F_n$ have distinct xs, and let $F_1^*, \ldots, F_n^*$ have distinct xs and be such that for each i $(1 \leqslant i \leqslant n)$, $\boldsymbol{\omega}_i = \boldsymbol{\omega}_i^*$. Now we can shift F_1 perpendicular to its face so that x_1 is carried into x_1^*, and all the other xs and all the $\boldsymbol{\omega}$s are left unchanged. After this motion T_1 the configuration of the F_i $(1 \leqslant i \leqslant n)$ is the same as ever it was, save that the point of intersection of the F_i has been moved; so by the nth-order stationarity of Z

$$g_n(F_1, F_2, \ldots, F_n) = g_n(F_1^*, F_2, \ldots, F_n). \tag{20}$$

The application of $(n-1)$ further motions $T_2, \ldots, T_n$ defined similarly to T_1 then yields the lemma.

Let N be the point-process on the line L induced by Z. Then N inherits the properties (1) and (2) of Z, so that N is nth-order stationary under translations of the line. From Lemma 1 we see, on integrating out the generalized angles, that

$$a_n \equiv \mathbf{E}\left\{\prod_{i=1}^{n} dN(x_i)\right\} \Big/ \prod_{i=1}^{n} dx_i \tag{21}$$

is not a function of the x_i $(1 \leqslant i \leqslant n)$, provided they are all distinct. By taking a section of Z by an r-space containing L $(1 \leqslant r < n)$, we obtain a process of flats in r-space which satisfies $(a)-(d)$ with r for n. Because this process $Z(r)$ induces the same N as does Z, we find that if $1 \leqslant r \leqslant n$,

$$a_r \equiv \mathbf{E}\left\{\prod_{i=1}^{r} dN(x_i)\right\} \Big/ \prod_{i=1}^{r} dx_i \tag{22}$$

is not a function of the x_i $(1 \leqslant i \leqslant r)$, provided they are all distinct. We define for completeness $a_0 \equiv 1$.

Suppose we go to the limit and consider processes of flats in infinite-dimensional space. This presents difficulties because of the vacuity of the idea of 'generalized angle' in infinite-dimensional space: we can, however, realize such a process by considering for all $n > 1$ a process Z_n of flats in n-space which satisfies $(a)-(d)$ and is a section of the process Z_{n+1} in $(n+1)$-space. In this case we get a process, Z say, of points on the line that is locally summable to all orders, is stationary to all orders under translations of the line, and satisfies equation (22) for all $r \geqslant 0$. Since a process Z on the line has these properties if and only if it has a complete sequence of a_r satisfying equation (22), the Z having these properties will be referred to as Z '*having an a-sequence*'.

Proposition 5. *If Z has a-sequence (a_r), then there is a non-negative random variable L such that a_r is the* r*th moment of L for all $r \geqslant 0$. Conversely, if L is a non-negative random variable having a complete sequence of moments (a_r), then there is a Z with a-sequence (a_r).*

Proof. For any interval I on the line let $c(I)$ be its length, and define

$$m_r(I) = \mathbf{E}(Z(I))^r \tag{23}$$

for each $r \geqslant 0$. Then if $c(I)$ is finite the same is true of m_r for every r. Let us, with Bartlett [1], define the coefficients d_{rs} $(0 \leqslant s \leqslant r < \infty)$ by the identities

$$n^r \equiv \sum_{s=0}^{r} d_{rs}\, n(n-1)\ldots(n-s+1) \tag{24}$$

for all non-negative integers n and r. We note that $d_{rr} = 1$ for all $r \geqslant 0$, but $d_{r0} = 0$ if $r > 0$. Bartlett shows that for all I such that $c(I)$ is finite

$$m_r(I) = \sum_{s=0}^{r} d_{rs}\, a_s (c(I))^s \tag{25}$$

for all $r \geqslant 0$; it follows from this that the a_r are the factorial moments of $Z(I_1)$, where I_1 is an interval of unit length.

The a_r will be the moment-sequence of a non-negative random variable if and only if (Widder [36]) for all $r \geqslant 0$ the $(r+1)\times(r+1)$ matrices $U(r)$ and $V(r)$ are non-negative definite, where

$$u_{i,j} = a_{i+j} \quad \text{and} \quad v_{i,j} = a_{i+j+1}.$$

On the other hand, this non-negative definiteness holds if and only if *all* the principal minors of each matrix $U(r)$ and $V(r)$ are non-negative (see Mirsky [26]). Consider first the matrices $U(r)$. A typical minor will be

$$M(\mathbf{a}) = \begin{vmatrix} a_{k_1+k_1} & a_{k_1+k_2} & \ldots & a_{k_1+k_t} \\ a_{k_2+k_1} & a_{k_2+k_2} & \ldots & a_{k_2+k_t} \\ \vdots & \vdots & \vdots & \vdots \\ a_{k_t+k_1} & a_{k_t+k_2} & \ldots & a_{k_t+k_t} \end{vmatrix};$$

M is a fixed function of the sequence $\mathbf{a} = (a_r)$. We know that for every I of finite length $M(\mathbf{m}(I)) \geqslant 0$, for by definition the $m_r(I)$ are the moments of a non-negative random variable. So using equation (25) to substitute for $(m_r(I))$ in $M(\mathbf{m}(I))$, for each p such that $1 \leqslant p \leqslant t$ dividing through the pth row and column of $M(\mathbf{m}(I))$ by $(c(I))^{k_p}$, and letting $c(I) \to \infty$, we see that $M(\mathbf{a}) \geqslant 0$. Since this holds for all M, we have that $U(r)$ is non-negative definite for each $r \geqslant 0$. Similar argument shows that $V(r)$ is non-negative definite for each $r \geqslant 0$, so the a_r are indeed the moments of a non-negative random variable L.

Conversely, suppose that we are given a non-negative random variable L which has moments of all orders. Define the process Z to be a mixed Poisson process with mixing variable L. We easily check that Z has an a-sequence such that a_r is the rth moment of L for each r; for the Poisson process with rate l has an a-sequence $a_r(l) = l^r$.

***Lemma* 6.** *If Z_n is a doubly stochastic Poisson process of flats in n-space with rate-process V that is square-summable, second-order stationary and continuous in mean square; then the point-process Z of its intersections with a fixed line enjoys the same properties.*

The proof is immediate from looking at the moment-generating functional of Z_n.

***Proposition* 6.** *Under the conditions of Lemma 6, Z is mixed Poisson.*

Proof. Let P be an arbitrary plane containing our line. Again by considering the moment-generating functional of Z_n, we find that the process Z_2 of lines in the plane formed by the intersections of the flats of Z_n with P is doubly stochastic Poisson with rate-process V that is square-summable, second-order stationary and continuous in mean square. This and Lemma 5 imply that (since Z is the process of intersections of the lines of Z_2 with our line)

$$a_2 = \frac{\mathbf{E}(dZ(x)\,dZ(y))}{dx\,dy}$$

does not depend on x or y if $x \neq y$. So using Lemma 6 we see that the covariance function of the rate-process V of Z is a constant; since V is continuous in mean square and second-order stationary, it follows that

$$\mathbf{E}(V(x)\,V(y)) = \mathbf{E}(V(x))^2 = \mathbf{E}(V(y))^2 < \infty$$

for *all* x and y. So

$$\mathbf{E}(V(x) - V(y))^2 = 0 \tag{26}$$

for all x and y. Take now some fixed x_0 and call the random variable $V(x_0)$ M: M is a.s. non-negative, and square-summable. Equation (26) shows now that for all positive integers k and all choice of $t_i \geqslant 0$ and x_i $(1 \leqslant i \leqslant k)$

$$\mathbf{E}\left\{\exp\left(-\sum_{i=1}^{k} t_i\,V(x_i)\right)\right\} = \mathbf{E}\left\{\exp\left(-M\sum_{i=1}^{k} t_i\right)\right\}. \tag{27}$$

Now the moment-generating functional of Z, which, as we observed in Section 2.1.2, characterizes Z uniquely, is given by

$$b_Z(f) = \mathbf{E}_V\left\{\exp\left(-\int_{-\infty}^{\infty} V(x)\,(1 - \mathrm{e}^{-f(x)})\,dx\right)\right\} \tag{28}$$

for all functions f that are uniform limits of non-negative step-functions of compact support. Approximating the integral in equation (28) by Darboux sums and using equation (27) and bounded convergence with respect to a probability measure, we find that Z is a mixed Poisson process with mixing random variable M.

If we strengthen the hypotheses of Proposition 5 we can improve on its analytic result with a stochastic conclusion, that in the circumstances of that proposition Z is actually a mixed Poisson process: I find that this result has also been got by McFadden [25]. Our method of proof involves the idea of exchangeability of a point-process, so we shall defer the statement and proof of this result till Section 2.1.8 on exchangeable point-processes.

2.1.5 EXCHANGEABLE STOCHASTIC POINT PROCESSES: THE CASE OF THE REAL LINE R

Let $W(V)$ be the set of all half-left-open (open) binary-rational-end-pointed intervals on the real line R. For $0 \leqslant i < \infty$ and $-\infty < n < \infty$ define

(29) $$S_{i,n} =]n2^{-i}, (n+1)2^{-i}] \quad \text{and} \quad T_{i,n} =]n2^{-i}, (n+1)2^{-i}[;$$

let

(30) $$W_i^* \equiv \{S_{i,n}\}_{n=-\infty}^{\infty} \quad \text{and} \quad V_i^* \equiv \{T_{i,n}\}_{n=-\infty}^{\infty},$$

so that

(31) $$W^* \equiv \bigcup_{i=0}^{\infty} W_i^* \subset W \quad \text{and} \quad V^* \equiv \bigcup_{i=0}^{\infty} V_i^* \subset V.$$

We shall at first be concerned exclusively with the $(W, S_{i,n})$ set-up. Let the Lebesgue measure of a measurable set S on R be written $m(S)$.

Theorem 1. *Let Z be a stochastic point process on R. Denote the number of its points in the measurable set S by $Z(S)$. Suppose that*

(a) $m(S) < \infty$ implies $Z(S) < \infty$ a.s.;

(b) for each i, the events $s_{i,n} \equiv \{Z(S_{i,n}) = 0\}$ are exchangeable. That is, if $n_1, n_2, \ldots, n_k$ are all different,

$$\mathbf{P}\left(\bigcap_{r=1}^{k} s_{i,n_r}\right)$$

is a function of i and k but not of the particular choice of the n_r $(1 \leqslant r \leqslant k)$.

Then Z is a mixed Poisson process (in the sense of Nawrotzki [28]).

Proof. We remark trivially that from their definition, for all i and n

(32) $$s_{i,n} \equiv s_{i+1,2n} \cap s_{i+1,2n+1}.$$

By the exchangeable-event theory of Kendall [13] we have that for each i there is a random variable x_i such that $0 \leqslant x_i \leqslant 1$ a.s. and, for all $k \geqslant 0$ and choices of $n_1, \ldots, n_k$ all different from each other

(33) $$\mathbf{P}\left(\bigcap_{r=1}^{k} s_{i,n_r} \,\middle|\, x_i\right) = x_i^k.$$

Let A_i be the minimal σ-field with respect to which all the events $s_{i,n}$ are measurable. Clearly A_i is a sub-σ-field of the σ-field of the process Z, and x_i is A_i-measurable. By equation (32) the A_i are increasing in i, so that x_i is A_j-measurable if $j \geqslant i$. In fact from equation (32) we have for all $k \geqslant 0$

(34) $$\mathbf{E}_{A_i}(x_{i+1})^{2k} = x_i^k,$$

where $\mathbf{E}_{A_i}$ means conditional expectation with respect to the σ-field A_i. The sequence

$$(A_i, (x_i)^{2i}) \quad (0 \leqslant i < \infty)$$

is thus a martingale; and the x_i are bounded between zero and unity, so there exists a limit random variable x defined on the σ-field

$$A \equiv \sigma\left(\bigcup_{i=0}^{\infty} A_i\right),$$

and

(35) $$\mathbf{E}_{A_r}(x)^{2-s} = (x_r)^{2r-s}$$

for all r and s such that $r \geqslant s \geqslant 0$. At once, if

(36) the sets S_{i_r,n_r} $(1 \leqslant r \leqslant k)$ in W^* are pairwise disjoint

then

(37) $$\mathbf{P}\left(\bigcap_{r=1}^{k} s_{i_r,n_r} \,\middle|\, x\right) = x^M,$$

where $M = \sum_{r=1}^{k} m(S_{i_r,n_r})$. Put $L = -\log(x)$. Since $0 \leqslant x \leqslant 1$ we have $0 \leqslant L \leqslant \infty$; let E be the event $L = \infty$. Then, conditional on E, $x = 0$. Consequently

(38) $$\mathbf{P}\left(\bigcap_{r=1}^{k} s_{i_r,n_r} \,\middle|\, E\right) = 0,$$

provided condition (36) holds. Hence conditional on E the points of Z are a.s. dense on R, contradicting our assumption (*a*) unless $\mathbf{P}(E) = 0$. So we must have $L < \infty$ a.s.

By equations (35) and (37) and the definition of L we have now

$$\mathbf{P}\left(\bigcap_{i=1}^{k}\{Z(S_i)=0\}\,|\,L\right)=\exp\left(-L\sum_{i=1}^{k}m(S_i)\right) \tag{39}$$

provided the S_i $(1\leqslant i\leqslant k)$ in W are pairwise disjoint. Now the S_i, being all in W, are commensurable: there exists a fixed binary rational $d>0$ such that each S_i $(1\leqslant i\leqslant k)$ is the disjoint union of h_i (say) subintervals in W each of which has length d. For each positive integer n of the form 2^l for some non-negative integer l define

$$n_i\equiv nh_i \quad (1\leqslant i\leqslant k). \tag{40}$$

Let $X_i(n_i)$ be the number of the n_i subintervals of S_i which contain points of Z. In view of equation (39), conditional on L the X_i are independent binomial random variables with (in the usual notation $B(n,p)$)

$$n=n_i \quad \text{and} \quad p=Lm(S_i)/n_i.$$

We define, for convenience,

$$g(a,r)=\mathrm{e}^{-a}a^r/r!$$

$(a\geqslant 0;\ r$ a non-negative integer). From the Poisson limit theorem we obtain

$$\lim_{n\to\infty}\mathbf{P}\left(\bigcap_{i=1}^{k}\{X_i(n_i)=r_i\}\,|\,L\right)=\prod_{i=1}^{k}g(Lm(S_i),r_i). \tag{41}$$

For all positive-integer-valued k-vectors $\mathbf{t}=(t_1,\ldots,t_k)$ define

$$C(\mathbf{t})\equiv\bigcap_{i=1}^{k}\left\{\begin{matrix}\text{The least distance between any two points of } Z \text{ in}\\ S_i \text{ lies in the range } [t_i^{-1}m(S_i),(t_i+1)^{-1}m(S_i)[\end{matrix}\right\}.$$

Then the $C(\mathbf{t})$ for different $\mathbf{t}$ are mutually exclusive; and in view of our assumption (a) their union is a.s. with respect to the σ-field of Z. So

$$\mathbf{P}\left(\bigcup_{\mathbf{t}}C(\mathbf{t})\,|\,L\right)=1 \text{ a.s. } (L);$$

whence it is permissible to (and we shall) take

$$\mathbf{P}\left(\bigcup_{\mathbf{t}}C(\mathbf{t})\,|\,L\right)=1.$$

We define further

$$\mathbf{P}_{n,\mathbf{t}}(\mathbf{r})=\mathbf{P}\left(C(\mathbf{t})\cap\bigcap_{i=1}^{k}\{X_i(n_i)=r_i\}\,|\,L\right) \quad \text{(see equation (40))},$$

and again

$$q_{\mathbf{t}}(\mathbf{r}) = \mathbf{P}\left(C(\mathbf{t}) \cap \bigcap_{i=1}^{k} \{Z(S_i) = r_i\} \,|\, L\right).$$

We observe that if $n_i \geqslant t_i$ for each i (which is bound to happen for all large enough n, by equation (40)) we have $\mathbf{P}_{n,\mathbf{t}}(\mathbf{r}) = q_{\mathbf{t}}(\mathbf{r})$. Hence

$$\mathbf{P}\left(\bigcap_{i=1}^{k} \{Z(S_i) = r_i\} \,|\, L\right) = \sum_{\mathbf{t}} q_{\mathbf{t}}(\mathbf{r}) = \sum_{\mathbf{t}} \lim_{n\to\infty} \mathbf{P}_{n,\mathbf{t}}(\mathbf{r})$$

by what we have just done

$$= \lim_{n\to\infty} \sum_{\mathbf{t}} \mathbf{P}_{n,\mathbf{t}}(\mathbf{r}) = \lim_{n\to\infty} \mathbf{P}\left(\bigcap_{i=1}^{k} \{X_i(n_i) = r_i\} \,|\, L\right).$$

This equality and equation (41) give us

$$\text{(42)} \qquad \mathbf{P}\left(\bigcap_{i=1}^{k} \{Z(S_i) = r_i\} \,|\, L\right) = \prod_{i=1}^{k} g(Lm(S_i), r_i).$$

Now let k be any positive integer, and let I_i $(1 \leqslant i \leqslant k)$ be *any* finite disjoint half-left-open intervals on R: we can suppose without loss of generality that I_i lies entirely to the left of I_j for every i and j such that $1 \leqslant i < j \leqslant k$. Let z be arbitrary such that $0 < z < 1$. Then there are $S_1, \ldots, S_k$ and $T_1, \ldots, T_k$, all of them in W, such that

$$\text{(43)} \qquad \left.\begin{array}{llll} S_i \supset I_i \supset T_i & \text{and} & m(T_i) \geqslant zm(S_i) & \text{if } 1 \leqslant i \leqslant k \text{ and } i \text{ is odd;} \\ S_i \subset I_i \subset T_i & \text{and} & m(S_i) \geqslant zm(T_i) & \text{if } 1 \leqslant i \leqslant k \text{ and } i \text{ is even.} \end{array}\right\}$$

For all measurable sets S the random variable Z is non-negative. So for all choices of non-negative integers $r_1, \ldots, r_k$ we have

$$\mathbf{P}\left(\bigcap_{\substack{1\leqslant i\leqslant k\\ i \text{ odd}}} \{Z(S_i) \geqslant r_i\} \cap \bigcap_{\substack{1\leqslant i\leqslant k\\ i \text{ even}}} \{Z(S_i) \leqslant r_i\} \,|\, L\right)$$

$$\geqslant \mathbf{P}\left(\bigcap_{\substack{1\leqslant i\leqslant k\\ i \text{ odd}}} \{Z(I_i) \geqslant r_i\} \cap \bigcap_{\substack{1\leqslant i\leqslant k\\ i \text{ even}}} \{Z(I_i) \leqslant r_i\} \,|\, L\right)$$

$$\text{(44)} \qquad \geqslant \mathbf{P}\left(\bigcap_{\substack{1\leqslant i\leqslant k\\ i \text{ odd}}} \{Z(T_i) \geqslant r_i\} \cap \bigcap_{\substack{1\leqslant i\leqslant k\\ i \text{ even}}} \{Z(T_i) \leqslant r_i\} \,|\, L\right).$$

Now all the S_i and T_i are in W, so equation (42) applies to the first and last probabilities in inequality (44). The right-hand side of equation (42) is continuous in the measures of the sets involved; so if we let $z \uparrow 1$ in the relations (43) we get from the relations (44) and (42)

$$\mathbf{P}\left(\bigcap_{i=1}^{k}\{Z(I_i)=r_i\}\,\Big|\,L\right)=\prod_{i=1}^{k}g(Lm(I_i),r_i) \tag{45}$$

for all positive integers k and all choices of disjoint half-left-open intervals $I_i\,(1\leqslant i\leqslant k)$. Thus conditional on L, Z can be identified with (the set of occurrence-times of) a Poisson process with parameter L, i.e. Z is mixed Poisson.

The converse *of the theorem is trivial.* That is, every mixed Poisson process with mixing random variable L such that $L<\infty$ a.s. satisfies equation (45) and so is exchangeable in the sense of the statement of the theorem.

2.1.6 A RESTATEMENT IN TERMS OF RANDOM CLOSED SETS

We retain the notation of equations (29)–(31). The theory of random closed sets has been developed by Kendall [14], and we introduce his terminology:

A *strong incidence function* is a map from V^* to the two-point set $\{0, 1\}$ such that if T in V^* is covered by any collection of $T_{i,n}$ with $Z(T_{i,n})=0$ for every i and n, then $Z(T)=0$ also.

The *random closed set* C puts a probability measure on the strong incidence functions Z, so that we can speak of a random Z; and we can interpret the statement $\{Z(T)=1\,(\text{or }0)\}$ as $\{C$ intersects (resp. does not intersect) $T\}$.

For each i and n such that $0\leqslant i<\infty$ and $-\infty<n<\infty$ let $t_{i,n}$ be the event $\{Z(T_{i,n})=0\}$; the measure of C will attach probability to all these. We want to show that if C is in some sense exchangeable, then C coincides with the set of occurrence-times of a mixed Poisson process. By analogy with (*a*) and (*b*) of the last section we might hope that

(a^*) $C\neq R$ a.s., and

(b^*) for each fixed i, the events $t_{i,n}(-\infty<n<\infty)$ are exchangeable,

would suffice. But this is not the case; for to a set C satisfying (a^*) and (b^*) we can adjoin the point $\{0\}$ with probability 1. The new set C' will still be closed and satisfy (a^*) and (b^*), but is not stationary and so cannot be the set of occurrence-times of a mixed Poisson process. We could get round this difficulty by introducing such a condition as (*c*) C a.s. contains no binary rational;

but I think it is better to proceed as follows.

Lemma 7. *For any S in W the event $s \equiv \{Z(S) = 0\}$ is C-measurable.*

Proof. For each non-negative integer i define T_i in V to be the interior of S translated to the right through 2^{-i}. Then

$$S = \bigcup_{r=1}^{\infty} \bigcap_{i=r}^{\infty} T_i.$$

Let us put $t_i \equiv \{Z(T_i) = 0\}$. I assert

$$s = \bigcup_{r=1}^{\infty} \bigcap_{i=r}^{\infty} t_i; \tag{46}$$

we now prove this. Suppose that s does not occur, that is, S has a point—P, say—in C. Then clearly T_i contains P for all large enough i; so for all r the intersection on the right-hand side of equation (46) is void. Hence the right-hand side itself is void.

Suppose contrariwise that S is void of points of C. Then each T_i is void of points of C to the left of, and including, the right-hand end-point —x, say—of S. Suppose if possible that there are infinitely many i such that T_i contains a point P_i (or several points) of C. Then there must be a sequence of such points converging to x. So by the closure of C x must be in C also, a contradiction. So if s occurs, t_i also occurs for all large enough i and so the right-hand side of equation (46) occurs. Thus equation (46) holds, which implies at once that s is C-measurable.

Theorem 1*. *Let C be a random closed set on R, such that*

(*a'*) *$C \neq R$ a.s., and*

(*b'*) *for each fixed i, the events $s_{i,n}$, measurable by the Lemma 7, are exchangeable* (*see* (*b*) *of Theorem* 1).

Then C can be identified with the set of occurrence-times of a mixed Poisson process.

Proof. As in Theorem 1 let A_i be the σ-field generated by the $s_{i,n}$, and let A be the minimal σ-field containing all the A_i, so that A is a sub-σ-field of the σ-field B of C. As in equation (39) we have that there is a random variable L which is A- (and so also B-) measurable such that $0 \leqslant L \leqslant \infty$ and

$$\mathbf{P}\left(\bigcap_{i=1}^{k} \{Z(S_i) = 0\} \,\middle|\, L\right) = \exp\left(-L \sum_{i=1}^{k} m(S_i)\right) \tag{47}$$

whenever the S_i $(1 \leqslant i \leqslant k)$ in W are pairwise disjoint. From (a') we see that $L < \infty$ a.s. Because C is closed and any T in V is a countable union of $S_{i,n}$ we conclude that there is a B-measurable random variable L such that $0 \leqslant L < \infty$ and

$$\mathbf{P}\left(\bigcap_{i=1}^{k}\{Z(T_i)=0\}\,\middle|\,L\right)=\exp\left(-L\sum_{i=1}^{k}m(T_i)\right) \tag{48}$$

for every positive integer k and all choices of $T_1, \ldots, T_k$ in V and pairwise disjoint.

If now we regard L as having been sampled we are in the position of Kendall [14] (in having a random closed set C possessing his properties of stationarity and of independence of increments) and can conclude that C can be identified with the set of occurrence-times of a Poisson process with rate L: unconditionally, C can be identified with the set of occurrence-times of a mixed Poisson process.

Note 1. It is appropriate to consider this exchangeability problem in the context of random closed sets, for the hypotheses of the problem are phrased (see (b) and (b')) in terms of the hitting of test sets by the process or random closed set.

Note 2. Earlier in this thesis† we said that we would give a probabilistic interpretation of the additive convexity of H, the pointwise closure of the set of infinitely divisible standard p-functions. Now Kendall [15] has shown that if we have a stationary random closed set C inducing a random strong incidence function Z, and if we put for all rational $t \geqslant 0$

$$U(t) = \mathbf{P}(Z(\,]0, t[\,) = 0),$$

then U is decreasing and convex in $t \geqslant 0$ (t rational), and so we can define U for all $t \geqslant 0$ by continuity. We have of course $U(0) = 1$, but U need not be continuous at $t = 0$. These conditions are not only necessary but also sufficient that U be the U-function of some stationary C. If we have C and C' with U-functions U, U', and if C, C' are independent, then evidently $V(t) = U(t)\,U'(t)$ is also a U-function—in fact, the U-function of $C \cup C'$. (These results are all due to Kendall.) Kendall further shows that if U is infinitely divisible in the sense that for each positive integer n there is a U_n such that $U = (U_n)^n$, then U is not just convex but logarithmically convex; and of course U is continuous save perhaps at $t = 0$. So U is infinitely divisible if and only if it is in H, the pointwise closure of the

† Davidson here refers to earlier sections of his Ph.D. Thesis which are not reproduced. See, however, his papers on Delphic Semigroups in the companion volume, *Stochastic Analysis*.

set of i.d. standard p-functions. Now the convex combination under addition of U-functions in H corresponds to mixing of the random closed sets involved; so we obtain a probabilistic interpretation of the convexity of H. It does not go very far, though, for the U-function does not in general specify the random closed set giving rise to it.

Note 3. Theorem 1* can now be rephrased: *If a random closed set C satisfies* (a') *and* (b'), *then its U-function is completely monotone* (from formula (42) with $k = 1$ therein). *The exchangeability then implies that C can be identified with the set of occurrence-times of a mixed Poisson process.*

2.1.7 WHAT HAPPENS WHEN THE CARRIER SPACE OF THE PROCESS IS COMPACT: THE CASE OF THE CIRCLE

Let the circle be of perimeter 1, and let W_n^* be the set of 2^n equal clockwise-half-open intervals, one of which has its open end-point at the angle $x \equiv 0$; the intervals to be of length 2^{-n} and disjoint. Call these intervals $S_{n,r}$ ($n \geqslant 0$; $0 \leqslant r \leqslant 2^n - 1$). Let the process of points on the circle be denoted by Z, and let $s_{n,r}$ be the event $\{Z(S_{n,r}) = 0\}$.

Theorem 1**. *Suppose that*

(a) $N \equiv Z(S_{0,0}) < \infty$ *a.s.;*

(b) *for each* $n \geqslant 0$, *the events* $s_{n,r}$ *are exchangeable.*

Then N can have any distribution (on the non-negative integers); but conditional on N, the process can be identified with the distribution of N points independently and uniformly round the circle.

Proof. We still have trivially for all $n \geqslant 0$ and $0 \leqslant r \leqslant 2^n - 1$

(49) $$s_{n,r} = s_{n+1,2r} \cap s_{n+1,2r+1}.$$

The important difference from Section 2.15 is that for each fixed n there are but a finite number of exchangeable events $s_{n,r}$. Let us write $m_n = 2^n$; and define N_n to be the number of the $S_{n,r}$ that contain points of Z. It is clear that

(50) $$N_n \uparrow N \text{ a.s.} \quad \text{as} \quad n \to \infty.$$

Define

$$W^* = \bigcup_{n=0}^{\infty} W_n^*;$$

suppose now that for $1 \leqslant i \leqslant k$ the sets S_i are in W^* and that they are pairwise disjoint. Obviously, for all large enough n each S_i is a union of sets pairwise disjoint in W_n^*. Let

$$m_n' = m_n \sum_{i=1}^{k} m(S_i)$$

(where $m(.)$ is Lebesgue measure on the circle) so that, for all large enough n, m_n' is an integer. Then by the exchangeable-event theory of Kendall [13],

$$\mathbf{P}\left(\bigcap_{i=1}^{k} \{Z(S_i) = 0\} \,|\, N_n\right) = \prod_{s=0}^{m_n'-1} \left\{\frac{m_n - N_n - s}{m_n - s}\right\}. \tag{51}$$

From equation (50) and the fact that N_n and N are discrete random variables,

(52)

$$\lim_{n\to\infty} \mathbf{P}\left(\bigcap_{i=1}^{k} \{Z(S_i) = 0\} \,|\, N_n\right) \quad \text{exists and is equal to} \quad \mathbf{P}\left(\bigcap_{i=1}^{k} \{Z(S_i) = 0\} \,|\, N\right).$$

If we put $M = \sum_{i=1}^{k} m(S_i)$, then equations (51) and (52) yield

$$\mathbf{P}\left(\bigcap_{i=1}^{k} \{Z(S_i) = 0\} \,|\, N\right) = \lim_{C} \prod_{s=1}^{m_n'} \left\{1 - \frac{N_n}{m_n - s + 1}\right\}, \tag{53}$$

where the condition C is $n \to \infty$; $N_n \uparrow N$; $m_n' = M m_n$.

Now if $|z| < \frac{1}{2}, |\log(1-z) + z| < |z|^2$, so provided $n > n_0(N)$

$$\left|\sum_{s=1}^{m_n'} \log\left(1 - \frac{N_n}{m_n - s + 1}\right) + N_n \sum_{s=1}^{m_n'} \frac{1}{m_n - s + 1}\right| \leqslant N_n^2 \sum_{s=1}^{m_n'} \left\{\frac{1}{(m_n - s + 1)^2}\right\}, \tag{54}$$

since $N_n \leqslant N$. The right-hand side of the inequality (54) converges to zero as $n \to \infty$ by the general principle of convergence; so since, if $M < 1$,

$$\lim_{C} \left\{-N_n \sum_{s=1}^{m_n'} \frac{1}{m_n - s + 1}\right\} \quad \text{exists and equals} \quad N \log(1 - M), \tag{55}$$

we have from inequality (54)

$$\lim_{C} \sum_{s=1}^{m_n'} \log\left(1 - \frac{N_n}{m_n - s + 1}\right) \quad \text{exists equal to } N \log(1 - M) \tag{56}$$

if $M < 1$. Taking the exponential of equation (56) and using equation (53) we get

$$\mathbf{P}\left(\bigcap_{i=1}^{k}\{Z(S_i)=0\}\,|\,N\right)=\left(1-m\left(\bigcup_{i=1}^{k}S_i\right)\right)^N \tag{57}$$

if $m(\bigcup_{i=1}^{k}S_i)<1$; this formula is trivial when

$$m\left(\bigcup_{i=1}^{k}S_i\right)=1,$$

provided we put $0^0=1$. By the methods of section 2.1.5 we can extend equation (57) to obtain

$$\mathbf{P}\left(\bigcap_{i=1}^{k}\{Z(S_i)=r_i\}\,|\,N\right) \tag{58}$$

for any choice of $k\geqslant 1$ and non-negative integers r_i $(1\leqslant i\leqslant k)$. In this case explicit calculation is difficult, but all we need here is that the probabilities (58) are determined uniquely by equation (57), even when the sets S_i in (58) are not restricted to be in W^* but may be any disjoint clockwise-half-open segments of the circle. In view of this uniqueness of (58), we can say that conditional on the value of N the points of Z are distributed independently and uniformly round the circle; and there are, by definition of N, N of these points.

On the other hand, if we take an arbitrary a.s. finite non-negative-integer valued random variable N, and put N points independently and uniformly on the circle, the resulting process Z is exchangeable.

Note. The non-negative-integer valued random variable N is a mixed Poisson random variable (with mixing random variable L) if and only if its probability-generating function $\mathbf{E}(z^N)$ satisfies

$$\mathbf{E}(z^N)=\int_0^\infty \exp\{L(z-1)\}\,dm(L),$$

where m is the probability measure on $[0,\infty[$ corresponding to the random variable L. It follows at once that if N is mixed Poisson, its probability-generating function cannot be a polynomial. So since, in Theorem 1**, N can have any distribution whatever, we cannot conclude in that theorem that Z is a mixed Poisson process.

Definition. Let (X,m) be a measure space and let U be a class of measurable sets $S\subset X$ such that $m(S)<\infty$ for each S in U. If Z is a point-process on X, we say that Z has the *CIP*(U) if

(*a*) whenever $m(S)<\infty$, $Z(S)<\infty$ a.s.;

(*b*) whenever S is in U, conditional on $Z(S)$ the points of Z in S are independently and uniformly distributed in S.

Nawrotzki [28] shows that when $X = R$, m is Lebesgue measure and U is the class of all left-half-open intervals, Z is mixed Poisson if and only if Z is stationary and has the $CIP(U)$. We deduce from our Theorem 1 that in this case, Z is stationary with the $CIP(U)$ if and only if Z is a.s. locally finite and exchangeable (conditions (*a*) and (*b*) in Section 2.1.5).

When X is the circle, m is Lebesgue measure and U is the set of clockwise-half-open intervals, it is an easy deduction from Theorem 1** that Z is stationary with the $CIP(U)$ if and only if Z is a.s. locally finite and exchangeable. Thus far the theorem agrees with the conclusion in the last paragraph; that we cannot extend Theorem 1** to conclude that under these circumstances Z is mixed Poisson just reflects the inferior strength of exchangeability on finite, as opposed to infinite, sets of events.

We note, on the other hand, that the proof of Theorem 1** does not use the fact that X is there a circle; the proof would apply equally to the case $X =]0, t] \subset R$ for any positive t. Thus we could obtain Theorem 1 from Theorem 1** and Nawrotzki's result quoted above.

We can—in a rather *ad hoc* way—carry over the results of the last three sections to more general spaces. Suppose we have a carrier space X: let it have a Borel field $\mathscr{B}$, and let $\mathscr{B}^*$ be the collection of Borel sets of finite measure (some Borel measure m being given). For our method of proof to work, the structure of X must be such that there is a family F of sets in $\mathscr{B}^*$ all the members of each generation of which are mutually congruent in some sense; further, F must generate $\mathscr{B}^*$.

Let $F = S_{i,a(i)}$ $(0 \leqslant i < \infty$, $a(i)$ in some index set $A(i))$. The 'generation' is i. The congruence must be such that any two members of the same generation have the same m-measure; we assume that $m(S_{i,a(i)}) > 0$ for all i. Then each S_i must split into the same finite number of S_{i+1}.

The assumptions on the process Z are that $Z(S)$ is an a.s. finite non-negative-integer-valued random variable for each S in $\mathscr{B}^*$; and that the events $S_{i,a(i)}$ ($a(i)$ in $A(i)$) are exchangeable for each fixed i.

Under these circumstances, if X is in $\mathscr{B}^*$ we can conclude that conditional on the total number of points of Z in X, those points are distributed independently and m-uniformly over X; if X is not in $\mathscr{B}^*$, then Z is a mixed Poisson process on X.

We can now generalize Theorems 1 and 1** to the case where the carrier space X is a product $R^r \times K^k$ of lines and circles. The sets $S_{i,a(i)}$ will be products of binary-rational-end-pointed left- and clockwise-half-open intervals of length 2^{-i}. Under the familiar assumptions Z will be a mixed Poisson process if $r>0$; if $r=0$, we can only deduce the conditional independence and uniformity properties.

2.1.8 APPLICATION OF THE THEORY TO PROPOSITION 5

***Proposition* 7.** *Let Z be a point-process on the line which is locally summable to all orders, is stationary to all orders under translations of the line and has product-moment densities of all orders such that for all non-negative integers n,*

$$a_n \equiv \mathbf{E}\left(\prod_{i=1}^{n} dZ(x_i)\right) \Big/ \prod_{i=1}^{n} dx_i \tag{59}$$

is independent of the x_i $(1 \leqslant i \leqslant n)$ provided they are all different.

Suppose further that

$$\sum_{n=0}^{\infty} a_n^{-1/n} \quad \textit{diverges.} \tag{60}$$

Then Z is exchangeable in the sense of Theorem 1.

Proof. Z, being locally summable to all orders, must be a.s. locally finite. So exchangeability will be proved if we show that

(61) If (S_r) is any sequence of disjoint half-left-open intervals on the line such that each S_r has a common length—c, say—then

$$\mathbf{P}\left(Z\left(\bigcup_{r=1}^{k} S_{i_r}\right) = 0\right)$$

depends on c and k but not on the indices i_r $(1 \leqslant r \leqslant k)$.

Consider the random variable

$$V = Z\left(\bigcup_{r=1}^{k} S_{i_r}\right):$$

it is of course non-negative-integer valued. Let T be some half-left-open interval covering all the S_{i_r}; let T be of (finite) length t and put $W = Z(T)$.

For any non-negative integers N and k write

$$N^{(k)} = N(N-1)\dots(N-k+1).$$

Then we have for all non-negative integers k

$$v_k \equiv \mathbf{E}(V^{(k)}) \leqslant \mathbf{E}(W^{(k)}) \equiv w_k. \tag{62}$$

On the other hand, it is immediate that $w_k = t^k a_k$, so that

$$\sum_{n=0}^{\infty} w_n^{-1/n} = \sum_{n=0}^{\infty} (t^n a_n)^{-1/n} = t^{-1} \sum_{n=0}^{\infty} a_n^{-1/n} = \infty.$$

From equation (62), then,

$$\sum_{n=0}^{\infty} v_n^{-1/n} \text{ diverges.} \tag{63}$$

Now for each n, v_n is the left-hand derivative at $x = 1$ of the probability-generating function $f_V(x) \equiv \mathbf{E}(x^V)$; and this function is of course absolutely monotone on the interval $[0, 1]$, so that

$$v_n = \text{lub}\{|f_V^{(n)}(x)| : 0 \leqslant x \leqslant 1\}. \tag{64}$$

From the theory of quasi-analytic functions of Carleman [3], equations (64) and (63) tell us that f_V is uniquely determined by its derivatives at the point $x = 1$: that is, the distribution of V is given uniquely by the sequence $(v_{k'}: 0 \leqslant k' < \infty)$. Hence in particular

$$\mathbf{P}\left(Z\left(\bigcup_{r=1}^{k} S_{i_r}\right) = 0\right) = \mathbf{P}(V = 0) \tag{65}$$

is determined by (i.e. is a function of) the moments of the random variable V; for, of course, the moments of V are determined as functions of its factorial moments $v_{k'}$, and vice versa.

By the additivity of Z we have

$$\mathbf{E}(V^r) = \mathbf{E}\{Z(S_{i_1}) + Z(S_{i_2}) + \ldots + Z(S_{i_k})\}^r \tag{66}$$

for all $r \geqslant 0$. But we have also, generalizing the formula of Bartlett [1, p. 84],

$$\mathbf{E}\prod_{j=1}^{k} \{Z(S_{i_j})\}^{r_j} = \sum_{s_1=0}^{r_1} \cdots \sum_{s_k=0}^{r_k} \left\{a_{\Sigma s_j} c^{\Sigma s_j} \prod_{j=1}^{k} d_{r_j s_j}\right\} \tag{67}$$

(where the d_{rs} are given by the identities (24)), since all the Ss have length c. This and equation (66) show that $\mathbf{E}(V^r)$ depends only on c, k and $a_0, \ldots, a_r$. So by relation (65)

$$\mathbf{P}\left(Z\left(\bigcup_{r=1}^{k} S_{i_r}\right) = 0\right)$$

depends only on c, k and the sequence (a_n); so we have obtained relation (61), proving the proposition.

Corollary. *Under the conditions of the proposition, Z is a mixed Poisson process; and the distribution of the mixing random variable L is uniquely determined by its moments a_n.*

Proof. That Z is mixed Poisson follows from Theorem 1. If Z has mixing random variable L, we have from the definition of a mixed Poisson process that

$$\mathbf{P}(Z(\,]0,t]\,)=0)=\mathbf{E}(e^{-tL}). \tag{68}$$

From the proposition just proved we know that the left-hand side of equation (68) is for all $t>0$ given uniquely in terms of the sequence (a_n). So the distribution of L, being determined by its Laplace transform, is also given uniquely by the sequence (a_n). That the a_n are the moments of L was proved in the course of Proposition 5.

The condition (60) in the proposition is obviously vital to our proof; but it is technical and detracts from the generality of the result. I wanted to state, with McFadden [25], the proposition which is now

Conjecture 1. *If Z satisfies just condition* (59), *then Z is mixed Poisson.*

This may be true; what is not true is that condition (59) is enough to define the process Z. McFadden's argument purports not only to show that Z is mixed Poisson but also to find the distribution of the mixing variable; and his proof of Conjecture 1 falls down as follows. He uses his formula 2.16 which (among other things) says (in our notation)

$$\mathbf{P}(Z(\,]0,t]\,)=0) \tag{69}$$

is determined by the sequence (a_n)—whether or not our condition (60) is satisfied.

But let (b_n) be the common moment-sequence of two non-negative random variables L and M which have distinct distributions (there exists such a triple $\{(b_n),L,M\}$: see Feller [8]). Now let Z and Z^* be mixed Poisson processes with mixing random variables L and M respectively. Because their Laplace transforms specify the distributions of L and M uniquely, we have from equation (68) that there is a $t>0$ such that

$$\mathbf{P}(Z(\,]0,t]\,)=0)\neq\mathbf{P}(Z^*(\,]0,t]\,)=0).$$

According, however, to the last sentence in the corollary above we have

$$a_n(Z)=a_n(Z^*)=b_n$$

for all non-negative integers n, so that (69) is contradicted.

In view of this lack of uniqueness we are led to make the more plausible

Conjecture 2. *If Z satisfies condition* (59), *and* (a_n), *a moment-sequence by Proposition* 5, *determines the distribution whose moment-sequence it is uniquely, then Z is exchangeable* (*and so is a mixed Poisson process*).

[***Note B.*** (This unfinished 'comment' was found among Davidson's papers.)

During the last year I have had many bad quarters of an hour because of the following problem. Let Z be a point-process on the line, and let the probability space associated with it be $(W, A, \mathbf{P})$. Let $s_{i,n}$ be the event that Z puts no points in the interval $]n2^{-i}, (n+1)2^{-i}]$. Let A_i be the σ-field generated by the exchangeable events $s_{i,n}$ for all integers n; and let B_i be the sub-σ-field of the events in A_i that are invariant under translations through 2^{-i}, so that for all $i \geqslant 0$, $B_i \subset A_i \subset A_{i+1}$.

We want to prove that for all positive integers r and non-negative integers i, $\mathbf{E}_{A_i}(x_{i+1})^2 = x_i$, where x_i is defined equivalently as (a.s.)

$$x_i = \mathbf{E}_{B_i}(x_{i,n}) \tag{70}$$

for any n ($x_{i,n}$ the indicator of $s_{i,n}$),

$$x_i = \lim_{n\to\infty}\left((1/n)\sum_{j=1}^{n} x_{i,j}\right), \tag{71}$$

by the exchangeable-event theory of Kendall [13].

Let then $\mathbf{P}_{A_i}$ be $\mathbf{P}$ restricted to A_i (any i). What we want to prove is that for all events G in A_i

$$\int_G x_i^r \, d\mathbf{P}_{A_i} = \int_G x_{i+1}^{2r} \, d\mathbf{P}_{A+1},$$

that is,

$$\mathbf{E}(x_i^r I_G) = \mathbf{E}(x_{i+1}^{2r} I_G). \tag{72}$$

Now by the definition of A_i, we can take G to be of the form

$$G = s_{i,n_1} \cap s_{i,n_2} \cap \ldots \cap s_{i,n_k}$$

for some positive integer k and some choice of $n_1, n_2, \ldots, n_k$, all distinct. In this case the left-hand side of equation (72) becomes

$$\begin{aligned}
\mathbf{E}(x_i^r x_{i,n_1} x_{i,n_2} \ldots x_{i,n_k}'), &= \mathbf{E}(\mathbf{E}_{B_i}(x_i^r x_{i,n_1} x_{i,n_2} \ldots x_{i,n_k})) \\
&= \mathbf{E}(x_i^r \, \mathbf{E}_{B_i}(x_{i,n_1} x_{i,n_2} \ldots x_{i,n_k})) \\
&= \mathbf{E}(x_i^r x_i^k) \\
&= \mathbf{E}(x_i^{r+k}).
\end{aligned}$$

For similar reasons, the right-hand side of equation (72) equals $\mathbf{E}(x_{i+1}^{2r+2k})$. But these two are equal by exchangeable-event theory and the fact that (for all non-negative integers i and all integers n)

$$s_{i,n} = s_{i+1,2n} \cap s_{i+1,2n+1},$$

thus concluding the proof.]

REFERENCES

[*Note C.* This is the list of references for the thesis as a whole, and not all the entries in it are cited in the present extract.]

1. M. S. Bartlett, *An Introduction to Stochastic Processes*, Cambridge (1966).
2. N. Bourbaki, *Éléments de Mathématique: Intégration*, Hermann, Paris (1956).
3. T. Carleman, *Les Fonctions Quasi Analytiques*, Collection Borel, Gauthier-Villars, Paris (1926).
4. E. Čech, *Topological Spaces*, Academia, Prague (1966).
5. H. Cramér, "On the factorization of probability distributions", *Arkiv for Matematik* **1** (1949), 61–65.
6. M. W. Crofton, "Probability", *Encyclopaedia Britannica*, IXth edition (1885).
7. D. Dugué, "Arithmetique des Lois de Probabilités", *Mémorial des Sciences Mathématiques* **137**, Gauthier-Villars, Paris (1957).
8. W. Feller, *An Introduction to Probability Theory and its Applications, Vol.* 2, Wiley (1966).
9. Revd. J. Hagen, "On Division of Series". *Am. J. Math.* **5** (1882), 236–237.
10. D. G. Kendall, "Delphic Semigroups, Infinitely Divisible Regenerative Phenomena, and the Arithmetic of *p*-functions (to appear). [2.2 of *Stochastic Analysis*, Wiley.]
11. ——, "Delphic Semigroups", *Bull. Am. math. Soc.* **73** (1) (1967), 120–121.
12. ——, "Renewal Sequences and their Arithmetic", *Proceedings of the Loutraki Symposium on Probabilistic Methods in Analysis*, Springer, Berlin (1967). [2.1 of *Stochastic Analysis*, Wiley.]
13. ——, "On finite and infinite sequences of exchangeable events", *Report* **64**, Department of Statistics, Johns Hopkins University (1966). [*Stud. Sci. Math. Hung.* **2** (1967), 319–327.]
14. ——, *A Theory of Random Sets* (unpublished). [Now partly published in this book.]
15. ——, *U-functions* (verbal communication). [See (14).]

15a. ——, [see 3.4 of the companion volume, *Stochastic Analysis*, Wiley.]

16. J. F. C. Kingman, "The stochastic theory of regenerative events". *Ztschr. Wahrsch'theorie & verw. Geb.* **2** (1964), 180–224.
17. ——, "Some further analytical results in the theory of regenerative events", *J. Math. Anal. Appl.* **11** (1965), 422–433.
19. P. Lévy, "L'addition des variables aléatoires definies sur une circonférence", *Bull. Soc. Math. France* **67** (1939), 1–41.
20. ——, "Sur les exponentielles de polynomes . . .", *Ann. Sci. de l'École Normale Supérieure, Série* 3 **54** (1937), 231–292.

20a. ——, "L'arithmétique des lois de probabilité", *J. Math. pures et appliquées* **103** (1938), 17–40.

21. Yu. V. Linnik, *Decomposition of Probability Distributions* Oliver and Boyd (1964).
22. M. Loève, *Probability Theory*, 3rd ed., Van Nostrand, Princeton (1963).
23. E. Lukacs, *Characteristic Functions*, Griffin, London (1960).
24. A. I. Markushevich, *Theory of Functions of a Complex Variable*, Vol. 1, Prentice-Hall, New Jersey (1965).

25. J. A. McFadden, "The mixed Poisson process", *Sankhyā* (*A*) **27** (1965), 83–92.
26. L. Mirsky, *An Introduction to Linear Algebra*, Oxford University Press (1955).
27. I. P. Natanson, *Theory of Functions of a Real Variable*, Vol. 1, Ungar, New York (1955).
28. K. Nawrotzki, "Ein Grenzwertsatze für homogene zufällige Punktfolgen (Verallgemeinerung eines Satzes von A. Rényi)", *Math. Nachrichten* **24** (1962), 201–217.
29. I. V. Ostrowski, "On factorizations of infinitely divisible laws without Gaussian components" (in Russian), *Doklady Academii Nauk SSSR* **161**(1) (1965), 48–52.
30. K. R. Parthasarathy, R. Ranga Rao and S. R. S. Varadhan, "On the category of indecomposable distributions on topological groups", *Trans. Am. math. Soc.* **102**(2) (1962), 200–217.
31. ——, "Probability distributions on locally compact abelian groups", *Ill. J. Math.* **7** (1963), 337–369.
32. R. R. Phelps, *Lectures on Choquet's Theorem*, Van Nostrand, Princeton (1966).
33. A. P. Robertson and W. J. Robertson, *Topological Vector Spaces*, Cambridge University Press (1964).
34. L. A. Santaló, "Introduction to integral geometry", *Actualités Scientifiques et Industrielles* **1198**, Hermann, Paris (1953).
35. E. C. Titchmarsh, *The Theory of Functions*, Oxford (1932).
36. D. V. Widder, *The Laplace Transform*, Princeton (1946).

[*Concluding note.* It is only fair to state that the second half of Davidson's Ph.D. thesis, reprinted here, was regarded by him as to some extent exploratory. Just before his death he had returned to these problems, and had he been able to publish a reconsidered version of this work, it is likely that much would have been altered and completed. In editing this material for publication now, I have only made the minimum of alterations dictated by the circumstances that this document is the second half of a longer one (the first half being concerned with Delphic semi-groups). I have *not* attempted to amend or complete the arguments as the author himself might have done. D. G. K.]

2.2
Exchangeable Point-processes

ROLLO DAVIDSON

In a recent paper Benczur [2] proved the

Theorem 1. *Let Z be a point-process on the line such that if $I_1, I_2, \ldots$ is any sequence of disjoint intervals all of the same length, the events $\{Z(I_i) = 0\}$ are exchangeable (= equivalent). Suppose also that for all $\varepsilon > 0$ there is $\delta > 0$ such that for all intervals I of the line with $0 < |I| < \delta$, $\mathbf{P}(Z(I) \geqslant 2) < \varepsilon |I|$. Then Z is a mixed Poisson process.*

In my doctoral thesis (Cambridge, 1968) I proved a similar theorem, stated here in its greatest generality. We consider point-processes on $R^r \times K^k$, where K is the circle of unit circumference. In this space, we let

$$S_{i;n_1\ldots n_r;m_1\ldots m_k} = \prod_{j=1}^{r}]n_j 2^{-i}, (n_j+1) 2^{-i}] \times \prod_{j'=1}^{k}]m_{j'} 2^{-i}, (m_{j'}+1) 2^{-i}]$$

be the typical product of left-half-open and clockwise-half-open intervals in the coordinate spaces, of equal length 2^{-i} in each space and with end-points whose coordinates are binary rationals with denominators not greater than 2^i. There are thus only countably many Ss.

Theorem 1'. *Let Z be a point-process on $R^r \times K^k$ such that*

(i) Z has a.s. no multiple points;

(ii) Z is a.s. locally finite (i.e. if I is any finite interval in $R^r \times K^k$, then $Z(I) < \infty$ a.s.);

(iii) for each $i \geqslant 0$, the events

$$Z(S_{i;n_1\ldots n_r;m_1\ldots m_k}) = 0 \quad (-\infty < n_j < \infty, 1 \leqslant j \leqslant r; 0 \leqslant m_{j'} < 2^i, 1 \leqslant j' \leqslant k)$$

are exchangeable.

Then (a) if $r>0$, Z is a mixed Poisson process;

(b) if $r=0$, K^k is compact; $Z(K^k)$ may have any honest distribution; and conditional on $Z(K^k)=N$, the N points of Z are distributed independently and uniformly over K^k.

The proof is omitted. [See Part II of the author's Ph.D. thesis.]

I became interested in processes of the Benczur type because of the following problem.

Let Z be a point-process on the line which may be represented, for each positive integer n, as the intersection of the given line with a process $Z^{(n)}$ of $(n-1)$-flats in Euclidean n-space, where $Z^{(n)}$ is stationary to the nth order under the rigid motions of n-space, possesses an nth-order product-moment density, and has a.s. no parallel flats. Then one may show that Z has product-moment densities of all orders which are constants: for all $n \geqslant 0$,

$$\frac{\mathbf{E}[dZ(x_1)\dots dZ(x_n)]}{dx_1 \dots dx_n} = a_n, \tag{1}$$

if all the x_i are different. (For a discussion of product-moment densities, see Bartlett [1].) It may also be shown that a process Z having such a representation as a section of $Z^{(n)}$ has a.s. no multiple points. So we become interested in processes Z on the line with a.s. no multiple points and satisfying (1) for all $n \geqslant 0$. If the a_n are a sequence proceeding in this way from a Z, one may show that $\{a_n : n \geqslant 0\}$ is the moment-sequence of a non-negative random variable L, say (see Theorem 3 below).

Now the mixed Poisson process with mixing random variable L has property (1) for each n (trivial), and it is tempting to conjecture that if a Z with a.s. no multiple points satisfies equation (1) for all n, then it is a mixed Poisson process. Indeed, McFadden [4] purports to prove this, but we shall see later that his proof breaks down. However, we may state

***Theorem* 2.** *Let Z be a point-process on the line having a.s. no multiple points, having product-moment densities of all orders, and satisfying equation* (1) *for all $n \geqslant 0$. Suppose also that*

$$\sum_{n=0}^{\infty} a_n^{-(1/n)} = \infty. \tag{2}$$

Then Z is exchangeable (i.e. satisfies (iii) of Theorem 1′); and so, applying Theorem 1′ to Z (and $r=1$, $k=0$ in $R^r \times K^k$), Z is mixed Poisson.

Proof. Since

$$\mathbf{E}(Z(I)) = \int_I \frac{\mathbf{E}dZ(x)}{dx}\,dx = a_1|I| < \infty,$$

Z is a.s. locally finite. Consequently, it is sufficient to show the truth of:

(3) if $\{S_r\}$ is any sequence of disjoint half-left-open intervals on the line such that $|S_r| = c$ (say) for each r, then $\mathbf{P}(Z(\bigcup_{r=1}^k S_{i_r}) = 0)$ depends on c and k but not on the choice of the indices i_r $(1 \leqslant r \leqslant k)$.

Consider the random variable

$$V = Z\left(\bigcup_{r=1}^{k} S_{i_r}\right).$$

It is of course non-negative-integer-valued. Let T be some left-half-open interval covering all the S_{i_r}, such that $|T| = t < \infty$, say. Put $W = Z(T)$.

For all non-negative integers N, k write

$$N^{(k)} = N(N-1)\ldots(N-k+1);\ N^{(0)} = 1. \tag{4}$$

Then we have for all non-negative k

$$v_k \equiv \mathbf{E}(V^{(k)}) \leqslant \mathbf{E}(W^{(k)}) \equiv w_k. \tag{5}$$

On the other hand, we have by the definition (Bartlett) of product-moment densities, $w_k = \mathbf{E}(W^{(k)}) = t^k a_k$, so that

$$\sum_{n=0}^{\infty} w_n^{-1/n} = \sum_{n=0}^{\infty} (t^n a_n)^{-1/n} = t^{-1} \sum_{n=0}^{\infty} a_n^{-1/n} = \infty,$$

by equation (2). From equation (5), then,

$$\sum_{n=0}^{\infty} v_n^{-1/n} = \infty \tag{6}$$

also.

Now, for each n, v_n is the left-hand derivative at $x = 1$ of the probability-generating function $f_V(x) = \mathbf{E}(x^V)$; and this function is of course absolutely monotone on $[0, 1]$, so that

$$v_n = \sup\{|f_V^{(n)}(x)| : 0 \leqslant x \leqslant 1\}. \tag{7}$$

From the theory of quasi-analytic functions of Carleman [3], equations (6) and (7) tell us that f_V is uniquely determined by its derivatives at $x = 1$; that is, the distribution of V is given uniquely by the sequence $v_{k'}: k' \geqslant 0$. Hence in particular

$$\mathbf{P}\left(Z\left(\bigcup_{r=1}^{k} S_{i_r}\right) = 0\right) \equiv \mathbf{P}(V = 0) \tag{8}$$

is determined by (i.e. is a function of) the moments of the random variable V, for of course the moments of V are determined as functions of its factorial moments $v_{k'}$, and vice versa. Now by the additivity of Z we have

$$\mathbf{E}(V^r) = \mathbf{E}(Z(S_{i_1})+Z(S_{i_2})+ \dots +Z(S_{i_k}))^r \tag{9}$$

for all $r \geqslant 0$. But we have also, generalizing the formula of Bartlett ([1], p. 84),

$$\mathbf{E}\prod_{j=1}^{k} \{Z(S_{i_j})\}^{r_j} = \sum_{s_1=0}^{r_1} \dots \sum_{s_k=0}^{r_k} a_{\Sigma s_j} c^{\Sigma s_j} \prod_{j=1}^{k} d_{r_j, s_j}, \tag{10}$$

where the $d_{r,s}$ $(r \geqslant s \geqslant 0)$ are given by the identities, for positive integers n,

$$n^r = \sum_{s=0}^{r} d_{r,s} n^{(s)}$$

(see [4]). In equation (10), we recall that the a_i are the members of the sequence of product-moment densities of Z, and c is the common length of the S_{i_j}.

But now equations (10) and (9) show that $\mathbf{E}(V^r)$ depends only on c, k and a_i $(1 \leqslant i \leqslant r)$. Thus by equation (8)

$$\mathbf{P}\left(Z\left(\bigcup_{r=1}^{k} S_{i_r}\right) = 0\right) \tag{11}$$

depends only on c, k and the sequence $\{a_n;\, n \geqslant 0\}$. So we have obtained condition (3) and the theorem.

Theorem 3. *Let Z be a point-process on the line with a.s. no multiple points and with product-moment densities of all orders satisfying equation* (1) *for all $n \geqslant 0$. Then $\mathbf{a} = \{a_n : n \geqslant 0\}$ is the moment-sequence of a non-negative random variable L, say; but, in general, the distributions of the point-process are not determined by the sequence* **a**.

Proof. For every interval I of the line put $m_r(I) = \mathbf{E}(Z(I))^r$ for integers $r \geqslant 0$. Then by the Bartlett formula (see equation (10))

$$m_r(I) = \sum_{s=0}^{r} d_{r,s}\, a_s\, |I|^s \tag{12}$$

for all $r \geqslant 0$, where the $d_{r,s}$ are given by equation (11). We note that $d_{r,r} = 1$ for all $r \geqslant 0$.

Now **a** will be the moment-sequence of a non-negative random variable if and only if (Widder [5]) for all $r \geqslant 0$ the $(r+1)\times(r+1)$ matrices $U(r)$, $V(r)$ are non-negative definite, where $U_{i,j}(r) = a_{i+j}$, $V_{i,j}(r) = a_{i+j+1}$. On the other hand, this non-negative definiteness holds if and only if *all* the principal minors of $U(r)$, $V(r)$ are non-negative. Now a typical principal minor of $U(r)$ will be $M(\mathbf{a}) = |b_{i,j}|$, say, where $b_{i,j} = a_{k_i+k_j}$ $(1 \leqslant i,j \leqslant t$, say). M is to be regarded as a fixed function of the sequence $\mathbf{a}$.

Now we know that $M(\mathbf{m}(I)) \geqslant 0$ for every interval I of finite length, because $\mathbf{m}(I)$ is the moment-sequence of the non-negative random variable $Z(I)$. So, first we use equation (12) to substitute for the $m_n(I)$ in $M(\mathbf{m}(I))$; then we divide, for $1 \leqslant p \leqslant t$, through the pth row and column of $M(\mathbf{m}(I))$ by $|I|^{k_p}$; finally, we let $|I| \to \infty$. Since $d_{r,r} = 1$ for all $r \geqslant 0$, we see that $M(\mathbf{a})$ is the limit (as $|I| \to \infty$) of $M(\mathbf{m}(I))$ and is accordingly non-negative.

So $U(r)$ is non-negative definite; a similar argument holds for $V(r)$; and thus $\mathbf{a}$ is indeed the moment-sequence of a non-negative random variable.

We turn to the second part of the theorem. It is trivial that if L is any non-negative random variable with all of its moments a_n (say) finite, then the mixed Poisson process Z_L (say) with mixing random variable L has a.s. no multiple points, and satisfies equation (1) for all n. Now we may choose the sequence $\{a_n\}$ such that there exist two different distributions on the non-negative reals, associated with random variables L and M, say, such that

$$\mathbf{E}(L^n) = \mathbf{E}(M^n) = a_n$$

for all $n \geqslant 0$. Let Z_L, Z_M be mixed Poisson processes with mixing random variables L, M. Then

$$(13) \quad \mathbf{P}(Z_L(\,]0,t]\,) = 0) = \mathbf{E}_L(e^{-Lt}); \; \mathbf{P}(Z_M(\,]0,t]\,) = 0) = \mathbf{E}_M(e^{-Mt}),$$

for every $t > 0$. But the distribution of a non-negative random variable is determined by its moment-generating function, so there must be a $t > 0$ such that $\mathbf{E}(e^{-Lt}) \neq \mathbf{E}(e^{-Mt})$, and then by equation (13)

$$\mathbf{P}(Z_L(\,]0,t]\,) = 0 \neq \mathbf{P}(Z_M(\,]0,t]\,) = 0).$$

McFadden's argument purports to deduce the distributions of Z from the a_n, with no conditions on the a_n save that they all be finite; but our proof above shows that this cannot be done, and, in fact, the argument leading to his theorem (our Theorem 2 without the condition (2)) is invalid because of an unjustifiable change of order of summation. The following problems are thus left open:

(*a*) Is Theorem 2 still true when the condition (2) is omitted?

(*b*) Is Theorem 2 still true when the condition (2) is weakened to

(2′) The Stieltjes moment problem for the sequence $\{a_n\}$ is determined?

REFERENCES

1. M. S. Bartlett, *An Introduction to Stochastic Processes*, Cambridge University Press (1966).
2. A. Benczur, "On sequences of equivalent events", *Studia Scient. Math. Hungarica* **3** (1968), 451–458.
3. T. Carleman, "Les fonctions quasi-analytiques", *Collection Borel* Gauthier-Villars, Paris (1926).
4. J. A. McFadden, "The mixed Poisson process", *Sankhyā* (*A*) **27** (1965), 83–92.
5. D. V. Widder, *The Laplace Transform*, Princeton University Press (1946).

2.3

Positive-definiteness of Product-moment Densities of Line-processes

ROLLO DAVIDSON

Let Z be a second-order stationary line-process in the plane, which is

(*a*) locally square-summable,

(*b*) the possessor of a second-order product-moment density, $f(w, w')$, say,

(*c*) a.s. free of pairs of parallel lines.

We assume that $f(.,.)$ is measurable; and that for any set S of lines

$$\mathbf{E}(Z(S)(Z(S)-1)) = \int_S \int_S f(w, w')\, dw\, dw',$$

which is what we have tacitly assumed all along.

Theorem 1. *Under these circumstances, the Fourier coefficients of f (which is a function of the angle between the lines) are all of them non-negative. If f is continuous, it is non-negative definite.*

Proof. Let W be the class of binary-rational end-pointed clockwise-half-open intervals on the circle: call such intervals $I, J, \ldots$. For I in W we define I^* to be the rectangle on the cylinder of height 1 and base I; and let T be the shift of this rectangle along the cylinder by a unit.

Consider the random variables $Z(T^n I^*)$ $(n \geq 0)$. We have

$$\mathbf{E}Z(T^n I^*) = k\,.\,mI,$$

m being the Lebesgue measure of I; and

$$\mathbf{E}(Z(T^n I^*)Z(T^m I^*)) = \oint_I \oint_I f(\theta-\theta')\,d\theta\,d\theta' \quad \text{if } m \neq n$$

$$= \oint_I \oint_I f(\theta-\theta')\,d\theta\,d\theta' + kmI \quad \text{if } m = n.$$

So the sequence $(Z(T^n I^*))$ is second-order stationary in $n \geqslant 0$. Accordingly there exists

$$Y(I) = \underset{n\to\infty}{\text{l.i.m.}}\, n^{-1} \sum_{r=0}^{n-1} Z(T^r I^*).$$

We have $Y(\varnothing) = 0$, $Y(.)$ is square-summable; and if $I \supset J$, $Y(I) \geqslant Y(J)$ a.s. (Z), because there exists a subsequence (n') of (n) such that

$$Y(J) = \lim_{n'\to\infty} (\text{a.s.})\, n'^{-1} \sum_{r=0}^{n'-1} Z(T^r J^*),$$

and then there exists a sub-subsequence (n'') of (n') such that

$$Y(I) = \lim_{n''\to\infty} (\text{a.s.})\, n''^{-1} \sum_{r=0}^{n''-1} Z(T^r I^*).$$

Now if $I \supset J$, $Z(T^r I^*) \geqslant Z(T^r J^*)$ surely; so the inequality holds as asserted.

Now

$$\mathbf{E}\, Y(I) = kmI.$$

Further,

$$\mathbf{E}(Y^2(I)) = \lim_{n\to\infty} (1/n^2)\, \mathbf{E}(Z(nI^*))^2 \quad \text{(in an obvious notation)}$$

$$= \int_I \int_I f(\theta-\theta')\,d\theta\,d\theta' + O(1/n) \quad \text{as } n \to \infty$$

$$= \int_I \int_I f(\theta-\theta')\,d\theta\,d\theta'.$$

Similarly

$$\mathbf{E}(Y(I)\,Y(J)) = \int_I \int_J f(\theta-\theta')\,d\theta\,d\theta'$$

for any I, J in W.

Let now g be any continuous function on the circle. We have, defined uniquely by Z, the random functional $Y(g)$: a limit of Riemann sums.

This functional is linear. We have

$$Y(g) = \lim_{r\to\infty} \sum_{s=0}^{2^r-1} g(x_s^r)\, Y(I_s^r),$$

where $I_s^r (0 \leqslant s \leqslant 2^r - 1)$ are the intervals in W of length $2\pi 2^{-r}$ and end-points x_s^r:

$$x_s^r = 2\pi s 2^{-r}.$$

Take g to be complex-valued: $g(x) = e^{inx}$. Then of course $\mathbf{E}(Y(g)\,\bar{Y}(g))$ is real and

$$0 \leqslant \mathbf{E}(Y(g)\,\bar{Y}(g)) < \infty.$$

The expectation in the middle is

$$\mathbf{E}\left(\lim_{r\to\infty} \sum_{s=0}^{2^r-1} e^{inx_s^r}\, Y(I_s^r) \lim_{r\to\infty} \sum_{s'=0}^{2^r-1} e^{-inx_{s'}^r}\, Y(I_{s'}^r)\right).$$

By dominated convergence, this is

$$\oint\oint f(\theta - \theta')\, e^{in(\theta-\theta')}\, d\theta\, d\theta',$$

implying that (since we know $f(x)$ is an even function of x)

$$\infty > \oint f(x) \cos nx\, dx \geqslant 0,$$

proving the first assertion of the theorem. If f is continuous, it is the uniform limit of its Fourier series, the coefficients in which are all non-negative; so f is non-negative definite.

In any case, if g is any continuous function of the angle x,

$$\infty > \oint\oint f(x-y)\, g(x)\, \bar{g}(y)\, dx\, dy \geqslant 0.$$

This implies that a mixed Poisson process has highest density of intersections (see Section 2.1.3 of Part II of my Ph.D. thesis [Reprinted without change of section numbering as 2.1 of this volume]) among *all* processes Z as above.

2.4

Construction of Line-processes: Second-order Properties

ROLLO DAVIDSON

2.4.1 SUMMARY

Let LP4 be the class of those random aggregates Z of lines in the Euclidean plane satisfying (*a*)–(*d*) below; let LP5 consist of those elements of LP4 which also satisfy (*e*).

(*a*) The number of lines of Z cutting any circle is a.s. finite.

(*b*) Z has a.s. no multiple lines.

(*c*) Z is locally square-summable (i.e. the mean-square number of lines of Z cutting any circle is finite) and second-order stationary under the translations and rotations of the plane.

(*d*) Z has a.s. no parallel (or antiparallel) lines.

(*e*) Z is strictly stationary under the translations and rotations of the plane.

A $Z \in$ LP4 is said to be a dsP (doubly-stochastic Poisson) process if it is got by putting a random Borel measure Λ on the manifold of lines, and then letting Z be an inhomogeneous Poisson process with rate measure Λ. The problem is: are all $Z \in$ LP5 dsP?

First we need a good description of the dsPZ; and it turns out that the Λ associated with any dsPZ is (in the (p, θ) representation of lines: p is the perpendicular distance of the line from the origin, and θ the angle between the perpendicular and a fixed axis) a product measure:

$$d\Lambda(p, \theta) = dp\, dY(\theta).$$

Now to show the existence of 'nice' non-dsP processes on the line or plane it suffices to show that the second-order behaviour of the dsP processes is not the most general possible for 'nice' processes. But alas, in our case, for each $Z^* \in$ LP4 there exists a dsP $Z \in$ LP5 with the same second-order behaviour (which is determined by that of the Y associated with Z; and we have an analytic description of that in terms of the 'squared' probability measures on the circle). However, it is possible to show that not all $Z \in$ LP4 are dsP.

We have tried two main ways of constructing non-dsP elements of LP5. The first (reminiscent of work on traffic distribution) is to take a point-process N on a fixed line and put lines through its points. This tends to break down because when we force stationarity on the resulting line-process it turns out (at least in the easy cases) to be dsP after all. But at least one can show that the second-order behaviour of N is that of a mixed Poisson process. The other main way is to associate lines with the points of a stationary point-process in the plane, but this violates (*c*).

At present, then, we think that everything in LP5 is dsP.

2.4.2 INTEGRAL GEOMETRY

Let w be an oriented line in the Euclidean plane R^2; then the standard coordinates of w are (p, θ), where $-\infty < p < \infty$ and θ is an angle. Here p is the perpendicular (signed) distance of w from some fixed origin O and θ is the angle made by this perpendicular with some fixed direction Ox. In Figure 1, $p > 0$; but if w' were parallel with w, the same distance from, and the other side of, O, we would have $p(w') = -p(w)$.

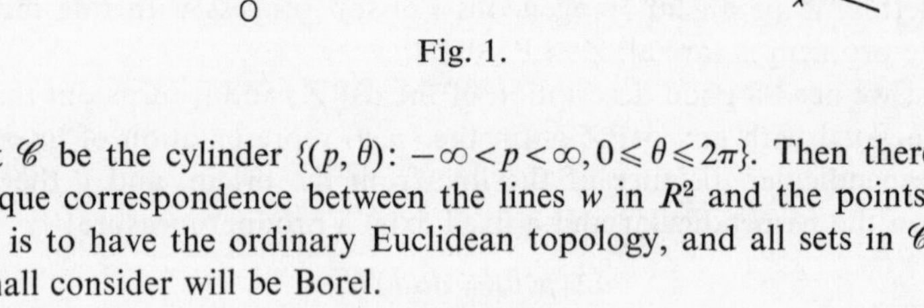

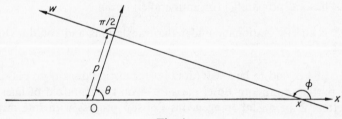

Fig. 1.

Let $\mathscr{C}$ be the cylinder $\{(p, \theta)\colon -\infty < p < \infty, 0 \leqslant \theta \leqslant 2\pi\}$. Then there is a biunique correspondence between the lines w in R^2 and the points w of $\mathscr{C}$. $\mathscr{C}$ is to have the ordinary Euclidean topology, and all sets in $\mathscr{C}$ that we shall consider will be Borel.

Let $\mathscr{M}^*$ be the group of rigid motions (translations and rotations) of R^2. Then each $T^* \in \mathscr{M}^*$ induces a motion T of $\mathscr{C}$. Let us write, for each positive q, $B_q = \{(p, \theta): |p| < q\}$.

Proposition 1. *Let $\mathscr{M}$ be the set of motions T of $\mathscr{C}$ as T^* runs through $\mathscr{M}^*$. Then $\mathscr{M}$ is the group generated by the motions*

$$R_\alpha: p \to p, \quad \theta \to \theta + \alpha \quad (0 \leqslant \alpha < 2\pi),$$
$$S_d: p \to p + d\cos\theta, \quad \theta \to \theta \quad (-\infty < d < \infty).$$

Proof. Immediate from the following observations:

(1) R_α corresponds to rotating Ox clockwise through the angle α;

(2) S_d corresponds to translating O a distance $-d$ along Ox;

(3) the two classes of motion above generate $\mathscr{M}^*$.

It is to be noted that R_α is a rotation of $\mathscr{C}$, and that S_d is a *parallel shear* of $\mathscr{C}$. That is, if $\mathscr{C}$ is slit along a generator and S_d is applied to the strip so obtained, the image of a line perpendicular to the axis of the strip is a sine curve.

Proposition 2. *No translation of $\mathscr{C}$ is a member of $\mathscr{M}$.*

Proof. Let the translation be $T = \{p \to p + q;\ \theta \to \theta\}\ (q \neq 0)$. Without loss of generality q is positive. Then consider the band $B_{q/2} = B$, say. It is clear that $B \cap T(B) = \emptyset$. On the other hand, B is the set of lines lying within the non-zero distance $q/2$ of O; so, under any $T' \in \mathscr{M}$, $T'(B)$ is the set of lines lying within distance $q/2$ of some point $\mathrm{O}' \in R^2$. Therefore, $B \cap T'(B) \neq \emptyset$, so that $T \neq T'$.

Proposition 3. (Crofton [4], Santaló [13]). *There is, up to positive factors, a unique positive Borel measure on $\mathscr{C}$ invariant under $\mathscr{M}$, and this measure (which we shall denote by m) can be taken to have the density $dp\,d\theta$.*

Returning to Figure 1, we now have, for almost all (w.r.t. m) w, the representation $w = (p, \theta) = (x, \varphi)$, where $x = p\sec\theta$, $\varphi = \theta + \pi/2$; x is the distance of the meet of w with Ox from O, and φ is the angle of intersection. We now have also $m(dw) = |\sin\varphi|\,dx\,d\varphi$.

2.4.3 LINE-PROCESSES

By a line-process Z we mean a non-negative integer-valued random Borel measure on $\mathscr{C}$, which satisfies

(*a*) For all positive q, $Z(B_q)$ is a.s. finite;

(*b*) Z has a.s. no atoms of mass greater than 1.

Thus Z corresponds to a random aggregate of lines in R^2, only finitely many of which cut any circle. We define further the conditions:

(*c*) $\mathbf{E}(Z^2(B_q))$ is finite for all finite positive q, and for all $T \in \mathscr{M}$,

$$\mathbf{E}(Z(A)) = \mathbf{E}(Z(TA)),$$
$$\mathbf{E}(Z(A)Z(B)) = \mathbf{E}(Z(TA)Z(TB)),$$

provided there is a q such that $A \cup B \subset \mathrm{B}_q$;

(*d*) Z has a.s. no parallel (or antiparallel) lines;

(*e*) the finite-dimensional distributions of Z are stationary under $\mathscr{M}$, that is,

$$\mathbf{P}\left(\bigcap_{i=1}^{k}(Z(A_i) = n_i)\right) = \mathbf{P}\left(\bigcap_{i=1}^{k}(Z(TA_i) = n_i)\right)$$

for all $k, n_1, \ldots, n_k, A_1, \ldots, A_k$ and $T \in \mathscr{M}$.

We shall study only those line-processes satisfying (*a*)–(*d*)—*these will form the class* LP4; *those members of* LP4 *satisfying also* (*e*) *will form the class* LP5.

We pause to look at (*d*). It will be clear throughout the paper that parallel lines are a pathology, essentially because we have a distance between them, so that we have all the complications of the theory of ordinary point-processes on the line. For LP4 the theory is quite different.

Proposition 4. *Any strictly stationary line-process Z (i.e. one that satisfies* (*e*)) *has a.s. no, or infinitely many, pairs of parallel lines.*

Proof. Let the event F = (there exists at least one pair of parallel lines) have positive probability. Let N be the process of points on the whole line Ox given by its intersections with those lines of Z which have at least one other line of Z parallel to them. Then N is strictly stationary (this may be verified by elementary calculations) and has at least two points in it. By Cauchy's inequality, $\mathbf{E}_F \exp - N([-2r, 2r[) \leqslant \mathbf{E}_F \exp - 2N([-r, r[)$ for any positive r. Using this doubling repeatedly, by bounded convergence we get

$$\mathbf{E}_F \exp - N(R) \leqslant \mathbf{P}(N([-r, r[) = 0 \mid F),$$

where R is the whole real line. Since $\mathbf{P}(.\mid F)$ is a probability measure, we obtain by continuity

$$\mathbf{E}_F \exp - N(R) \leqslant \mathbf{P}(N(R) = 0 \mid F) = 0,$$

by definition of F. Thus $N(R)$ is a.s. infinite given F.

Note. Ryll-Nardzewski [12] states effectively this result, but the method used here appears to be useful later.

The obvious example of an LP5 is the Poisson line-process, studied e.g. by Miles [8], [9], which has, for disjoint $A_1, \ldots, A_k$ on $\mathscr{C}$,

$$\mathbf{P}\left(\bigcap_{i=1}^{k} (Z(A_i) = n_i)\right) = \prod_{i=1}^{k} \left\{ e^{-\lambda m(A_i)} (\lambda m(A_i))^{n_i} / n_i! \right\},$$

where λ is a non-negative constant.

We may generalize this somewhat as follows. Let Λ be any random Borel measure on $\mathscr{C}$, satisfying (*a*) and (*c*) (with therein Λ for Z), and (*e*) if we desire to construct an LP5 rather than an LP4. Then, having sampled Λ, we put on an inhomogeneous Poisson process with local rate $\Lambda(dw)$, so that

$$\mathbf{P}\left(\bigcap_{i=1}^{k} (Z(A_i) = n_i)\right) = \mathbf{E}_{\Lambda} \prod_{i=1}^{k} \left\{ e^{-\Lambda(A_i)} (\Lambda(A_i))^{n_i} / n_i! \right\}.$$

We shall, of course, require the satisfaction of (*b*) and (*d*) for the resulting line-process Z; and it turns out that this imposes heavy conditions on Λ. Processes of the type just described are called *doubly stochastic Poisson* processes, and *we say that they form the class* dsP. The fundamental, and as yet unsolved, question is, *do there exist members of* LP5 − dsP? We shall return to this question in Section 2.4.6; meanwhile we shall investigate the second-order properties of the members of LP4 with this question in mind.

2.4.4 SECOND-ORDER PROPERTIES

In this section, Z is arbitrary in LP4. Define, for arbitrary bounded Borel sets A, B of $\mathscr{C}$, $\mu(A \times B) = \mathbf{E}Z(A)Z(B)$.

Proposition 5. (Krickeberg). *μ has a unique extension to a Borel measure on $\mathscr{C} \times \mathscr{C}$.*

Proof by countable additivity and square-summability of Z.

Theorem 1. *Every $Z \in$ LP4 is second-order stationary under reflections of R^2 (equivalently, under translations of $\mathscr{C}$).*

Note. This theorem, for the special case where Z possesses a second-order product-moment density, appears in my Cambridge Ph.D thesis (1967). It was proved in full generality by Krickeberg (1969). His proof (which proceeds by disintegration of μ) is much more sophisticated and elegant than ours here, but ours is thematic.

Proof of the theorem. It is clearly only necessary to consider the particular reflection in Ox. This, T_0 say, is given by $T_0(p, \theta) = (-p, \pi - \theta)$. It is required to prove that if A, B are bounded Borel sets, then

$$\mathbf{E}Z(A)Z(B) = \mathbf{E}Z(TA)Z(TB);$$

for it is easy to see that the pseudo-Haar measure on $\mathscr{C}$, and hence the first-order moment measure of Z, are invariant under T_0 (and translation of the cylinder, of course). By Proposition 5, it suffices only to consider the case where A and B are congruent rectangular shields on $\mathscr{C}$. That is, A has the form $[p, p+l[\times[\gamma, \delta[$ and B has a congruent form. It will be noticed that we are taking A and B to be half-open. By a rotation we may take

$$A = [p, p+l[\times[-\alpha-\beta, -\alpha[, \quad B = [q, q+l[\times[\alpha, \alpha+\beta[,$$

where $0 < \beta < \pi$. We suppose at first that there is no generator of $\mathscr{C}$ common to $\overline{A}$ and $\overline{\beta+\pi}$; equivalently, that $\max(|\alpha|, |\alpha+\beta|) < \pi/2$.

Let Q_d be the translation of $\mathscr{C}$ through d: $(p, \theta) \to (p+d, \theta)$. Then we at once verify that if we put $d = -(p+q+1)$, then $T_0 A^0 = Q_d B^0$ and $T_0 B^0 = Q_d A^0$, where the superscript ^{0} denotes that the interior is taken. Now what we shall demonstrate is stationarity under Q_d. Since the open rectangles generate the Borel sets, second-order stationarity under T_0 will follow from Proposition 5. Also it is clear that we could deduce stationarity under Q_d from that under T_0.

The idea of the proof is to approximate the effect of translation on A and B by splitting them up lengthwise and applying suitable shears to the pairs of split pieces. Define the distortion of the shear

$$(p, \theta) \to (p + d\cos(\theta - \alpha), \theta)$$

as $|d|$; this is the maximum displacement under the shear. For any shield C (that does not encircle $\mathscr{C}$), let $P(C)$ be its point furthest clockwise and with least (algebraic) p-value. For convenience, put $T = Q_d$.

Divide A and B each into $n = 2^m$ $(m \to \infty)$ congruent shields A_i, B_j, each of which has the same length, and $1/n$th the angular width, of the original shield. Let S_{ij} be the (unique) shear such that $P(S_{ij}TA_i) = P(B_j)$ and $P(S_{ij}TB_j) = P(A_i)$. This shear exists and is unique because the angular interval subtended by $\overline{A \cup B}$ is less than π. For the same reason, there is m so large that we could adjoin shields congruent to A_i, B_j on each side of A and B, and the angular interval subtended by the union of the augmented A and B would still be less than π. Let the augmenting, extreme shields be, without loss of generality, A_0 and B_{n+1}; and let $S_{0,n+1}$ (which again exists by the small-angular-interval argument) be defined similarly to S_{ij}. Let the distortion of $S_{0,n+1}$ be δ; it is easy to see that this is at least as large as the distortion of each S_{ij}.

Now we have to consider the sets $S_{ij}\,TA_i\,\Delta B_j$ ($= B_{ij}$, say) and $S_{ij}\,TB_j\,\Delta A_i$ ($= A_{ij}$, say). For we have

$$|\mu(TA\times TB)-\mu(A\times B)| = |\mathbf{E}Z(A)Z(B)-\sum_i\sum_j \mathbf{E}Z(TA_i)Z(TB_j)|$$

$$= |\sum_i\sum_j \mathbf{E}\{Z(A_i)Z(B_j)-Z(S_{ij}\,TB_j)Z(S_{ij}\,TA_i)\}|$$

by stationarity

$$\leqslant \sum_i\sum_j \mu(A_{ij}\times B_{ij}).$$

Now $A_{ij}\subset A'_i$, the disjoint union of two shields with least p-values those of the ends of A_i, comprising the same generators as A_i, half-open similarly to A_i (and so to A itself), and of length (each) $\beta\delta/2^m$. So

$$|\mu(TA\times TB)-\mu(A\times B)|\leqslant \sum_i\sum_j \mu(A'_i\times B'_j),$$

where the B'_j are similarly defined for B. Let

$$A' = \bigcup_{i=1}^{n} A'_i,\quad B' = \bigcup_{j=1}^{n} B'_j.$$

Then $|\mu(TA\times TB)-\mu(A\times B)|\leqslant\mu(A'\times B')$. But each of A' and B' decreases to the empty set as $m\to\infty$, so the right-hand side decreases to zero. Therefore the left-hand side must vanish, and we have stationarity in the special case where $\bar{A}$ and $\overline{B+\pi}$ have no generator in common.

In the general case, we again divide the shields A and B into n shields congruent to each other and of the same length as A and B. Then, in the previous notation,

$$\mathbf{E}Z(A)Z(B) = \mathbf{E}\sum_i\sum_j Z(A_i)Z(B_j) = \mathbf{E}(\Sigma+\Sigma'),$$

say, where Σ is taken over those i and j such that $\bar{A}_i$ and $\overline{B_j+\pi}$ have a generator in common, and Σ' is the remainder. Now consider Σ. Since the sets

$$\bigcup_{i,j}(A_i\times B_j)$$

involved in Σ decrease to the line segment

$$\{(w,w')\colon \theta' = \theta+\pi;\ p,p',\theta,\theta' \text{ lie within bounds fixed by } A \text{ and } B\}$$

in the product subspace $A\times B$ of $\mathscr{C}\times\mathscr{C}$; and since this line segment is a.s.

not charged by Z (because it has a.s. no parallel lines), $\sum \to 0$ a.s. Therefore,

$$\sum_i \sum_j Z(A_i) Z(B_j) - \Sigma'$$

decreases to zero a.s. as $n = 2^m \to \infty$. Taking expectations and using the previous result, we have the theorem.

Example. *If Z may charge the line $\{(w, w')\colon \theta' = \theta + \pi\}$, that is, Z admits antiparallel lines, Z need not be stationary under reflections. For let Z_1 be a Poisson process of oriented lines. For each $w \in Z_1$ introduce w' antiparallel to, and to the right of, w, at a distance d (fixed positive) from w. Let Z be the whole process so obtained*—the railway-line process. *Then Z satisfies all of* (a)–(e) *except* (d). *But if we reflect Z we get pairs of antiparallel lines which lie always on each other's left, instead of on the right; and it follows from this that Z is not stationary under reflections, either strictly or to the second order.*

We now associate to every $Z \in$ LP4 a Y-process, which will turn out to be a non-atomic random measure on the circle, second-order stationary under its rotations and strictly stationary if Z is.

Definition. Let $\mathscr{K}^*$ be the class of binary-rational endpointed, clockwise half-open intervals I of the circle K. For $I \in \mathscr{K}^*$, let $I^* \subset \mathscr{C}$ be the shield of base I and height 1; we suppose that I^* is closed below and open above. Let T be the translation of the cylinder through height 1. Define

$$Y(I) = \lim_{n\to\infty} \sum_{r=0}^{n} Z(T^r I^*)/(n+1).$$

This is a reasonable definition because, by Theorem 1, the sequence $Z(T^r I^*)$ is second-order stationary. By the assumption (c) on Z, we have that $\mathsf{E}Y^2(K)$ is finite; and it is clear, by taking suitable subsequences of $\{n \to \infty\}$, that $Y(.)$ is a monotone square-summable non-negative random set function on $\mathscr{K}^*$. It is then a trivial matter to deduce, via the Riesz–Markov theorem, that we can extend Y in a unique manner to a random Borel measure on K. It is also trivial that Y inherits the stationarity (under rotations) of Z.

Proposition 6. *Y has a.s. no atoms.*

Proof. There exists a sequence $\{n_k \to \infty\}$ such that for all $I \in \mathscr{K}^*$, Y is the a.s. limit of

$$\sum_{r=0}^{n} Z(T^r I^*)/(n+1);$$

and so we can restrict attention to those realizations of Z for which all the a.s. limits exist (there are only countably many of them). Y can still be extended to a Borel measure on K. Suppose that Y has an atom at θ_0. It is clear that the atoms of Z not lying on the generator $\{\theta = \theta_0\}$ do not contribute, via the a.s. $(C, 1)$ limits, to our atom of Y. Therefore the only way there can arise an atom at θ_0 is that there be infinitely many atoms of Z on the generator $\{\theta = \theta_0\}$, but this would mean that there was in Z a whole sheet of parallel lines, which contradicts (*d*).

We define the *intensity*, λ say, of Z, to be the mean number of points of Z in any set of m-measure 1; or equivalently, $\lambda = \mathbf{E}Z(B_1)/4\pi$.

***Theorem* 2.** *Given* λ, *the covariance measure* μ *of* Z *is determined by that,* ν *say, of* Y.

Proof. It suffices to determine μ for congruent shields (as usual half-open) A and B of common length a, in three cases:

(*i*) $\overline{I(A)} \cap \overline{I(B)} = \emptyset$;

(*ii*) $I(A) = I(B)$ or they are contiguous, but $A \cap B = \emptyset$;

(*iii*) $A = B$.

We may also, and do, demand that a be a binary rational.

In the first case we may, by shears and a translation and the limiting process used in the first part of Theorem 1, move A and B so that they coincide with the shields $a(I(A))^*$ and $a(I(B))^*$. In view of the definition of Y, and the fact that a is a binary rational, we have at once that

$$\mathbf{E}\{Z(A)\,Z(B)\} = a^2\,\mathbf{E}\{Y(I(A))\;Y(I(B))\}.$$

Using this, we may apply, in cases (*ii*) and (*iii*), the approximation method of the second part of the proof of Theorem 1, to obtain, in case (*ii*), $\mathbf{E}\{Z(A)\,Z(B)\} = a^2\,\mathbf{E}\{Y(I(A))\;Y(I(B))\}$; while in case (*iii*),

$$\mathbf{E}Z^2(A) = a^2\,\mathbf{E}Y^2(I(A)) + a\mathbf{E}Y(I(A)) = a^2\,\mathbf{E}Y^2(I(A)) + a\lambda|I(A)|,$$

where $|\,.\,|$ is Lebesgue measure on the circle of unit radius. It follows that once we know λ, the covariance measure of Z is determined by that of its Y.

Corollary. *Given any* $Z \in \mathrm{LP4}$, *there exists* $Z^* \in \mathrm{LP4} \cap \mathrm{dsP}$ *with the same intensity and covariance measure.*

Proof. Given Z we have Y on K. Given Y on K construct Z^* on $\mathscr{C}$ as follows: the rate measure of Z^* is to be $\Lambda = Y \times l$, where l is Lebesgue measure on the line; and Z^* is to be an inhomogeneous Poisson process with this rate measure. It is at once clear that Z^* satisfies (*a*)–(*c*). Further,

Z^* gives rise to the same Y that we (and Z) started with, so the covariance measures of Z and Z^* are identical (by Theorem 2 and the fact that their intensities are obviously the same). Now since Y is continuous so is the rate measure of Z^*; which means that a.s. there will be no parallel lines in Z^* (contrariwise, there would be if Y did possess atoms, but it doesn't). So $Z^* \in$ LP4 (and LP5 if Z is).

2.4.5 EXISTENCE OF $Z \in$ LP4 − dsP

***Proposition** 7. Let N be a pseudo-Poisson process on the line. Through each point of N put a line; the lines to have orientations independent of each other and N, and common density proportional to* $|\sin\varphi|$ *(see Figure* 1). *Let Z be the line-process so obtained. Then* $Z \in$ LP4 − dsP.

Proof. By a pseudo-Poisson process we mean a strictly stationary point-process on the line with stationary uncorrelated (but not independent) increments. Such processes have been constructed by Lee [6], Rényi [11] and Shepp [5]. Using the (x, φ) representation of the lines of Z, it is easy to calculate the intensity and covariance measure of Z: on $\mathscr{C}$, with the (p, φ) representation, these are

$$\mathbf{E}Z(A) = \lambda m(A), \quad \mathbf{E}\{Z(A)\,Z(B)\} = \lambda m(A \cap B) + \lambda^2 m(A)\,m(B),$$

where, if λ' is the expected number of points of N in an interval of length 1, $\lambda' = 4\lambda$.

It is immediately clear that $Z \in$ LP4. Suppose $Z \in$ dsP. Then by Theorem 3 below, N is mixed Poisson. But a mixed Poisson process is ergodic if and only if it is the Poisson process; and the pseudo-Poisson process N is ergodic but is, by construction, not Poisson.

Note. For reasons which will appear in Section 2.4.6, the Z constructed here does not belong to LP5.

***Theorem** 3. Let* $Z \in$ LP4 $\cap$ dsP. (*i*) *Let Z have rate measure* Λ. *Then there is a version of* Λ *which is a product measure:* $\Lambda = Y \times l$, *where l is Lebesgue measure on the line.* (*ii*) *If w is any line in the plane, then* $N = Z \cap w$ *is a mixed Poisson process; that is, there is a non-negative random variable v such that conditional on v, N is Poisson with rate v.*

Proof. We have from Theorem 2 that if A and B are congruent shields, then $\mathbf{E}\{Z(A)\,Z(B)\} = a^2\,\mathbf{E}\{Y(I(A))\;Y(I(B))\} + \mathbf{E}Z(A \cap B)$, where a is the common length of A and B. From this, and elementary calculations of the relation between the covariance measures of a doubly stochastic Z and its rate measure Λ, we find that $\mathbf{E}\{\Lambda(A)\,\Lambda(B)\} = a^2\,\mathbf{E}\{Y(I(A))\;Y(I(B))\}$. It follows at once that if A and B occupy the same generators, then

$$\mathbf{E}\{\Lambda(A)\,\Lambda(B)\} = \mathbf{E}\Lambda^2(A).$$

Immediately, since their mean values are also the same, we have

$$\Lambda(A) = \Lambda(B) \text{ a.s.}$$

Since Λ is continuous in mean square, this implies that we may choose a version of Λ that is a product measure of the form described. This proves (*i*).

Turning to (*ii*), we first show that $Z \in \text{dsP}$ implies $N \in \text{dsP}$ (for the line). Now we may represent—with probability 1, by stationarity—Z on $\mathscr{C}'$, the (x, φ) strip $\{-\infty < x < \infty, 0 < \varphi < \pi \text{ or } \pi < \varphi < 2\pi\}$. Further, Z is still constructed on $\mathscr{C}'$ by putting a random square-summable σ-finite Borel measure on $\mathscr{C}'$ and then putting on an inhomogeneous Poisson process with our random meaure as rate. Then N is obtained from Z by mapping the point (x, φ) down onto x. So we have Figure 2.

$$\begin{array}{ccc} \Lambda & \xrightarrow{\gamma} & Z \\ \zeta\big\downarrow & & \big\downarrow\zeta \\ P & \xrightarrow{\gamma'} & N \end{array}$$

Fig. 2.

Here Λ is the rate measure of Z. γ and γ' are the operations of taking the inhomogeneous Poisson process, and ζ is the operation of integrating over φ. P is then a square-summable σ-finite Borel measure on the line; and, by considering rectangles on the strip, we see easily that the diagram is (not a.s., but in distribution) commutative. Thus N is indeed doubly stochastic Poisson.

Clearly, N inherits the stationarity of Z; and the material motions of Z are

(1) $(x, \varphi) \to (x + d \tan\varphi, \varphi)$

(from translation of w perpendicular to its length);

(2) $(x, \varphi) \to (x + d, \varphi)$ (from translation of w along its length).

One shows, by methods similar to those of Theorem 1, that

(*i*) If A and B are two rectangles of C' such that there is no generator common to A and B, then for all real x we have

$$\mathbf{E}\{Z(A)\,Z(B)\} = \mathbf{E}\{Z(A)\,Z(B + x)\}.$$

(*ii*) If A and B are two bands (sections transverse to its generators) of C', then for all real x we have

$$\mathbf{E}\{Z(A)Z(B)\} - \lambda m(A \cap B) = \mathbf{E}\{Z(A)Z(B+x)\} - \lambda m(A \cap (B+x)).$$

(*iii*) If I and J are any two intervals of the line, then for all real x we have

$$\mathbf{E}\{N(I)N(J)\} - \lambda' |I \cap J| = \mathbf{E}\{N(I)N(J+x)\} - \lambda' |I \cap (J+x)|.$$

It is clear that (*iii*) is an immediate deduction from (*ii*), and that (*ii*) will follow, under (*d*), from (*i*) by the approximation method of the last part of the proof of Theorem 1. Again (*i*) may be proved using the motions (1) and (2) and the methods of the first part of the proof of Theorem 1.

Now that we have (*iii*), we may apply the proof of (*i*) of the present theorem to conclude that the measure P is such that for any two congruent intervals I and J on the line, $P(I) = P(J)$ a.s. Consequently, P is a random multiple of Lebesgue measure, which is equivalent to saying that N is mixed Poisson.

After this analysis there are two problems outstanding:

(1) *To find what restrictions are imposed on a random measure (second-order stationarity on the line, say) by niceness of its covariance measure.*

(2) *To characterize analytically the covariance measures of* LP4s (*see Theorem* 2).

The first problem is of considerable interest in its own right but is not on the theme of this paper; we treat the second. First we observe that this problem is equivalent, by Theorem 2, to characterizing covariance measures of square-summable second-order stationary non-atomic random measures Y on the circle K.

Theorem 4. (*i*) *μ (on K) is the kernel measure of the covariance measure of such a Y if and only if $\mu = \alpha\mu'$, $\alpha \geqslant 0$ a constant, where μ' has no atoms and lies in the convex set D' of probability measures generated by the set* Sqp′ *of squared probability measures without atoms on K. (A probability measure is called squared if it is of the form $\nu * \nu^*$, where $\nu^*(I) = \nu(-I)$ and ν is itself a probability measure.) (ii) Let $Z \in$ LP4. Then there exists $Z^* \in$ LP5 $\cap$ dsP having the same intensity and covariance measure, and Z^* is got as follows:*

(*a*) *take a certain random non-atomic probability measure on K;*

(*b*) *put an independent uniform rotation on this measure—call the result Y_0;*

(*c*) *take a random variable Λ, independent of the previous choices—Λ may take the values 0 and Λ_0, with probabilities q and $p = 1-q$;*

(*d*) *set* $Y = \Lambda Y_0$, *and let* Z^* *be the doubly stochastic Poisson process with rate measure* $Y \times 1$.

Before going through the proof, we observe that this theorem raises the question, 'When is a probability measure on K in D'?' and we naturally push this question back one stage, by asking 'When is a probability measure squared?' Neither of these questions has been answered (so far as I know), even for the more famous (but possibly more difficult) case where the measures are on the line. For that case, we may make the following remarks. Let $\mathbf{P}$ be a probability measure on the line and let f be its characteristic function. That $\mathbf{P}$ lie in the convex hull of the set of squared probability measures it is necessary that f be real and non-negative; and if $\mathbf{P}$ possesses atoms, one of them must be at the origin. $\mathbf{P}$ will lie in the convex hull in question if f has the form $f = \frac{2}{3} + \frac{1}{3}$ (any real ch.f.), and $\mathbf{P}$ will be a squared measure if f is real and infinitely divisible, or if f is the square of a Pólya characteristic function.

Proof of the theorem. We start with a Z and so also with its Y, which has a random Fourier sequence (F.S.) $\{b_n\}$, say,

$$b_n = \int e^{in\theta} Y(d\theta).$$

We see at once that $\mathbf{E}(b_0) = \lambda$, the intensity of the process, whereas $\mathbf{E}(b_n) = 0$ for $n \neq 0$. Let M be the covariance measure of Y. Then by second-order stationarity of Y we may disintegrate M as $M = \mu \times \kappa$, where κ is proportional to Lebesgue measure on K and μ is the kernel measure, also on K. Now we define the Fourier double sequence $\{a_{m,n}\}$ of M,

$$\begin{aligned} a_{m,n} &= \int\int e^{in\theta - im\varphi} M(d(\theta, \varphi)) \\ &= \int\int e^{in\theta - im\varphi} Y(d\theta)\, Y(d\varphi) \\ &= \mathbf{E}(b_n \bar{b}_m). \end{aligned}$$

Then, using the disintegration of M, it is clear that $a_{m,n}$ vanishes unless $m = n$, when we have $a_{n,n} = \mathbf{E}(|b_n|^2) = a_n(\mu)$, say. For by the disintegration we have also

$$a_{n,n} = \int e^{in\psi} \mu(d\psi).$$

We now have obvious conditions on λ, viz., that $\lambda^2 \leqslant a_0$ and $\lambda > 0$ if $a_0 > 0$. We shall see later that these are the only relations between λ and the F.S. of μ.

Let L be the set of all totally finite non-negative measures on K; let Lp be the set of those of L whose total mass is 1. Let $\mathrm{Sq} \subset L$ be the set of squared measures in L (the 'square roots' also lying in L), and let

$$\mathrm{Sqp} = \mathrm{Sq} \cap \mathrm{Lp}.$$

We give L the topology of ordinary convergence of the F.S.s. To put a prime (′) on any of these spaces is to restrict attention to the continuous measures in it. Let D be the closed convex hull of Sqp in L, and let D' be the closed convex hull of Sqp′ in L′ with the relative topology.

First we observe that $\mu \in L'$. For a.s. $Y \in L'$. Consequently, by the standard criterion for continuity of a measure on K, we have that

$$(C, 1) \lim |b_n|^2 = 0 \text{ a.s.}$$

Since Y is square-summable we may take expectations and interchange the limit and the expectation, to get $(C, 1) \lim a_n(\mu) = 0$. Since all the as are non-negative, we find that $(C, 1) \lim |a_n(\mu)|^2 = 0$, which means that μ is continuous.

We may suppose that μ is not the zero measure (which corresponds to a null process, so that the theorem is trivial). Let then $\mu' = \mu/a_0(\mu)$; then $\mu' \in \mathrm{Lp}'$. Also, by the representation of the F.S. of μ, we have that

$$\mu' = \int_{m \in \mathrm{Sq}'} m \mathbf{P}_0(dm),$$

where $\mathbf{P}_0$ is some probability measure on the Borel subsets of Sq′ (which is itself a Borel subset of $[0, \infty[\times \mathrm{Sqp}$, the latter being a compact metric space); and

$$a_0(\mu') = \int_{m \in \mathrm{Sq}'} a_0(m)\, \mathbf{P}_0(dm) = 1.$$

Now $\mathbf{P}_0$ is not concentrated on the null measure z. Let its atom there then be a, $0 \leqslant a < 1$. Define $\mathbf{P}_1$ on the Borel sets of Sq′ by

$$\mathbf{P}_1(A) = \mathbf{P}_0((1-a)A - \{z\})/(1-a).$$

Then we have

$$\mu' = \int_{m \in \mathrm{Sq}' - \{z\}} m \mathbf{P}_1(dm),$$

and

$$\int_{\mathrm{Sq}'-\{z\}} a_0(m)\,\mathbf{P}_1(dm) = (1-a)^{-1}\int_{\mathrm{Sq}'-\{z\}} a_0(m)\,\mathbf{P}_0(d(1-a)m)$$
$$= (1-a)^{-1}(1-a) = 1.$$

Now define $\mathbf{P}$ on Sqp′ by, if $A \subset \mathrm{Sqp}'$ (so that $A \times]0,\infty[\subset \mathrm{Sq}'$),

$$\mathbf{P}(A) = \int_{A\times]0,\infty[} a_0(m)\,\mathbf{P}_1(dm).$$

It is clear from the properties of $\mathbf{P}_1$ just proved that $\mathbf{P}$ is a probability measure. We assert that

$$\mu' = \int_{\mathrm{Sqp}'} m^*\,\mathbf{P}(dm^*).$$

To show this we have to prove that for all n,

$$a_n(\mu') = \int_{\mathrm{Sqp}'} a_n(m^*)\,\mathbf{P}(dm^*).$$

But the right-hand side of this

$$= \int_{\mathrm{Sq}'} a_n(m^*)\,a_0(m)\,\mathbf{P}_1(dm),$$

where m^* is m scaled to a probability,

$$= \int_{\mathrm{Sq}'} a_n(m)\,\mathbf{P}_1(dm),$$

by the definition of F.S. and the relations of m^* and m,

$$= a_n(\mu')$$

by properties of $\mathbf{P}_1$. So we have exhibited μ' as a probability mixture of elements of Sqp′. It follows at once that $\mu' \in D'$.

We have now proved the first part of Theorem 4(*i*). We shall now prove its (*ii*) and the last part of (*i*) simultaneously. By the first part of (*i*), starting from any $Z \in \mathrm{LP4}$ we end up with a λ, an a_0 and a μ', such that λ and a_0 satisfy the conditions given earlier. So now let us start with λ and a_0 satisfying those conditions, and $\mu' \in D'$.

Since $\mu' \in D'$, μ' also lies in D, the closed convex hull of Sqp in Lp, so that D is compact. Therefore (see e.g. Phelps [10]) there exists a probability measure h on Sqp representing μ'. But if h puts positive mass on Sqp − Sqp′, μ' would have an atom at $\theta = 0$. Thus h must be concentrated on Sqp′, that is, μ' is a probability mixture of elements $\nu * \nu^*$ of Sqp′.

Thus we may sample, with respect to h, a probability measure $\nu \in \mathrm{Lp}'$. We may then rotate it uniformly round K, obtaining a random strictly stationary probability measure Y_0 on K.

Now we have to deal with λ and a_0. Consider a random variable Λ taking the two values Λ_0 and 0 with probabilities p and $q = 1-p$ respectively. Then $\mathbf{E}\Lambda = p\Lambda_0$, $\mathbf{E}\Lambda^2 = p\Lambda_0^2$. So if we dilate Y_0 by Λ, we obtain $\mathbf{E}(\Lambda Y_0(K)) = p\Lambda_0$, $\mathbf{E}(\Lambda Y_0(K))^2 = p\Lambda_0^2$, since Y_0 has to be a probability measure and we assume that Λ and Y_0 are independent. Because of the conditions $\lambda^2 \leqslant a_0$, $\lambda > 0$ if $a_0 > 0$, we can solve the equations

$$p\Lambda_0 = \lambda, \; p\Lambda_0^2 = a_0$$

for $\Lambda_0 \geqslant 0$ and the probability p, so that $\Lambda_0 = a_0/\lambda$ and $p = \lambda^2/a_0$. (If a_0 or λ vanishes they both do and we may take $\Lambda_0 = 0, p = 1$.) Then if we put $Y = \Lambda Y_0$, we have $\mathbf{E}Y(K) = \lambda$, $\mathbf{E}Y^2(K) = a_0$; and in fact the kernel measure of the covariance measure of Y is, as desired, $a_0\,\mu' = \mu$. For the kernel measure of the covariance measure of Y_0, given ν, is computed as $\nu * \nu^*$, and then we take averages. Now if we put on Z^* as the doubly stochastic Poisson process with rate $Y \times 1$, $Z^* \in \mathrm{LP5}$ and has the same intensity and covariance measure as Z.

2.4.6 THE BIG PROBLEM

Do there exist elements of LP5 − dsP?

The relevance of the previous work to this is that it might have been possible to prove that the covariance measures of the class dsP did not exhaust those of LP5. However, as we have shown, this is not the case (it is the case for point-processes on the line: see, e.g., Bartlett [2]).

How might one try to construct elements of LP5?

(1) By taking a point-process on a fixed line and putting lines through its points.

(2) By taking a stationary point-process in the plane and putting lines through its points.

(3) By tinkering with a Poisson line-process.

We first look at (1). Clearly this is the general method of construction; but how are we to put the lines through the points of the point-process (which we call N, and which has, of course, to be strictly stationary under shifts of the line)?

Proposition 8. *Let $Z \in \mathrm{LP5}$ be constructed by method* (1), *where the lines are put through the points with orientations independent of each other and N,* *and with a common continuous density. Then N is mixed Poisson and* $Z \in \mathrm{dsP}$.

Proof. That N is mixed Poisson may be deduced from the work of Breiman [3], as strengthened by Thedéen [14]. It is easy to modify their work to show the following: let N be a spatially strictly stationary summable process of cars (points) on a road (line), with velocities independent of each other and the positions of the cars and having a common density which is a.e. continuous and bounded on compacta. Then letting N_t be the process of cars as observed at time $t \geqslant 0$ ($N_0 = N$), N_t converges in laws of finite-dimensional distributions to a mixed Poisson process as $t \to \infty$.

Applying this to our N, and observing that Z has to be stationary under translations of the fixed line through distances t perpendicular to its length, N (which is summable because Z is) must be mixed Poisson. By construction, then, $Z \in \mathrm{dsP}$.

It is thus difficult to see how we should assign the orientations to the points of N to obtain a $Z \in \mathrm{LP5} - \mathrm{dsP}$; and this is as far as I have been able to take method (1). But in any case, we have the condition on N given in

Proposition 9. *N has a second-order product-moment density which is a constant.*

Proof. The second-order product-moment density is defined (see Bartlett [1]) by $g(x,y) = \lim \mathbf{E}N(I)\,N(J)/|I|\,|J|$, where I and J are congruent intervals with centres, and shrinking down to, x and y respectively ($x \neq y$). (Of course if g turns out to exist and be a constant it can be defined for $x = y$ by continuity.) But from Theorem 3(*ii*) we know that so soon as I and J are so small that they are disjoint, the numerator of the limit in the definition of $g(x,y)$ is a constant times the product of the lengths of I and J. It is thus clear that $g(x,y)$ exists for N and is a constant. Now admittedly Theorem 3(*ii*) only applies to $Z \in \mathrm{dsP}$, but by Theorem 2 the covariance behaviour of these Z exhausts that of all $Z \in \mathrm{LP4}$, and so also that of all $Z \in \mathrm{LP5}$.

We now turn to method (2). Let $(P, T) = \{x_i, \theta_i\}$ be a marked point-process (in the sense of Matthes [7]) in the plane, the x_i being the points of P and the θ_i being the orientations assigned to them. We identify $Z \in \mathrm{LP5}$ with (P, T) by constructing Z with lines going through the x_i with the orientations θ_i. We assume that (P, T) is strictly stationary under the rigid motions of the plane, in which case P is also stationary under these motions. Now P has to be locally square-summable, otherwise Z would certainly not be square-summable (consider those lines whose parent points lie in a convex compact region of non-square-summability of P). Consequently P is well distributed in the sense of Goldman [5]. We assume throughout that P has a.s. infinitely many points.

Proposition 10. *If T is a process of uniform orientations independent of each other and P, then Z puts a.s. infinitely many lines through each circle.*

Proof. Let the circle be of radius $r>0$ and centre the origin O. Let the points of P lie at distances $r_1 \leqslant r_2 \leqslant r_3 \leqslant \dots$ from O. Then, the Ps being well distributed, there exist positive constants (conditional on P) h and $k<h$ such that for all $n, |n^{-\frac{1}{2}} r_n - h| < k$.

Now the probability that the line whose parent point lies at a distance $d>r$ from O will pass through our circle is $(1/\pi)\sin^{-1}(r/d) \sim (r/\pi d)$ as $d \to \infty$. But the incidences of different lines on our circle are independent; so we may apply the divergence case of the Borel–Cantelli lemmas, with

$$\sum_{n=1}^{\infty} \mathbf{P}_n = \sum_{n=1}^{\infty} (1/\pi)\sin^{-1}(r/r_n) \geqslant H \sum_{n=1}^{\infty} n^{-\frac{1}{2}} \quad \text{for some } H = H(h,k) > 0$$

$$= \infty,$$

so that infinitely many of the lines of Z hit our circle.

We now look at the general case. It is clear, by simple addition, that the mean number of lines cutting any circle is infinite, so that method (2) cannot yield an LP4. But the problem of whether the actual number cutting any circle is infinite (even with positive probability) is of independent interest. We might proceed as follows.

The problem for the circle is clearly equivalent to that for a finite non-empty line segment, J say, forming part of some line w (say) in the plane. We divide the process Z up into subprocesses Z_n consisting of those lines of Z whose parent points lie between the distances n (inclusive) and $n+1$ (exclusive) from w. Those lines of Z whose parent points lie below (at negative distances from) w are from now on disregarded; if we can do without them so much the better. Then each Z_n is a marked point-process $\{x_i, y_i, \theta_i\}$ stationary under shifts of x, where x_i is the abscissa along w, y_i is the ordinate (lying in $[0,1[$), and θ_i is the orientation assigned. Further, the Z_n have identical (but of course not independent) distributions. Let X, X_n be the Z-, Z_n-charges of J; let

$$Y_n = \sum_{r=0}^{n-1} X_r.$$

Then for all $t>0$

$$\mathbf{E}e^{-tX} = \lim_{n\to\infty} \mathbf{E}e^{-tY_n} = \lim_{n\to\infty} \mathbf{E}e^{-t\sum_{r=0}^{n-1} X_r} \leqslant \lim_{n\to\infty} \left(\prod_{r=0}^{n-1} \mathbf{E}e^{-ntX_r}\right)^{1/n},$$

by repeated use of Hölder's inequality.

Now since all the Xs are non-negative integers, we have

$$\mathbf{P}(X_r = 0) = \mathbf{P}_r \text{ (say)} \leqslant \mathbf{E}e^{-ntX_r} \leqslant \mathbf{P}_r + e^{-nt}.$$

So

$$\mathbf{E}e^{-tX} \leqslant \lim_{n\to\infty} \left(\prod_{r=0}^{n-1} (\mathbf{P}_r + e^{-nt})\right)^{1/n}.$$

Suppose that (and here is the gap) we have $\mathbf{P}_r \leqslant 1-c$ $(c>0)$ uniformly in r; then the formula above gives

$$\mathbf{E}e^{-tX} \leqslant \lim_{n\to\infty} ((1-c+e^{-nt})^n)^{1/n}$$

$$\to 1-c, \quad \text{independent of } t>0.$$

It would follow at once, from the discontinuity of the Laplace-Stieltjes transform at the origin, that X was infinite with the positive probability c.

Now observe the point where there was the gap. In fact we do not there need that $\mathbf{P}_r$ should be bounded away from unity; it is sufficient that $\{\mathbf{P}_r\} \not\to 0$ as $r\to\infty$, that is, that the X_r do not converge to zero in probability. Of course in the case discussed in Proposition 8 we actually have convergence of the $\mathbf{P}_r$ to a non-zero limit; and there are other cases, e.g. when equal orientations are permitted, where we can bridge the gap. Even in the general case the result required appears likely enough; even more so when we consider a very slightly modified version of the problem in different terms. We omit the ordinates y_i from the marked point-processes Z_n. Then Z_0 may be regarded as an array of cars on a road at time zero, spatially stationary and with a stationary array of speeds which remain constant in time. Z_n is then the same process of cars observed at time n, at least in distribution, and X_n is the number of cars in a fixed interval J of road at time n. Or we may take X_n to be the number of cars that pass a fixed observer between times n and $n+1$; one does not thereby change the problem of whether X_n converges to zero in probability.

Turning now to method (3), an obvious way of constructing non-doubly stochastic Poisson processes in R^1 or R^2 is to take a Poisson process and then modify it in some way. For example, one may attach new points to the old ones or to selected clusters of old ones, or subtract points from clusters, or change the geometry of clusters, and so on. Now the first of these methods only works because of the compactness of the group of rotations about the 'old point', so that we can make a coordinate-free stationary assignment of the new points to the old. With lines that have to be skew, this cannot be done. For clusters, the situation is vaguer, but the general trouble is that any line will appear in infinitely many clusters, so that if we are deleting lines independently from each cluster they will all disappear. On the other hand, if we are adding lines, it seems that clusters may be associated with points of R^2, and then—since these points will form a stationary process—the troubles of method (2) arise.

Thus we think at the moment that dsP does exhaust LP5. If this is so, of course, we may use Miles' results on distributions associated with the Poisson process to get expressions for the same distributions associated with any $Z \in \text{LP5}$. For example, let ρ be the experimental intensity of Z:

$$\rho = \lim_{r \to \infty} Z(B_r)/4\pi r.$$

Let δ be the diameter of the incircle of a random polygon; then $\delta\rho$ has an exponential distribution with mean 2 (see [8]).

I am greatly indebted to Professor D. G. Kendall for proposing to me the topic of line-processes; and to Professor K. Krickeberg for his interest in the problems discussed here, leading to the decisive first proof of Theorem 1. Also I am very grateful to them and to Dr. F. Papangelou for many fruitful conversations.

REFERENCES

1. M. S. Bartlett, *An Introduction to Stochastic Processes*, Cambridge University Press (1966).
2. ——, "The spectral analysis of point processes", *J. R. statist. Soc.* B **25** (1963), 264–296.
3. L. Breiman, "The Poisson tendency in traffic distribution", *Ann. math. Statist.* **34**, (1963), 308–311.
4. M. W. Crofton, "Probability", *Encyclopaedia Britannica*, IX edition (1885), vol. XXI.
5. J. R. Goldman, "Stochastic point processes: limit theorems", *Ann. math. Statist.* **38** (1967), 771–779; appendix by L. Shepp.
6. P. M. Lee, "Some examples of infinitely divisible point processes", *Studia Sci., Math. Hungarica* **3** (1968), 219–224.
7. K. Matthes, "Stationäre zufällige Punktfolgen I", *Jahresbericht der D.M.V.* **66** (1963), 66–79.
8. R. E. Miles, "Random polygons determined by random lines in a plane, I, II", *Proc. Nat. Acad. Sci.* **52** (1964), (4) 901–907, (5) 1157–1160.
9. ——, "Poisson flats in Euclidean space", to appear in *Advances in Applied Probability*.

10. R. R. Phelps, *Lectures on Choquet's Theorem*, Van Nostrand, Princeton (1966).
11. A. Rényi, "Remarks on the Poisson process", *Studia Sci., Math. Hungarica* **2** (1967), 119–123.
12. C. Ryll-Nardzewski, "Remarks on processes of calls", *Proc. 4th Berkeley Symposium*, vol. 2, University of California Press (1961).
13. L. A. Santaló, "Introduction to integral geometry", *Act. Sci. Ind.* 1198, Hermann, Paris (1953).
14. T. Thedéen, "A note on the Poisson tendency in traffic distribution", *Ann. math. Statist.* **35** (1964), 1823–1824.

2.5

Invariance Properties of the Correlation Measure of Line-processes

KLAUS KRICKEBERG

2.5.1 INTRODUCTION

The invariance under reflections of the correlation measure of a second-order stationary line process in the plane with no parallel or antiparallel lines was proved by R. Davidson in his Ph.D. thesis under the assumption that the correlation measure has a density relative to the relevant integral geometric measure outside the diagonal ([5]; see also [6]). The purpose of this note is to show how this result can be obtained in full generality from the familiar theory of the disintegration of measures. The method obtained here works in similar cases as well.

I am very much indebted to R. Davidson for making available to me his thesis and for many stimulating discussions, to D. G. Kendall and the British Council for arranging our first encounter, and to U. Krause for a remark which influenced the organization of the paper.

2.5.2 IMAGES OF DISINTEGRATED MEASURES

Let Y be a locally compact space which has a countable base, and ν a positive Radon measure on Y. Consider an equivalence relation $\sim$ in Y and assume that $\sim$ is ν-measurable. We can then select a locally compact space Γ with a countable base and a ν-measurable map r of Y onto Γ such that $x \sim y$ if and only if $r(x) = r(y)$; in fact, the existence of such a pair Γ, r is necessary and sufficient for the ν-measurability of $\sim$ [3, § 3, No. 4]. For any $\gamma \in \Gamma$ we denote the equivalence class in Y described by γ, i.e. $r^{-1}\{\gamma\}$, by Y_γ. We know [3, § 3, No. 2] that there is a pseudo-image of ν under r; by definition this is a positive Radon measure κ on Γ such that a subset Δ of Γ is a κ-null set if and only if $r^{-1}\Delta$ is a ν-null set. For fixed κ there exists by [3, §3, No. 3] a scalarly κ-integrable family of positive measures $(\nu_\gamma)_{\gamma\in\Gamma}$ on Y with the property that ν_γ is carried by Y_γ for any $\gamma\in\Gamma$, and

$$\nu = \int_\Gamma \nu_\gamma\, \kappa(d\gamma). \tag{1}$$

If κ' stands for any other pseudo-image of ν under r and $(\nu'_\gamma)_{\gamma\in\Gamma}$ for any scalarly κ'-integrable family of positive Radon measures on Y such that ν'_γ is also carried by Y_γ and

$$\nu = \int_\Gamma \nu'_\gamma\, \kappa'(d\gamma),$$

then we have $\nu_\gamma = \rho(\gamma)\,\nu'_\gamma$ for κ-almost all γ where $\rho = d\kappa'/d\kappa$ denotes the Radon–Nikodym derivative of κ' with respect to κ.

Next let V be a map of Y into itself which is measurable and proper with respect to ν. This amounts to the fact that $V^{-1}K$ is ν-integrable for every compact subset K of Y, and it suffices to require this for all elements K of a base of Y consisting of compact sets [2, §6, No. 1]. For fixed K the set $V^{-1}K$ will be ν_γ-integrable for κ-almost all γ [2, §3, No. 3], hence for κ-almost all γ the map V will be measurable and proper with respect to ν_γ, too. We now assume that there is a κ-measurable and κ-proper bijective map v of Γ onto itself with the property that $VY_\gamma \subseteq Y_{v(\gamma)}$ for κ-almost all γ, and that the image κ' of κ under v has the same null sets as κ. Thus κ' is a pseudo-image of ν under r.

Let ν' be the image of ν and $\nu'_{v(\gamma)}$ the image of ν_γ under V. Taking any continuous function f on Y with a compact carrier we have by equation (1):

$$\nu'(f) = \nu(f\circ V) = \int_\Gamma \nu_\gamma(f\circ V)\,\kappa(d\gamma) = \int_\Gamma \nu'_{v(\gamma)}(f)\,\kappa(d\gamma)$$

$$= \int_\Gamma \nu'_\gamma(f)\,\kappa'(d\gamma),$$

hence

$$\nu' = \int_\Gamma \nu'_\gamma\,\kappa'(d\gamma). \tag{2}$$

Here, $(\nu'_\gamma)_{\gamma\in\Gamma}$ is a scalarly κ'-integrable family of Radon measures on Y, and ν'_γ is carried by Y_γ.

Consider an analogous ν-measurable and ν-proper map W of Y into itself and a κ-measurable and κ-proper bijective map w of Γ onto itself with the properties that $WY_\gamma \subseteq Y_{w(\gamma)}$ for κ-almost all γ, and that the image κ''

of κ under w has the same null sets as κ. Let ν'' be the image of ν and $\nu''_{w(\gamma)}$ the image of ν_γ under W, and set $\rho = d\kappa'/d\kappa''$. Then equation (2) implies immediately

Theorem 1. *The images ν' and ν'' of ν under V and W, respectively, coincide if and only if $\nu''_\gamma = \rho(\gamma)\,\nu'_\gamma$ for κ-almost all γ.*

Recall that ν is said to be invariant under V if $\nu' = \nu$. Taking for W the identity we obtain

Corollary 1. *Under the assumptions made on V before, ν is invariant under V if and only if $\nu_\gamma = \rho(\gamma)\,\nu'_\gamma$ for κ-almost all γ where $\rho = d\kappa'/d\kappa$. In particular, assuming one of the conditions '$\nu_\gamma = \nu'_\gamma$ for κ-almost all γ' and '$\kappa = \kappa'$' to be satisfied, the other one is necessary and sufficient for the invariance of ν under V.*

Specializing once more to the case where v is the identical map of Γ we get

Corollary 2. *Suppose that $VY_\gamma \subseteq Y_\gamma$ for κ-almost all γ. Then ν is invariant under V if and only if ν_γ is invariant under V for κ-almost all γ.*

We now turn to the case where the equivalence relation is defined by a group $\mathscr{L}$ of transformations of Y, i.e. $x \sim y$ if and only if $Tx = y$ for some $T \in \mathscr{L}$. We are also going to change slightly our point of view by directing our attention not only to a single measure ν but rather to the set of all measures invariant under $\mathscr{L}$. Thus let $\mathscr{L}$ be a locally compact group with a countable base which acts continuously on Y [1, §2, No. 4].

We assume that there exists a locally compact space Γ with a countable base and a map r of Y onto Γ such that $x \sim y$ if and only if $r(x) = r(y)$, and such that a subset Δ of Γ is Borelian if and only if $r^{-1}(\Delta)$ is Borelian in Y. Note that if the quotient space $Y/\!\sim$ is separated in the quotient topology, we may take this space for Γ and the canonical mapping of Y onto Γ for r [1, §2, No. 4 and §4, No. 5]; however, in the main application we have in mind the quotient topology will not be separated. Under the preceding assumption, given any positive Radon measure ν on Y, the map r is ν-measurable, and every $T \in \mathscr{L}$, being a homeomorphism of Y, is ν-measurable and ν-proper. Each Y_γ is a Borel set in Y.

A positive Radon measure ν on Y is said to be $\mathscr{L}$-invariant if it is invariant under every $T \in \mathscr{L}$. By Corollary 2 of Theorem 1, every $\mathscr{L}$-invariant measure ν has a representation (1) with some positive Radon measure κ on Γ where, given $T \in \mathscr{L}$, the measure ν_γ is concentrated on Y_γ and invariant under T for κ-almost all γ. This amounts to $\nu_\gamma f = \nu_\gamma(f \circ T)$ for κ-almost all γ and every continuous function f on Y with a compact carrier. Since $\nu_\gamma(f \circ T)$ is, for fixed f of this kind, a continuous function of $T \in \mathscr{L}$ [4, § 1, No. 1] and $\mathscr{L}$ contains a countable dense set, we find that, for κ-almost all γ, the measure ν_γ is invariant under $\mathscr{L}$. Conversely, having selected any positive Radon measure κ on Γ and a scalarly κ-integrable family $(\nu_\gamma)_{\gamma \in \Gamma}$ of measures on Y such that, for κ-almost all γ, ν_γ is concentrated on Y_γ and invariant under $\mathscr{L}$, the measure ν defined by (1) is invariant under $\mathscr{L}$.

We have thus obtained a kind of survey on all $\mathscr{L}$-invariant measures on Y. This survey will turn out especially simple and useful if we make the following assumption: for every γ there is, up to a non-negative factor, one and only one $\mathscr{L}$-invariant positive Radon measure τ_γ concentrated on Y_γ. Then, unless $\tau_\gamma = 0$, every non-empty subset of Y_γ which is open in Y_γ has a positive τ_γ-measure.

We assume that there exists a non-negative bounded Borelian function h on Y with the following properties: for any $\gamma \in \Gamma$, the set $Y_\gamma \cap \{h > 0\}$ contains a non-empty subset which is open in Y_γ; for every compact subset Δ of Γ, the set $r^{-1}(\Delta) \cap$ carrier (h) is relatively compact. This assumption will obviously be satisfied in the later examples. Then $0 < \tau_\gamma(h) < \infty$ for every γ for which $\tau_\gamma \neq 0$, and by renormalizing τ_γ we can assume that for every γ we have $\tau_\gamma(h) = 1$ or $\tau_\gamma(h) = 0$. The set $\{x\colon x \in Y, \tau_{r(x)} = 0\}$ is null for every $\mathscr{L}$-invariant measure on Y, and could as well be discarded.

Now let ν be an $\mathscr{L}$-invariant positive Radon measure on Y, and κ a pseudo-image of ν under r. Starting with the decomposition (1) we find for κ-almost all γ a number $\rho(\gamma) \geqslant 0$ such that $\nu_\gamma = \rho(\gamma)\,\tau_\gamma$. Given any compact subset Δ of Γ the function $g(x) = h(x)\,1_\Delta(r(x))$ is ν-integrable with

$$\nu(g) = \int_\Gamma \nu_\gamma(g)\,\kappa(d\gamma) = \int_\Gamma \tau_\gamma(h)\,\rho(\gamma)\,\kappa(d\gamma),$$

where $\tau_\gamma(h) = 1$ for κ-almost all γ. Hence ρ is a locally κ-integrable function on Γ, and the measure $\kappa' = \rho\kappa$ is a pseudo-image of ν under r. Writing again κ instead of κ' we obtain the decomposition

$$\nu = \int_\Gamma \tau_\gamma\,\kappa(d\gamma), \tag{3}$$

where κ is unique. Conversely, given any positive Radon measure κ on Γ such that the family $(\tau_\gamma)_{\gamma \in \Gamma}$ is scalarly κ-integrable, the measure ν defined by (3) is $\mathscr{L}$-invariant. Thus we have

Theorem 2. *Under the assumptions made on $\mathscr{L}$ before, the formula* (3) *yields a one-to-one correspondence between $\mathscr{L}$-invariant measures ν on Y and measures κ on Γ such that $(\tau_\gamma)_{\gamma \in \Gamma}$ is scalarly κ-integrable.*

Combining this with Corollary 1 of Theorem 1 we get the

Corollary. *Suppose that the preceding assumptions on* $\mathscr{L}$ *are satisfied. Let* V *be a homeomorphism of* Y *which induces a bijective transformation* v *of* Γ *such that* $VY_\gamma = Y_{v(\gamma)}$ *for every* γ, *or in other words* $r \circ V = v \circ r$. *Then* v *and* v^{-1} *are Borelian. Suppose in addition that for every* γ, *the measure* $\tau_{v(\gamma)}$ *is the image of* τ_γ *under* V. *Then an* $\mathscr{L}$*-invariant measure* ν *on* Y *represented in the form* (3) *is invariant under* V *if and only if the corresponding measure* κ *on* Γ *is invariant under* v.

In fact, the assertion that v and v^{-1} are Borelian follows immediately from our assumptions on r; hence v and v^{-1} are κ-measurable for every Radon measure κ on Γ. To apply the Corollary 1 of Theorem 1 in the case where ν is invariant under $\mathscr{L}$ we need to know that v is κ-proper and preserves κ-null sets. The first statement results easily from the formula $\kappa(\Delta) = \nu(h.1_{r^{-1}(\Delta)})$ where Δ is any Borel set in Γ; this leads, by the way, directly to the ν-invariance of κ. The second statement holds even if V is only assumed to preserve ν-null sets.

An important type of mappings to which the preceding corollary applies is the following one: let V be a homeomorphism of Y such that $VTV^{-1} \in \mathscr{L}$ for every $T \in \mathscr{L}$. Then $y \sim y'$ entails indeed $Vy \sim Vy'$, and hence there is a bijective transformation v of Γ such that $r \circ V = v \circ r$.

Consider next the particular case where V commutes with every T of $\mathscr{L}$, let ν_γ be any $\mathscr{L}$-invariant positive Radon measure concentrated on Y_γ, and $\nu'_{v(\gamma)}$ its image under V. Then for every $T \in \mathscr{L}$ and every continuous function f on Y with a compact carrier we have

$$\nu'_{v(\gamma)}(f \circ T) = \nu_\gamma(f \circ T \circ V) = \nu_\gamma(f \circ V \circ T) = \nu_\gamma(f \circ V) = \nu'_{v(\gamma)}(f),$$

i.e. $\nu'_{v(\gamma)}$ is also $\mathscr{L}$-invariant. Thus, assuming that for every γ there exists, up to a factor, only one $\mathscr{L}$-invariant positive Radon measure concentrated on Y_γ we find that $\nu'_{v(\gamma)}$ is a multiple of $\nu_{v(\gamma)}$. This shows that our hypothesis $\tau'_{v(\gamma)} = \tau_{v(\gamma)}$ is not an unrealistic one; it is, in fact, only a hypothesis regarding the normalization of the τ'_γs. If the normalizing function h is itself V-invariant, we get

$$\tau'_{v(\gamma)}(h) = \tau_\gamma(h \circ V) = \tau_\gamma(h) = 1$$

unless $\tau_\gamma = 0$, hence automatically $\tau'_{v(\gamma)} = \tau_{v(\gamma)}$. Given a finite group of homeomorphisms $V_1, \ldots, V_n$ of Y which commute with every $T \in \mathscr{L}$, we can always choose h so as to be invariant under every V_i by replacing it by

$$\sum_{i=1}^{n} h \circ V_i.$$

In particular, h can be chosen V-invariant if V is periodic.

2.5.3 A PARTICULAR CASE: A DIAGONAL GROUP IN A PRODUCT SPACE

Let X be a locally compact space with a countable base and $\mathscr{M}$ a locally compact group with a countable base which acts continuously on X. We fix a positive integer k and define Y to be the product space X^k. For $\mathscr{L}$ we take the group of all transformations of the form

$$(x_1, \ldots, x_k) \to (T_0 x_1, \ldots, T_0 x_k)$$

where $T_0 \in \mathscr{M}$. Then $\mathscr{L}$ acts continuously on Y and has a countable base, the topology of $\mathscr{L}$ being given in the obvious way by that of $\mathscr{M}$.

As before we assume that we have a locally compact space Γ with a countable base and a map r of Y onto Γ which 'represent' $Y/\sim$ in the sense previously defined.

An example of a homeomorphism V which commutes with every $T \in \mathscr{L}$ is given by

$$V(x_1, \ldots, x_k) = (x_{i_1}, \ldots, x_{i_k}),$$

where $i_1, \ldots, i_k$ is a fixed permutation of the numbers $1, \ldots, k$. We can choose the normalizing function h to be invariant under every mapping of this kind.

Another type of mappings are the diagonal ones:

$$V(x_1, \ldots, x_k) = (V_0 x_1, \ldots, V_0 x_k)$$

where V_0 is a homeomorphism of X such that $T_0 \in \mathscr{M}$ implies $V_0 T_0 V_0^{-1} \in \mathscr{M}$. Obviously $T \in \mathscr{L}$ implies $VTV^{-1} \in \mathscr{L}$. If V_0 commutes with every $T_0 \in \mathscr{M}$, V commutes with every $T \in \mathscr{L}$, and if V_0 is periodic, V is periodic.

Applying the corollary of Theorem 2 we obtain

Proposition 1. *Let V be any transformation of Y of the kind described before, and v the corresponding transformation of Γ. Suppose that for every γ there is, up to a factor, one and only one $\mathscr{L}$-invariant positive Radon measure τ_γ concentrated on Y_γ, and that $\tau_{v(\gamma)}$ is the image of τ_γ under V. Then an $\mathscr{L}$-invariant measure ν on Y represented by equation* (3) *is invariant under V if and only if the corresponding measure κ on Γ is invariant under v.*

2.5.4 MIXED MOMENTS OF RANDOM MEASURES

We start with a space X, a positive integer k, and a group $\mathscr{M}$ as in Section 2.5.3. We denote by $\mathscr{B}(X)$ the sigma-ring of all Borel sets, and by $\mathscr{B}_0(X)$ the set of all relatively compact Borel sets of X. In addition we take a probability space $(\Omega, \mathscr{F}, \mathbf{P})$ to be kept fixed once for all, and let $\mathbf{E}$ stand for 'expectation'. As usual we mean by a positive random measure on X a map z of $\mathscr{B}(X)$ into the set of all non-negative extended real-valued random variables with the following property: if (A_n) is a sequence of disjoint sets of $\mathscr{B}(X)$ and

$$A = \bigcup_n A_n,$$

then we have

$$z(A) = \sum_n z(A_n)$$

almost surely. Since $z(A_n) \geqslant 0$ it suffices to require only stochastic convergence.

The random measure z is said to be locally finite if $z(A) < \infty$ almost surely for any $A \in \mathscr{B}_0(X)$. It is called locally of kth order if $\mathbf{E}(z(A)^k) < \infty$ whenever $A \in \mathscr{B}_0(X)$. Assuming this the mixed moments of kth order of the random variables $z(A_1), \ldots, z(A_k)$ exist for arbitrary sets $A_1, \ldots, A_k$ in $\mathscr{B}_0(X)$. The function

$$\nu(A_1 \times \ldots \times A_k) = \mathbf{E}\left(\prod_{i=1}^{k} z(A_i)\right) \tag{4}$$

defined on the semiring $\mathscr{S}$ of all sets of the form $A_1 \times \ldots \times A_k$ where $A_i \in \mathscr{B}_0(X)$ is additive, and it follows from Fatou's lemma that

$$\nu(B) = \lim_{n \to \infty} \nu(B_n)$$

for any increasing or decreasing sequence (B_n) in $\mathscr{S}$ which converges to a set B in $\mathscr{S}$. We can then show in the usual way [7, p.56] that

$$\nu(B) = \sum_n \nu(B_n)$$

if (B_n) is a sequence of disjoint sets in $\mathscr{S}$, and

$$B = \bigcup_n B_n \in \mathscr{S}.$$

Hence [7, §2] ν can be uniquely extended to a positive Radon measure in $Y = X^k$ which will again be denoted by ν.

The measure ν is symmetric, i.e. invariant under any 'permutation of the axes' of the type W studied in Section 2.5.3. Of course, not every symmetric measure arises in this way, as shown by the measure in the space $Y = \{0, 1\}^2$ which consists of two unit masses at the points $(0, 1)$ and $(1, 0)$.

The random measure z is called stationary to the kth order if it is locally of kth order, and

$$\mathbf{E}\left(\prod_{i=1}^{l} z(TA_i)\right) = \mathbf{E}\left(\prod_{i=1}^{l} z(A_i)\right)$$

for any $l \leqslant k$, any sets $A_1, \ldots, A_l \in \mathscr{B}_0(X)$ and any $T \in \mathscr{M}$. In this case, ν is invariant under $\mathscr{L}$ as well as under any permutation of the axes, and is thus subject to Proposition 1.

Given a random measure z of kth order and a set $B \in \mathscr{B}(Y)$ we agree to say that almost surely there are no k-tuples in B if $\nu(B) = 0$. This can be justified as follows. Suppose we are given a version of z. By this we mean a non-negative function of $A \in \mathscr{B}(X)$ and $\omega \in \Omega$ which coincides, for each fixed A, almost surely with $z(A)$, and which is a measure on $\mathscr{B}(X)$ as a function of A for almost all fixed ω. To save letters we are going to denote a particular version again by z so that $z(A;\, \omega)$ is the value of that version for the set A and the outcome ω. For fixed ω, let $\nu\,(.\,;\, \omega)$ be the product measure in Y defined by $\nu(A_1 \times \ldots \times A_k;\, \omega) = z(A_1;\, \omega) \ldots z(A_k;\, \omega)$ if $A_1, \ldots, A_k \in \mathscr{B}_0(X)$. Then for every $B \in \mathscr{B}(Y)$ we have

(5) $$\nu(B) = \int_\Omega \nu(B;\, \omega)\, \mathbf{P}(d\omega).$$

In fact by equation (4) this formula is true if B has the form $A_1 \times \ldots \times A_k$ where $A_i \in \mathscr{B}_0(X)$, and hence is generally true. Loosely speaking, ν is the 'mixture', with respect to $\mathbf{P}$, of all measures $\nu\,(.\,;\, \omega)$ where ω runs through Ω. It follows that $\nu(B) = 0$ if and only if $\nu(B;\, \omega) = 0$ for almost all ω.

By a point-process in X we mean a positive random measure z with the property that, for every $A \in \mathscr{B}_0(X)$, $z(A)$ takes almost surely only integral values. In terms of a particular version, the phrase 'there are almost surely no k-tuples in B' can then be described in the following way which makes it intuitively clear: for almost all ω, there is no $(x_1, \ldots, x_k) \in B$ such that $z(\{x_1\};\, \omega) > 0, \ldots, z(\{x_k\};\, \omega) > 0$.

2.5.5 AN EXAMPLE

Take for X the set of all oriented lines in the oriented Euclidean plane R^2. We parametrize X in the usual way [8]: having chosen a fixed origin O and a fixed axis w_0 through O, an oriented line w will cut w_0 at a point s and at an angle $\vartheta + \frac{1}{2}\pi$. We then describe w by the pair (p, ϑ) where $p = s \cos \vartheta$ is the positive or negative distance of w from 0. If w is parallel or antiparallel to w_0 we have $\vartheta = \frac{3}{2}\pi$ or $\vartheta = \frac{1}{2}\pi$, respectively, and the sign of the distance p is determined by continuity, e.g. positive if w is antiparallel to w_0 and lies on the 'left bank' of w_0. Thus $-\infty < p < \infty$ and $0 \leqslant \vartheta \leqslant 2\pi$; topologically, X is simply the infinite two-dimensional cylinder $R \times S$, where S is the circle.

Let $\mathscr{M}$ be the group acting on X induced by the Euclidean motions of the plane. Explicitly, if S is the motion

$$(\xi, \eta) \to (\xi \cos \vartheta_0 - \eta \sin \vartheta_0 + \xi_0,\ \xi \sin \vartheta_0 + \eta \cos \vartheta_0 + \eta_0)$$

of R^2, taking for 0 the origin and for w_0 the first axis of coordinates the induced map of X becomes

$$(p, \vartheta) \to (p + \xi_0 \cos(\vartheta + \vartheta_0) + \eta_0 \sin(\vartheta + \vartheta_0), \vartheta + \vartheta_0),$$

and determines S uniquely. Under its natural topology obtained in this way, $\mathscr{M}$ is homeomorphic to the group of the Euclidean motions of R^2, and to $R \times R \times S$. Its action on X is continuous.

Denote by Y the product space X^2, and accordingly by $\mathscr{L}$ the group of all transformations of Y of the form $(w_1, w_2) \to (Tw_1, Tw_2)$, where $T \in \mathscr{M}$. There are then five types of equivalence classes of Y with respect to $\mathscr{L}$ which we are going to describe in terms of a partition of Γ into five sets $\{\delta\}$, $\{\alpha\}$, Γ_+, Γ_- and Γ_0.

$\{\delta\}$: The set of all double lines (w, w), i.e. the diagonal in Y, forms one equivalence class which we will represent by a point δ of Γ.

$\{\alpha\}$: Likewise, the set of all pairs of coincident lines with opposite directions $((p, \vartheta), (-p, \vartheta + \pi))$ forms one equivalence class which we will represent by a point α of Γ.

Γ_+: Suppose that w_1 and w_2 are parallel and unequal, say $w_1 = (p, \vartheta)$ and $w_2 = (p + \gamma, \vartheta)$ where $\gamma \neq 0$. Note that 'parallel' in this sense implies 'equidirected', and the distance γ between w_1 and w_2 is taken as positive if w_2 lies on the right bank of w_1. Then $(w_1, w_2) \sim (w_1', w_2')$ if and only if w_1' and w_2' are also parallel with the same distance γ, i.e. $w_1' = (p', \vartheta')$ and $w_2' = (p' + \gamma, \vartheta')$. Hence we may take γ to represent the equivalence class which contains (w_1, w_2), and the part Γ_+ of Γ corresponding to the classes of pairs of unequal parallel lines is the real line minus the origin $R - \{0\}$.

Γ_-: Suppose that w_1 and w_2 are antiparallel and non-coincident, say $w_1 = (p, \vartheta)$ and $w_2 = (\gamma - p, \vartheta + \pi)$ where $\gamma \neq 0$. Then $(w_1, w_2) \sim (w_1', w_2')$ if and only if w_1' and w_2' are also antiparallel with the same distance γ. Hence we may again take γ to represent the equivalence class which contains (w_1, w_2), and the part Γ_- of Γ corresponding to the classes of pairs of antiparallel and non-coincident lines is equal to $R - \{0\}$.

Γ_0: Suppose that w_1 and w_2 are neither parallel nor antiparallel and form the angle γ, say $w_1 = (p_1, \vartheta)$ and $w_2 = (p_2, \vartheta + \gamma)$, where γ is not an integral multiple of π. Then $(w_1, w_2) \sim (w_1', w_2')$ if and only if w_1' and w_2' form the same angle γ, i.e. $w_1' = (p_1', \vartheta')$ and $w_2' = (p_2', \vartheta' + \gamma)$. Hence we may take γ to represent the equivalence class which contains (w_1, w_2), and the part Γ_0 corresponding to the classes of pairs of neither parallel nor antiparallel lines is equal to the union of the two disjoint open intervals $0 < \gamma < \pi$ and $\pi < \gamma < 2\pi$.

The representation space $\Gamma = \{\delta\} \cup \{\alpha\} \cup \Gamma_+ \cup \Gamma_- \cup \Gamma_0$ is not separated in its quotient topology because the intersection of Γ_0 with any two neighbourhoods of any two points of Γ_+, or any two points of Γ_-, is never empty. We will, however, endow Γ with a finer topology, to be called the 'natural' one, which makes it locally compact with a countable base. To this end we use the ordinary topologies of Γ_+, Γ_- and Γ_0 given by their description in terms of $R-\{0\}, R-\{0\}$ and $]0,\pi[\cup]\pi,2\pi[$, respectively, and let Γ_+, Γ_- and Γ_0 be open in Γ. A basis of neighbourhoods of δ is to consist of the following sets: any union of δ with intervals $]-\varepsilon,0[$ and $]0,\varepsilon[$ of Γ_+ and intervals $]0,\varepsilon[$ and $]2\pi-\varepsilon,2\pi[$ of Γ_0 where $\varepsilon>0$. Similarly a typical neighbourhood of α consists of a union of α with intervals $]-\varepsilon,0[$ and $]0,\varepsilon[$ of Γ_- and intervals $]\pi-\varepsilon,\pi[$ and $]\pi,\pi+\varepsilon[$ of Γ_0.

Briefly, Γ is a circle with two lines welded on to it at opposite points.

Although the canonical map r of Y onto Γ is no longer continuous with respect to the natural topology of Γ, it has a continuous restriction to each of the five Borel sets $r^{-1}\{\delta\}$, $r^{-1}\{\alpha\}$, $r^{-1}(\Gamma_+)$, $r^{-1}(\Gamma_-)$ and $r^{-1}(\Gamma_0)$. Hence r is Borelian, i.e. $r^{-1}(\Delta)$ is a Borel set for every Borel set Δ in Γ. On the other hand, r is open in the natural topology since this topology is finer than the quotient topology. It follows that Δ is a Borel set if $r^{-1}(\Delta)$ is.

Next we look at the action of $\mathscr{L}$ in the various equivalence classes Y_γ. Each class Y_γ with $\gamma=\delta$, $\gamma=\alpha$, $\gamma\in\Gamma_+$ or $\gamma\in\Gamma_-$ can be mapped homeomorphically onto X by the projection $(w_1,w_2)\rightarrow w_1$. The action of $\mathscr{L}$ on Y_γ is hereby carried into the action of $\mathscr{M}$ on X. It is well known [8, §2] that there exists one and, up to a positive factor, only one non-vanishing positive Radon measure on X which is invariant under $\mathscr{M}$, namely, $|dp\,d\vartheta|$. The corresponding measure on Y concentrated on Y_γ will be denoted by τ_γ.

Suppose that $\gamma\in\Gamma_0$. In this case the map $((p_1,\vartheta),(p_2,\vartheta+\gamma))\rightarrow(\xi,\eta,\vartheta)$ where

$$\xi = (\sin\gamma)^{-1}(p_1\sin(\vartheta+\gamma)-p_2\sin\vartheta),$$

$$\eta = (\sin\gamma)^{-1}(p_2\cos\vartheta-p_1\cos(\vartheta+\gamma)),$$

is a homeomorphism of Y_γ onto the space $R\times R\times S$; the point (ξ,η) is the intersection of the lines (p_1,ϑ) and $(p_2,\vartheta+\gamma)$. Hence the action of $\mathscr{L}$ on Y_γ is carried by this map into the action on $R\times R\times S$ of the group of all transformations

$$(\xi,\eta,\vartheta)\rightarrow(\xi\cos\vartheta_0-\eta\sin\vartheta_0+\xi_0,\ \xi\sin\vartheta_0+\eta\cos\vartheta_0+\eta_0,\ \vartheta+\vartheta_0),$$

where $(\xi_0,\eta_0,\vartheta_0)$ runs through $R\times R\times S$. Again it is well known [8, §5] that there exists one and, up to a positive factor, only one non-vanishing positive Radon measure on $R \times R \times S$ which is invariant under this group, namely, the so-called kinematic measure $|d\xi\, d\eta\, d\vartheta| = |\sin\gamma|^{-1} |dp_1\, dp_2\, d\vartheta|$. The corresponding measure on Y concentrated on Y_γ will be denoted by τ_γ.

It follows now from Theorem 2 that a positive Radon measure ν in Y is invariant under $\mathscr{L}$ if and only if it admits a disintegration (3) where κ is a positive Radon measure on Γ.

Next we consider various other transformations of Y. As before, let W be the permutation of the axes: $(w_1, w_2) \to (w_2, w_1)$. The subsequent maps have the form

$$V_i: \qquad (w_1, w_2) \to (V_{i0}\, w_1, V_{i0}\, w_2),$$

where the transformation V_{i0} of X for $i = 1, 2, 3$ does not belong to $\mathscr{M}$.

V_{10}: The transformation of X induced by a reflection at a fixed line in the plane R^2, for example by the reflection at the axis w_0, i.e.

$$(p, \vartheta) \to (-p, \pi - \vartheta).$$

V_{20}: Change of the orientation of all lines, i.e. $(p, \vartheta) \to (-p, \vartheta + \pi)$.

V_{30}: Translation of the cylinder X, i.e. $(p, \vartheta) \to (p + p_0, \vartheta), p_0$ fixed, $\neq 0$.

An elementary geometric reasoning shows that the group of transformations of X generated by $\mathscr{M}$ and any maps of two given types V_{i0} does not contain a map of the remaining type. An alternative proof will be given below.

Obviously $T_0 \in \mathscr{M}$ entails $V_{i0}\, T_0\, V_{i0}^{-1} \in \mathscr{M}$. Accordingly, let w and v_i $(i = 1, 2, 3)$ be the transformations of Γ with the property that $r \circ W = w \circ r$ and $r \circ V_i = v_i \circ r$ for $i = 1, 2, 3$. It is easy to describe w and every v_i explicitly if we represent Γ_+, Γ_- and Γ_0 as above by the spaces $R - \{0\}$, $R - \{0\}$ and $]0, \pi[\cup]\pi, 2\pi[$, respectively:

$$w(\delta) = \delta; \quad w(\alpha) = \alpha, \quad w(\gamma) = -\gamma \quad \text{if } \gamma \in \Gamma_+,$$
$$w(\gamma) = \gamma \quad \text{if } \gamma \in \Gamma_-, \quad w(\gamma) = 2\pi - \gamma \quad \text{if } \gamma \in \Gamma_0.$$

$$v_1(\delta) = \delta; \quad v_1(\alpha) = \alpha, \quad v_1(\gamma) = -\gamma \quad \text{if } \gamma \in \Gamma_+,$$
$$v_1(\gamma) = -\gamma \quad \text{if } \gamma \in \Gamma_-, \quad v_1(\gamma) = 2\pi - \gamma \quad \text{if } \gamma \in \Gamma_0.$$

$$v_2(\delta) = \delta; \quad v_2(\alpha) = \alpha, \quad v_2(\gamma) = -\gamma \quad \text{if } \gamma \in \Gamma_+,$$
$$v_2(\gamma) = -\gamma \quad \text{if } \gamma \in \Gamma_-, \quad v_2(\gamma) = \gamma \quad \text{if } \gamma \in \Gamma_0.$$

$$v_3(\delta) = \delta; \quad v_3(\alpha) = 2p_0 \quad \text{in } \Gamma_-, \quad v_3(\gamma) = \gamma \quad \text{if } \gamma \in \Gamma_+,$$
$$v_3(\gamma) = \gamma + 2p_0 \quad \text{if } \gamma \in \Gamma_- \text{ and } \gamma + 2p_0 \neq 0,$$
$$v_3(\gamma) = \alpha \quad \text{if } \gamma = -2p_0 \in \Gamma_-,$$
$$v_3(\gamma) = \gamma \quad \text{if } \gamma \in \Gamma_0.$$

By looking at the changes of the orientation of $\Gamma_+ \cup \{\delta\}$, $\Gamma_- \cup \{\alpha\}$ and $\Gamma_0 \cup \{\delta\} \cup \{\alpha\}$ under these maps we see in the first place that the group of transformation of Y generated by $\mathscr{L}$ and any maps of three given types out of the types W, V_1, V_2 and V_3 does not contain a map of the remaining type.

Moreover, by the definition of τ_γ the image of τ_γ under W or V_i is the measure $\tau_{w(\gamma)}$ or $\tau_{v_i(\gamma)}$, respectively. Hence by the corollary of Theorem 2, an $\mathscr{L}$-invariant measure ν on Y written in the form (3) is invariant under W or V_i if and only if the corresponding measure κ is invariant under w or v_i, respectively. In applying this we will make use of the following trivial remark: suppose we have two bijective Borelian transformations v and v' of Γ with a Borelian inverse and a decomposition of Γ into two sets Γ_e and Γ_g which are invariant under both v and v' and such that v' is the identity on Γ_e and coincides κ-almost everywhere with v on Γ_g. Then if κ is invariant under v, it is also invariant under v'.

Finally, let z be a positive random measure on X which is stationary to the second order, let ν be its second moment measure defined by

$$\nu(A_1 \times A_2) = \mathbf{E}(z(A_1)\, z(A_2)),$$

and let κ be the corresponding measure on Γ. Then ν is invariant under W, hence κ is invariant under w. Therefore, using the terminology introduced at the end of the preceding section and recalling that $\kappa(\Delta) = 0$ is tantamount to $\nu(r^{-1}\Delta) = 0$ we find that

ν is invariant under reflections at fixed lines in the plane and under the change of the orientation of all lines if almost surely there are no pairs of antiparallel, non-coincident lines, i.e. $\kappa(\Gamma_-) = 0$;

ν is invariant under translations of the cylinder X if almost surely there are no antiparallel (coincident or non-coincident) lines, i.e. $\kappa(\Gamma_- \cup \{\alpha\}) = 0$.

REFERENCES

1. N. Bourbaki, *Topologie Générale*, chap. 3: Groupes topologiques (*Actualités scientifiques et industrielles* **1143**), 3ième éd., Hermann, Paris (1960).
2. ——, *Intégration*, chap. 5: Intégration des mesures. (*Actualités scientifiques et industrielles* **1244**), 2ième éd., Hermann, Paris (1967).

3. N. Bourbaki, *Intégration*, Chap. 6: Intégration vectorielle. (*Actualités scientifique et industrielles* **1281**), Hermann, Paris (1960).
4. ——, *Intégration*, Chap. 7: Mesure de Haar (*Actualités scientifiques et industrielles* **1306**), Hermann, Paris (1963).
5. R. Davidson, *Stochastic processes of flats and exchangeability*, Ph.D. thesis, Part 2, University of Cambridge, England (1967). (2.1 of this book.)
6. ——, "Construction of line-processes: second-order properties". (2.4 of this book).
7. K. Krickeberg, *Wahrscheinlichkeitstheorie*, Teubner, Stuttgart (1963).
8. L. A. Santaló, *Introduction to Integral Geometry* (Actual. scientifiques et industrielles 1198), Hermann, Paris (1953).

2.6

Moments of Point-processes

KLAUS KRICKEBERG

2.6.1 INTRODUCTION

The present article grew out of a series of lectures given by the author in the spring of 1970 in the joint seminar on probability theory of McMaster University and the Université de Montréal, and at various other Canadian Universities.† The subject of these talks has been the theory of the correlation measure of second-order stationary line-processes as given by R. Davidson [3] and the author [5]. In comparison, the article was expanded to treat also higher moments in a systematic way. To reduce invariance properties of point-processes and their moment measures to well-known properties of measures invariant under certain groups which arise in geometry may be regarded as its main theme. This allows us to derive in full generality propositions like the corollary to Theorem 2 which had been obtained in particular cases by R. Davidson using elementary methods, and Theorem 6 which, in the case $k = n = 2$, had been conjectured by the author and then proved by Davidson.

While writing these notes the author learned of the untimely death of Rollo Davidson. During the short time he had worked in this domain of intriguing problems on the border line of geometry and probability theory, his great imagination contributed a wealth of new insights of which the present article is only one testimony among many.

† The author is greatly indebted to many Canadian colleagues, in particular Professor M. Behara, for organizing this seminar, and to the National Research Council and various Canadian Universities for financing it. The article was written while the author was a visiting professor at the University of Buenos Aires under its multi-national programme where he had the benefit of stimulating discussions with Professor L. A. Santaló.

2.6.2 DISINTEGRATION OF INVARIANT MEASURES

Let Y be a locally compact space and $\mathscr{H}$ a locally compact group which acts continuously on Y [2, §2, No. 4]. We denote by $\sim$ the equivalence relation determined by $\mathscr{H}$, that is, $\xi \sim \eta$ if and only if $\eta = H\xi$ for some $H \in \mathscr{H}$. We make the following assumptions: Y and $\mathscr{H}$ have countable bases; there exists a Borel representation of $\sim$, that is, a locally compact space Γ with a countable base and a map r of Y onto Γ such that $\xi \sim \eta$ is equivalent to $r(\xi) = r(\eta)$ and a subset Δ of Γ is Borelian if and only if $r^{-1}(\Delta)$ is Borelian in Y. In particular, each equivalence class $Y_\gamma = r^{-1}\{\gamma\}$ will be a Borel set.

Note that in general Γ cannot be the quotient space $Y/\sim$ endowed with the quotient topology because the latter may not be separated. It would be interesting to know whether a Borel representation always exists in the present context.

Next we assume that there is a non-negative bounded Baire function b on Y with the following properties: for every $\gamma \in \Gamma$, the set $Y_\gamma \cap \{b > 0\}$ contains a non-empty subset which is open in Y_γ; for every compact subset Δ of Γ, the set $r^{-1}(\Delta) \cap \text{carrier}\,(b)$ is relatively compact. Again it may be that this is always true.

In the following the term 'measure' will always mean 'positive Radon measure'.

Our final, and crucial, assumption bears on the action of $\mathscr{H}$ in the various equivalence classes: for every γ there is an $\mathscr{H}$-invariant measure τ_γ in Y concentrated on Y_γ, and only one, up to a non-negative factor. We can then normalize τ_γ by requiring that $\tau_\gamma(b)$ be a Baire function of γ, bounded on every compact set, and $\tau_\gamma(b) > 0$ unless $\tau_\gamma = 0$.

We are now in a position to describe $\mathscr{H}$-invariant measures in Y in terms of measures in Γ.

Theorem 1. [5, §1.] *A measure ν in Y is $\mathscr{H}$-invariant if and only if there exists a measure κ in Γ such that the family $(\tau_\gamma)_{\gamma\in\Gamma}$ becomes scalarly κ-integrable, and*

$$\nu = \int_\Gamma \tau_\gamma \, \kappa(d\gamma). \tag{1}$$

κ is uniquely determined by ν.

Recall that equation (1) amounts to

$$\nu(f) = \int_\Gamma \tau_\gamma(f)\, \kappa(d\gamma)$$

for every ν-integrable function f.

The first application of this disintegration of an $\mathscr{H}$-invariant measure ν concerns its invariance under maps not in $\mathscr{H}$.

Theorem 2. [5, § 1.] *Let F be a homeomorphism of Y which induces a bijective transformation Φ of Γ such that $F(Y_\gamma) = Y_{\Phi(\gamma)}$ for every γ, or in other words*

(2) $$r \circ F = \Phi \circ r.$$

Suppose in addition that for every γ the measure $\tau_{\Phi(\gamma)}$ is the image of τ_γ under F. Then an $\mathscr{H}$-invariant measure ν in Y represented in the form (1) *is invariant under F if and only if the corresponding measure κ in Γ is invariant under Φ.*

Note that Φ and Φ^{-1} are necessarily Borel maps. The existence of a transformation Φ which satisfies equation (2) is assured, for example, if $FHF^{-1} \in \mathscr{H}$ for every $H \in \mathscr{H}$.

In the case where each τ_γ is finite we may, of course, use a constant normalization, more precisely

(3) $$\tau_\gamma(Y) = \beta$$

with some $\beta > 0$ for all γ such that $\tau_\gamma \neq 0$. Then if g denotes a κ-integrable function on Γ, it follows from (1) and (3) that

(4) $$\kappa(g) = \beta^{-1} \nu(g \circ r),$$

in particular

$$\kappa(\Delta) = \beta^{-1} \nu(r^{-1} \Delta)$$

for every Borel subset Δ of Γ.

2.6.3 FACTORIZED INVARIANT MEASURES

We pass to the particular case

(5) $$Y = Q \times T$$

where Q and T, too, are locally compact spaces with a countable base. Consider a Borelian subset Γ^0 of Γ with the following property:

$$(q, \theta) \sim (q', \theta)$$

for all $q, q' \in Q$ such that $r(q, \theta) \in \Gamma^0$. Let T^0 be the set of all $\theta \in T$ which satisfy $r(q, \theta) \in \Gamma^0$ for one, and hence for all $q \in Q$. Clearly

$$r^{-1}(\Gamma^0) = Q \times T^0$$

is a Borel set in Y, and therefore T^0 is Borelian in T. Moreover, $Q \times T^0$ is invariant under $\mathscr{H}$.

Let $\theta, \theta' \in T^0$. Then we have

$$(q, \theta) \sim (q', \theta') \tag{6}$$

for one pair $q, q' \in Q$ if and only if this holds for all pairs q, q'. In fact (6) implies $(p, \theta) \sim (q, \theta) \sim (q', \theta') \sim (p', \theta')$ for all $p, p' \in Q$. In this way (6) defines an equivalence relation in T^0 to be denoted again by $\sim$ for the sake of simplicity. Thus $(q, \theta) \sim (q', \theta')$ and $\theta \in T^0$ imply $\theta' \in T^0$ and $\theta \sim \theta'$ and, conversely, $\theta, \theta' \in T^0$ and $\theta \sim \theta'$ entail $(q, \theta) \sim (q', \theta')$. Hence each equivalence class Y_γ with $\gamma \in \Gamma^0$ induces an equivalence class T_γ in T^0 which in turn determines Y_γ by $Y_\gamma = Q \times T_\gamma$.

Next assume that for every $\gamma \in \Gamma^0$ the measure τ_γ can be factorized in the form

$$\tau_\gamma = \rho \otimes \sigma_\gamma,$$

where ρ is some fixed measure in Q and σ_γ a measure in T carried by T_γ. Then the representation (1) of an $\mathscr{H}$-invariant measure ν in Y for factorized functions

$$(f \otimes g)(q, \theta) = f(q) g(\theta)$$

takes the form

$$\nu(f \otimes g) = \rho(f) \int_{\Gamma^0} \sigma_\gamma(g)\, \kappa(d\gamma) + \int_{\Gamma - \Gamma^0} \tau_\gamma(f \otimes g)\, \kappa(d\gamma),$$

which is valid, for example, for any bounded Baire functions f on Q and g on T with compact carriers.

In particular, if $\nu(Q \times (T - T^0)) = 0$, or equivalently $\kappa(\Gamma - \Gamma^0) = 0$, we get

$$\nu(f \otimes g) = \rho(f) \int_{\Gamma^0} \sigma_\gamma(g)\, \kappa(d\gamma). \tag{7}$$

2.6.4 PRODUCTS OF IDENTICAL MEASURES

We start from a locally compact space X with a countable base, and denote by $\mathscr{K}(X)$ the space of all continuous functions on X with a compact carrier, and by $\mathscr{B}_0(X)$ the class of all relatively compact Borel subsets of X. Let μ be a measure in X. We employ as before the notation $\mu(f)$ for $f \in \mathscr{K}(X)$ as well as for more general μ-integrable functions, and $\mu(A)$ for $A \in \mathscr{B}_0(X)$ as well as for a general μ-measurable A. The measure μ is called diffuse if $\mu\{\xi\} = 0$ for every single point ξ of X, and a point measure if it is carried by a finite or countable set, assigning measure 1 to each point of that set.

Let k be a positive integer and $Y = X^k$. We write $\mu^{[k]}$ for the kth power of μ. Thus $\mu^{[k]}$ is the measure in Y defined by

$$\mu^{[k]}(f_1 \otimes \dots \otimes f_k) = \mu(f_1) \dots \mu(f_k) \quad \text{for } f_1, \dots, f_k \in \mathscr{K}(X).$$

Therefore, $\mu^{[k]}$ is symmetric, that is, invariant under all permutations of the axes of X^k.

Clearly, if μ is diffuse or a point measure, $\mu^{[k]}$ also is.

Consider a partition $\mathscr{J} = \{J_1, \dots, J_m\}$ of the set $\{1, \dots, k\}$ into m disjoint non-empty subsets $J_1, \dots, J_m$. We define the '$\mathscr{J}$-diagonal' $D_{\mathscr{J}}$ of Y to be the set of all $(\xi_1, \dots, \xi_k)$ of Y such that, for every $j = 1, \dots, m$, we have $\xi_i = \xi_{i'}$ for all $i, i' \in J_j$. By the projection $\Pi_{\mathscr{J}}$ of $D_{\mathscr{J}}$ onto the space X^m we mean the map $\Pi_{\mathscr{J}}(\xi_1, \dots, \xi_k) = (\eta_1, \dots, \eta_m)$ where $\eta_j = \xi_i$ for all $i \in J_j$. This definition of $\Pi_{\mathscr{J}}$ is, of course, dependent on the fact that we have arranged the sets $J_1, \dots, J_m$ in a definite order, whereas $D_{\mathscr{J}}$ is not, but any other order of the J_js will only amount to a permutation of the axes of X^m. Therefore we may be excused for the sloppy notation.

Obviously, $\Pi_{\mathscr{J}}$ is bijective, and a subset C of $D_{\mathscr{J}}$ is Borelian if and only if $\Pi_{\mathscr{J}} C$ is.

To simplify the notations in the following computation we will treat the case where $j < j'$ and $i \in J_j$, $i' \in J_{j'}$ imply $i < i'$; the general case can be reduced to this one by a permutation of the axes of X^k. Let l_j be the number of elements of J_j. Given any sets $B_{ji} \in \mathscr{B}_0(X)$ for $j = 1, \dots, m$ and $i = 1, \dots, l_j$ we have

(8)
$$\begin{aligned} &D_{\mathscr{J}} \cap (B_{11} \times \dots \times B_{1l_1} \times \dots \times B_{m1} \times \dots \times B_{ml_m}) \\ &= D_{\mathscr{J}} \cap ((B_{11} \cap \dots \cap B_{1l_1})^{l_1} \times \dots \times (B_{m1} \cap \dots \cap B_{ml_m})^{l_m}). \end{aligned}$$

Hence the part concentrated on $D_{\mathscr{J}}$ of any measure in Y is completely determined by the values of this measure for sets C of the form

(9)
$$C = D_{\mathscr{J}} \cap (B_1^{l_1} \times \dots \times B_m^{l_m})$$

with $B_j \in \mathscr{B}_0(X)$ which have the projection

(10)
$$\Pi_{\mathscr{J}} C = B_1 \times \dots \times B_m.$$

To compute the part of $\mu^{[k]}$ carried by $D_{\mathscr{J}}$ we make use of Fubini's theorem. In the first step we get, for a set of the type (9)

$$\mu^{[k]}(C) = \prod_{j=1}^{m} \mu^{[l_j]}(\tilde{B}_j),$$

where $\tilde{B}_j = \{(\xi_1, \dots, \xi_{l_j}) : \xi_1 = \dots = \xi_{l_j} \in B_j\}$. To evaluate further the jth factor of this product we have to distinguish the cases $l_j=1$ and $l_j>1$. In the first case, this factor is clearly equal to $\mu(B_j)$. In the second case we find, again by Fubini's theorem,

$$\mu^{[l_j]}(\tilde{B}_j)=\int_{B_j}\mu\{\xi\}^{l_j-1}\,\mu(d\xi)=\sum_{\xi\in B_j}\mu\{\xi\}^{l_j},$$

where it suffices, of course, to extend the sum over all points $\xi\in B_j$ such that $\mu\{\xi\}>0$. Thus

$$\mu^{[k]}(C)=\prod_{j:l_j=1}\mu(B_j)\prod_{j:l_j>1}\left(\sum_{\xi\in B_j}\mu\{\xi\}^{l_j}\right). \tag{11}$$

Theorem 3. *Suppose that $m<k$. Then μ is diffuse if and only if $\mu^{[k]}(D_{\mathscr{J}})=0$. It is a point measure if and only if*

$$\mu^{[k]}(C)=\mu^{[m]}(\Pi_{\mathscr{J}}\,C) \tag{12}$$

for every Borel subset C of $D_{\mathscr{J}}$.

Proof. The assumption $m<k$ is equivalent to $l_j>1$ for at least one j. Therefore the first assertion follows immediately from equation (11).

Next, suppose that μ is a point measure. Then $\mu\{\xi\}=1$ for every $\xi\in X$ such that $\mu\{\xi\}>0$, hence equation (12) follows from equations (10) and (11).

Conversely, suppose that equation (12) is true. The case $\mu=0$ being trivial we select a set A with $\mu(A)>0$. Let B be any set in $\mathscr{B}_0(X)$, and define $B_j=A$ if $l_j=1$ and $B_j=B$ if $l_j>1$. Applying equation (12) with the set C given by equation (9) we obtain on account of equations (10) and (11):

$$\mu(B)^l=\prod_{j:\,l_j>1}\left(\sum_{\xi\in B}\mu\{\xi\}^{l_j}\right), \tag{13}$$

where

$$l=\sum_{j:\,l_j>1}1.$$

Taking for B a one-point set $\{\eta\}$ with any $\eta\in X$ we find that $\mu\{\eta\}^l=\mu\{\eta\}^{l'}$, where

$$l'=\sum_{j:\,l_j>1}l_j>l,$$

hence $\mu\{\eta\}=1$ if $\mu\{\eta\}>0$. Therefore by equation (13), if B is any set in $\mathscr{B}_0(X)$, we have

$$\mu(B)^l=\left(\sum_{\xi\in B}\mu\{\xi\}\right)^l$$

which implies

$$\mu(B) = \sum_{\xi \in B} \mu\{\xi\}.$$

Since every $\mu\{\xi\}$ is equal to 0 or 1 we see that μ is in fact a point measure.

We remark that equation (12) is trivially true for every measure μ if $m = k$.

The intuitive meaning of equation (12) is, of course, that the measure $\mu^{[k]}$ restricted to $D_{\mathscr{J}}$ is essentially equal to $\mu^{[m]}$, more precisely, $\mu^{[m]}$ is the image under $\Pi_{\mathscr{J}}$ of the restriction of $\mu^{[k]}$ to $D_{\mathscr{J}}$. There are various other ways to express this fact, for example

$$\mu^{[m]}(B_1 \times \ldots \times B_m) = \mu^{[k]}(D_{\mathscr{J}} \cap (B_1^{l_1} \times \ldots \times B_m^{l_m})) \tag{14}$$

for all $B_j \in \mathscr{B}_0(X)$, or on the basis of equation (8),

$$\mu^{[k]}\left(D_{\mathscr{J}} \cap \bigotimes_{j=1}^{m} \bigotimes_{i=1}^{l_j} B_{ji} \right) = \mu^{[m]}\left(\bigotimes_{j=1}^{m} \bigcap_{i=1}^{l_j} B_{ji} \right), \tag{14'}$$

and in terms of functions instead of sets,

$$\mu^{[k]}\left(1_{D_{\mathscr{J}}} \bigotimes_{j=1}^{m} \bigotimes_{i=1}^{l_j} f_{ji} \right) = \mu^{[m]}\left(\bigotimes_{j=1}^{m} \prod_{i=1}^{l_j} f_{ji} \right) \tag{15}$$

for all $f_{ji} \in \mathscr{K}(X)$.

We observe that, the measures $\mu^{[k]}$ and $\mu^{[m]}$ being symmetric, the formulas (12), (14), (14′) and (15) hold as well for any other order of the elements of the J_js.

We denote by $\operatorname{cd}(\mathscr{J})$ the number of sets of the partition $\mathscr{J}$ of $\{1, \ldots, k\}$. Given two partitions $\mathscr{J}$ and $\mathscr{J}'$ we write $\mathscr{J}' \prec \mathscr{J}$ if $\mathscr{J}$ is a subpartition of $\mathscr{J}'$. Clearly $\mathscr{J}' \prec \mathscr{J}$ implies $D_{\mathscr{J}'} \subseteq D_{\mathscr{J}}$. To avoid trivialities we will henceforth assume that X contains at least two points. Then the converse statement holds: $\mathscr{J}' \prec \mathscr{J}$ if $D_{\mathscr{J}'} \subseteq D_{\mathscr{J}}$, and in this case $D_{\mathscr{J}'} = D_{\mathscr{J}}$ if and only if $\mathscr{J}' = \mathscr{J}$, that is, $\operatorname{cd}(\mathscr{J}') = \operatorname{cd}(\mathscr{J})$.

If $\operatorname{cd}(\mathscr{J}) = k$ or $\operatorname{cd}(\mathscr{J}) = 1$, there is only one partition

$$\mathscr{J}_{\max} = \{\{1\}, \ldots, \{k\}\}$$

or

$$\mathscr{J}_{\min} = \{1, \ldots, k\},$$

respectively. We have $D_{\mathscr{J}_{\max}} = X^k$ and, arranging the sets $\{j\}$ in their natural order, $\Pi_{\mathscr{J}_{\max}} = id$, whereas $D_{\mathscr{J}_{\min}}$ is the usual diagonal and $\Pi_{\mathscr{J}_{\min}}(\xi, \ldots, \xi) = \xi$. Clearly $\mathscr{J}_{\min} \prec \mathscr{J} \prec \mathscr{J}_{\max}$ for any partition $\mathscr{J}$.

The intersection of any two diagonals $D_{\mathscr{J}}$ and $D_{\mathscr{J}'}$ being again a diagonal we can write $D_{\mathscr{J}} \cap D_{\mathscr{J}'} = D_{\mathscr{J}^*}$ where $\mathscr{J}^* \prec \mathscr{J}$ and $\mathscr{J}^* \prec \mathscr{J}'$, and we have $\mathscr{J}^* = \mathscr{J}'$ if and only if $D_{\mathscr{J}'} \subseteq D_{\mathscr{J}}$.

Next we define

$$E_{\mathscr{J}} = D_{\mathscr{J}} - \bigcup_{\substack{\mathscr{J}':\, \mathscr{J}' \prec \mathscr{J} \\ \mathscr{J}' \neq \mathscr{J}}} D_{\mathscr{J}'}. \tag{16}$$

In concrete terms, this is the set of all $(\xi_1, \ldots, \xi_k) \in D_{\mathscr{J}}$ such that $i \in J_j$, $i' \in J_{j'}$ and $j \neq j'$ implies $\xi_i \neq \xi_{i'}$. Then $\bigcup_{\mathscr{J}} E_{\mathscr{J}} = Y$ since $D_{\mathscr{J}_{\max}} = Y$. Moreover, the sets $E_{\mathscr{J}}$ are mutually disjoint. In fact, suppose that $\mathscr{J} \nprec \mathscr{J}'$. Then $D_{\mathscr{J}} \cap D_{\mathscr{J}'} = D_{\mathscr{J}^*}$, where $\mathscr{J}^* \prec \mathscr{J}$ and $\mathscr{J}^* \neq \mathscr{J}$, hence $E_{\mathscr{J}} \cap D_{\mathscr{J}^*} = \varnothing$ by equation (16) and $E_{\mathscr{J}} \subseteq D_{\mathscr{J}}$, thus $E_{\mathscr{J}} \cap D_{\mathscr{J}'} = \varnothing$ and *a fortiori*

$$E_{\mathscr{J}} \cap E_{\mathscr{J}'} = \varnothing.$$

Note that, by equation (16), $E_{\mathscr{J}_{\min}} = D_{\mathscr{J}_{\min}}$.

Theorem 4. *Let ν be any measure in $Y = X^k$. Then there exists a unique decomposition of ν of the form*

$$\nu = \sum_{\mathscr{J}} \nu_{\mathscr{J}}, \tag{17}$$

where $\nu_{\mathscr{J}}$ is a measure carried by $D_{\mathscr{J}}$, and $\nu_{\mathscr{J}}(D_{\mathscr{J}'}) = 0$ if $\mathscr{J}' \prec \mathscr{J}$ and $\mathscr{J}' \neq \mathscr{J}$.

Proof. To prove the uniqueness it suffices to show that the measure $\nu_{\mathscr{J}}$ in any such decomposition is carried by $E_{\mathscr{J}}$, since the sets $E_{\mathscr{J}}$ are mutually disjoint. This, however, follows immediately from equation (16).

To prove the existence, we define

$$\nu_{\mathscr{J}}(C) = \nu(E_{\mathscr{J}} \cap C). \tag{18}$$

Clearly, $\nu_{\mathscr{J}}$ is carried by $D_{\mathscr{J}}$. Moreover, if $\mathscr{J}' \prec \mathscr{J}$ and $\mathscr{J}' \neq \mathscr{J}$, we have $D_{\mathscr{J}'} \cap E_{\mathscr{J}} = \varnothing$ by equation (16), hence $\nu_{\mathscr{J}}(D_{\mathscr{J}'}) = 0$ by equation (18).

Combining Theorem 4 with Theorem 3 we see that in the case of a diffuse measure μ the decomposition, equation (17), of $\nu = \mu^{[k]}$ is the trivial one, namely, $\mu^{[k]}_{\mathscr{J}_{\max}} = \mu^{[k]}$ and $\mu^{[k]}_{\mathscr{J}} = 0$ for all other $\mathscr{J}$. In the case of a point measure μ, by equation (12), the definition (18) takes the form

$$\mu^{[k]}_{\mathscr{J}}(C) = \mu^{[m]}(\Pi_{\mathscr{J}}(E_{\mathscr{J}} \cap C)) \tag{19}$$

where $m = \operatorname{cd}(\mathscr{J})$, in particular

$$\mu^{[k]}_{\mathscr{J}_{\min}}(C) = \mu(\Pi_{\mathscr{J}_{\min}}(D_{\mathscr{J}_{\min}} \cap C)).$$

2.6.5 MOMENTS OF RANDOM MEASURES

Let $(\Omega, \mathscr{F}, \mathbf{P})$ be some probability space. A random measure z in X is a function on $\mathscr{K}(X) \times \Omega$ with the property that $z(f, \omega)$ is a measure as a function of f for fixed ω and a random variable as a function of ω for fixed f. The former will be denoted by $z^I(\omega)$ and the latter by $z(f)$, that is

$$z^I(\omega) = (f \to z(f, \omega))$$

and

$$z(f) = (\omega \to z(f, \omega)).$$

By the distribution of z we mean the family of the distributions of all random vectors $(z(f_1), \ldots, z(f_n))$ with $f_1, \ldots, f_n \in \mathscr{K}(X)$ or, equivalently, of all random vectors $(z(A_1), \ldots, z(A_n))$ with $A_1 \ldots, A_n \in \mathscr{B}_0(X)$. Note that we may describe the distribution of $(z(f_1), \ldots, z(f_n))$ by the expectations

$$\mathbf{E}h(z(f_1), \ldots, z(f_n)) = \int_\Omega h(z(f_1, \omega), \ldots, z(f_n, \omega))\, \mathbf{P}(d\omega)$$

where h runs, for example, through all functions in $\mathscr{K}(R^n)$, or all bounded Baire functions on R^n.

The random measure z is said to be diffuse or a point-process if $z^I(\omega)$ is diffuse or a point measure, respectively, for $\mathbf{P}$-almost all ω. Intuitively, a point-process consists in throwing at random a finite or countable set of points into the space X in such a way that only finitely many of them fall into any given relatively compact set. If ω describes this realization of the random phenomenon in question, these points carry the point measure $z^I(\omega)$.

Let k be a positive integer. For each fixed $\omega \in \Omega$ we can form the kth power $(z^I(\omega))^{[k]}$ of the measure $z^I(\omega)$, to be denoted by $f \to z^{[k]}(f, \omega)$ which amounts to $z^{[k]I} = z^{I[k]}$. It is clear that $z^{[k]}$ represents a random measure in X^k. If z is diffuse or a point measure, $z^{[k]}$ has the same property. In the latter case, using the intuitive picture of a point-process we may say that $z^{[k]}(C)$ is the number of k-tuples of points that fall into the set $C \in \mathscr{B}_0(X^k)$.

The random measure z is called of kth order if the expectation $\mathbf{E}z^{[k]}(f)$ exists and is finite for every $f \in \mathscr{K}(X^k)$. This amounts, of course, to $z^{[k]}$ being of first order. A necessary and sufficient condition for this to happen is $\mathbf{E}(z(A)^k) < \infty$ for every $A \in \mathscr{B}_0(X)$, and z will then also be of lth order for every $l \leqslant k$.

If z is of kth order, the functional $f \to \mathbf{E}z^{[k]}(f)$ with $f \in \mathscr{K}(X^k)$ obviously represents a measure in X^k. We will denote this measure by ν_z^k and term it the kth moment measure of z. Thus $\nu_z^k = \nu_{z^{[k]}}^1$, and explicitly for functions in the form of a product:

$$\nu_z^k(f_1 \otimes \ldots \otimes f_k) = \mathbf{E}(z(f_1) \ldots z(f_k)).$$

It follows that ν_z^k shares with the measures $z^{[k]I}(\omega)$ the property of being symmetric.

Suppose that z is of kth order and $\mathscr{J}$ is a partition of $\{1, \ldots, k\}$ into disjoint non-empty sets with $m = \mathrm{cd}(\mathscr{J}) < k$. Then Theorem 3 has the following corollary.

Theorem 3, Corollary 1. *Under the preceding assumptions, z is diffuse if and only if $\nu_z^k(D_{\mathscr{J}}) = 0$.*

We can also derive

Theorem 3, Corollary 2. *In addition to the preceding assumptions, suppose that for almost all ω and every $A \in \mathscr{B}_0(X)$, the number $z(A, \omega)$ is an integer. Then z is a point-process if and only if*

$$\nu_z^k(C) = \nu_z^m(\Pi_{\mathscr{J}}\, C) \tag{20}$$

for every Borel set $C \subseteq D_{\mathscr{J}}$.

In fact, Theorem 3 shows that equation (20) is necessary for z to be a point-process. On the other hand, by equation (11) we have almost surely $z^{[k]}(C, \omega) \geqslant z^{[m]}(\Pi_{\mathscr{J}}\, C, \omega)$ for every C of the form of equation (9), and hence for every Borel set $C \subseteq D_{\mathscr{J}}$. Therefore equation (20) implies $z^{[k]}(C, \omega) = z^{[m]}(\Pi_{\mathscr{J}}\, C, \omega)$ for almost all ω where the exceptional set of ωs may now depend on C. However, since X has a countable base, an exceptional set of probability 0 may be chosen to serve for all C, thus $z^I(\omega)$ is a point measure for almost all ω by Theorem 3, that is, z is a point-process.

The assumption that almost all $z(A, \omega)$ are integers cannot be discarded as shown by the example of a one-point set X with $k = 2$, $m = 1$ and $z(X)$ uniformly distributed in the interval $[0, \frac{3}{2}]$.

By equation (20) the mth moment measure of a point-process is the image under $\Pi_{\mathscr{J}}$ of the restriction of its kth moment measure to $D_{\mathscr{J}}$ for any $\mathscr{J}$ such that $\mathrm{cd}(\mathscr{J}) = m$. Hence

Theorem 3, Corollary 3. *The moment measures of order less than k of a point-process are completely determined by its moment measure of order k.*

Of course, equation (20) can be rewritten in ways analogous to the previous transformations (14), (14′) and (15) of equation (12), in particular

$$\nu_z^m(B_1 \times \ldots \times B_m) = \nu_z^k(D_{\mathscr{J}} \cap (B_1^{l_1} \times \ldots \times B_m^{l_m})) \tag{21}$$

$$\nu_z^k\left(1_{D_{\mathscr{J}}} \bigotimes_{j=1}^{m} \bigotimes_{i=1}^{l_j} f_{ji}\right) = \nu_z^m\left(\bigotimes_{j=1}^{m} \prod_{i=1}^{l_j} f_{ji}\right). \tag{22}$$

Let z again be any random measure in X of kth order. For fixed ω, we have the decomposition of $z^{[k]I}(\omega)$ given by equations (17) and (18), that is,

$$z^{[k]}(C) = \sum_{\mathscr{J}} z^{[k]}(E_{\mathscr{J}} \cap C)$$

for every $C \in \mathscr{B}_0(X^k)$. Taking expectations we get

$$\nu_z^k(C) = \sum_{\mathscr{J}} \nu_z^k(E_{\mathscr{J}} \cap C),$$

and since the measure $C \to \nu_z^k(E_{\mathscr{J}} \cap C)$ is carried by $E_{\mathscr{J}}$, this is the decomposition of the measure $\nu = \nu_z^k$ defined by Theorem 4. In the case of a point-process, on account of equation (20), it takes the form

$$\nu_z^k(C) = \sum_{\mathscr{J}} \nu_z^{\mathrm{cd}(\mathscr{J})}(\Pi_{\mathscr{J}}(E_{\mathscr{J}} \cap C)). \tag{23}$$

2.6.6 THE DOUBLY STOCHASTIC POISSON PROCESS

Let μ be a fixed measure in X. Then [6, 9] there exists a random measure z in X with the following properties:

(*a*) If $A_1, \ldots, A_n$ are mutually disjoint sets in $\mathscr{B}_0(X)$, the random variables $z(A_1), \ldots, z(A_n)$ are independent.

(*b*) For every $A \in \mathscr{B}_0(X)$ the random variable $z(A)$ has a Poisson distribution with parameter $\mu(A)$, that is,

$$\mathbf{P}\{z(A) = m\} = \frac{1}{m!}\mu(A)^m \exp(-\mu(A)), \quad m = 0, 1, 2, \ldots. \tag{24}$$

Any random measure z with these properties is called a Poisson process with mean number of points μ.

It follows from (*b*) that

$$\nu_z^1 = \mu. \tag{25}$$

On the other hand, the distribution of z is completely determined by μ. Moreover, z is a point-process if and only if μ is diffuse. Given any point-process z which satisfies (*a*) and equation (25) where μ is diffuse, we can derive the condition (*b*); this is just a slight generalization of the classical Poisson limit theorem.

Next, consider an arbitrary random measure u in X. A random measure z in X is called a doubly stochastic Poisson process with mean number of points u if its distribution is the mixture with respect to $\mathbf{P}(d\omega)$ of the distributions of the various Poisson processes corresponding to the various measures $u^I(\omega)$ with $\omega \in \Omega$. More precisely, denoting for fixed ω by z_ω a Poisson process with mean number of points $u^I(\omega)$ and by

$\mathbf{E}_{\omega; f_1,\ldots,f_n}$ the expectation with respect to the distribution of the random vector $(z_\omega(f_1), \ldots, z_\omega(f_n))$, we should have

$$\mathbf{E}h(z(f_1), \ldots, z(f_n)) = \int_\Omega \mathbf{E}_{\omega; f_1,\ldots,f_n}(h)\, \mathbf{P}(d\omega) \tag{26}$$

for every $h \in \mathscr{K}(R^n)$, and hence for every bounded Baire function h on R^n. In particular, from equation (24),

$$\mathbf{P}\{z(A) = m\} = \frac{1}{m!}\int_\Omega u(A, \omega)^m \exp(-u(A, \omega))\, \mathbf{P}(d\omega).$$

It can be proved [6] that such a process z always exists and that the distributions of z and of u determine each other completely. Clearly any random variable $z(A)$ with $A \in \mathscr{B}_0(X)$ takes only integral values. Moreover, z is a point-process if and only if u is diffuse. Finally, z is of kth order if and only if u is, and the moment measures up to the kth order of z and of u can be computed from each other. We have

$$\nu_z^k(f_1 \otimes \ldots \otimes f_k) = \sum_{m=1}^{k} \sum_{\{J_1,\ldots,J_m\}} \nu_u^m\left(\bigotimes_{j=1}^{m} \prod_{i \in J_j} f_i\right), \tag{27}$$

where $\{J_1, \ldots, J_m\}$ runs through all partitions of $\{1, \ldots, k\}$ into m disjoint non-empty subsets. In particular,

$$\nu_z^1(f) = \nu_u^1(f),$$
$$\nu_z^2(f_1 \otimes f_2) = \nu_u^2(f_1 \otimes f_2) + \nu_u^1(f_1 f_2).$$

Consider a fixed partition $\mathscr{J} = \{J_1, \ldots, J_m\}$, and assume that u is diffuse, hence z is a point-process. For every $\omega \in \Omega$ the measure in X^k defined by

$$u^*(f_1 \otimes \ldots \otimes f_k, \omega) = u^{[m]}\left(\bigotimes_{j=1}^{m} \prod_{i \in J_j} f_i, \omega\right)$$

is carried by $D_{\mathscr{J}}$. From the first part of Theorem 3 it follows that

$$u^*(D_{\mathscr{J}'}, \omega) = 0 \text{ if } \mathscr{J}' \prec \mathscr{J} \text{ and } \mathscr{J}' \neq \mathscr{J}.$$

Taking expectations we see that the measure

$$f_1 \otimes \ldots \otimes f_k \to \nu_u^m\left(\bigotimes_{j=1}^{m} \prod_{i \in J_j} f_i\right)$$

has the same properties. Therefore, by Theorem 4, the decomposition (27) is identical with the decomposition (17) of the measure $\nu = \nu_z^k$. Hence, comparing equation (27) with equation (23) we get

$$\nu_u^m(A) = \nu_z^m(\Pi_{\mathscr{J}} E_{\mathscr{J}} \cap A) \tag{28}$$

for every $A \in \mathscr{B}_0(X^m)$; observe that $\Pi_{\mathscr{J}} E_{\mathscr{J}}$ is the set of all $(\eta_1, \ldots, \eta_m)$ with distinct components η_j. This is, in a sense, an inverse formula to equation (27).

A doubly stochastic Poisson process z is called a mixed Poisson process if its mean number of points u has the form $u(A, \omega) = \mu(A)\, a(\omega)$ with a fixed measure μ and a random variable a.

2.6.7 INVARIANCE OF RANDOM MEASURES

Suppose a locally compact group $\mathscr{G}$ with a countable base acts continuously in X. A random measure z in X is called strictly stationary if its distribution is invariant under $\mathscr{G}$. By this we mean that for any $f_1, \ldots, f_n \in \mathscr{K}(X)$ and any $G \in \mathscr{G}$ the random vectors $(z(f_1), \ldots, z(f_n))$ and $(z(f_1 \circ G), \ldots, z(f_n \circ G))$ have the same distribution.

Let k be a positive integer. We denote by $\mathscr{H}$ the diagonal group acting in $Y = X^k$ generated by $\mathscr{G}$, that is, the group of all transformations of the form

$$(\xi_1, \ldots, \xi_k) \to (G\xi_1, \ldots, G\xi_k)$$

with $G \in \mathscr{G}$. We make the assumptions of section 2.6.2. For simplicity, an $\mathscr{H}$-invariant measure or set in Y will also be termed $\mathscr{G}$-invariant.

Clearly, any diagonal $D_{\mathscr{J}}$ of Y is invariant under $\mathscr{G}$ and can thus be written as

(29) $$D_{\mathscr{J}} = r^{-1}(\Delta_{\mathscr{J}})$$

where $\Delta_{\mathscr{J}}$ is a Borelian subset of the representation space Γ introduced in Section 2.6.2. It follows that $\mathscr{J}' \prec \mathscr{J}$ if and only if $\Delta_{\mathscr{J}'} \subseteq \Delta_{\mathscr{J}}$. Defining

(30) $$\Gamma_{\mathscr{J}} = \Delta_{\mathscr{J}} - \bigcup_{\substack{\mathscr{J}' : \mathscr{J}' \prec \mathscr{J} \\ \mathscr{J}' \neq \mathscr{J}}} \Delta_{\mathscr{J}'}$$

we have, by equations (29) and (16),

(31) $$E_{\mathscr{J}} = r^{-1}(\Gamma_{\mathscr{J}}),$$

and the sets $\Gamma_{\mathscr{J}}$ form a partition of Γ. Therefore, the disintegration (1) of an $\mathscr{H}$-invariant measure ν in Y can be further decomposed into

(32) $$\nu = \sum_{\mathscr{J}} \int_{\Gamma_{\mathscr{J}}} \tau_\gamma\, \kappa(d\gamma),$$

and by equations (18) and (31), this decomposition coincides with the decomposition (17) of ν.

Consider a fixed partition $\mathscr{J}$ with $\operatorname{cd}(\mathscr{J}) = m$. The projection $\Pi_{\mathscr{J}}$ maps $D_{\mathscr{J}}$ in a one-to-one way onto X^m, and by the definition of the diagonal group, the equivalence classes with respect to $\mathcal{H}$ contained in $D_{\mathcal{J}}$ correspond via $\Pi_{\mathcal{J}}$ to the equivalence classes of X^m with respect to the diagonal group acting in X^m. The $\mathcal{G}$-invariant measure τ_γ with $\gamma \in \Delta_{\mathcal{J}}$ is transformed by $\Pi_{\mathcal{J}}$ into a $\mathcal{G}$-invariant measure $\Pi_{\mathcal{J}}\tau_\gamma$ in X^m carried by the corresponding equivalence class. Hence, making use of equation (30) we see that we know all the relevant measures τ_γ appearing in decompositions of the type (32) for the space X^m if we know them for the space X^k. In fact, the general $\mathcal{G}$-invariant measure ν' on X^m is given by

$$\nu' = \sum_{\mathcal{J}' : \mathcal{J}' \prec \mathcal{J}} \int_{\Gamma_{\mathcal{J}'}} \Pi_{\mathcal{J}} \tau_\gamma \kappa'(d\gamma),$$

where the sets $\Gamma_{\mathcal{J}'}$ and the measures τ_γ are the same as in equation (32) and κ' is some measure in

$$\bigcup_{\mathcal{J}' : \mathcal{J}' \prec \mathcal{J}} \Gamma_{\mathcal{J}'}.$$

In the sequel we will always assume that $\mathcal{G}$ acts transitively on X. This is tantamount to requiring that the ordinary diagonal $D_{\mathcal{J}\min}$ consists of exactly one equivalence class, and we will denote the representing point of Γ by δ so that

$$D_{\mathcal{J}\min} = E_{\mathcal{J}\min} = Y_\delta, \quad \Delta_{\mathcal{J}\min} = \Gamma_{\mathcal{J}\min} = \{\delta\}.$$

The image of τ by $\Pi_{\mathcal{J}\min}$ is then a $\mathcal{G}$-invariant measure in X, and the only one, up to a factor.

A random measure z in X is said to be stationary up to the kth order if it is of kth order, and if for every integer $m \leqslant k$ the moment measure ν_z^m in X^m is $\mathcal{G}$-invariant. We can then represent its moment measure $\nu = \nu_z^k$ in the form of equation (32).

Let z be a point-process. The moment measures ν_z^m with $m < k$ being given in terms of ν_z^k by equation (21), it suffices to require the $\mathcal{G}$-invariance, that is, the $\mathcal{H}$-invariance, of ν_z^k to ensure kth-order stationarity. Moreover, writing

$$(33) \qquad \nu_z^k = \sum_{\mathcal{J}} \int_{\Gamma_{\mathcal{J}}} \tau_\gamma \kappa(d\gamma)$$

it follows from equation (20) and the preceding discussion that the corresponding decomposition of ν_z^m has the form

$$\nu_z^m = \sum_{\mathcal{J}' : \mathcal{J}' \prec \mathcal{J}} \int_{\Gamma_{\mathcal{J}'}} \Pi_{\mathcal{J}} \tau_\gamma \kappa(d\gamma)$$

with the same measure κ where $\mathcal{J}$ may be any partition of $\{1, \ldots, k\}$ such that $\mathrm{cd}(\mathcal{J}) = m$. In particular

$$\nu_z^1 = \kappa\{\delta\} \tau.$$

Next, consider the case of a doubly stochastic Poisson process with mean number of points u. Clearly, z is strictly stationary if and only if u is, and likewise z is stationary up to the kth order if and only if u is. In the case where u is diffuse and therefore z a point-process, the decompositions (27) and (33) are identical, hence

$$\nu_u^m = \int_{\Gamma_{\mathscr{I}}} \Pi_{\mathscr{I}} \tau_\gamma \kappa(d\gamma)$$

in accordance with equation (28). In particular,

(34) $$\nu_u^k = \int_{\Gamma_{\mathscr{I}_{\max}}} \tau_\gamma \kappa(d\gamma).$$

Finally, suppose that X is compact and, therefore, Y too. Then the image μ^{*k} of the kth power $\mu^{[k]}$ of any measure μ in X under the map r is well defined, and

(35) $$\mu^{*k}(g) = \mu^{[k]}(g \circ r)$$

for every bounded Baire function g on Γ. Let z be any random measure in X. Applying equation (35) to all measures $\mu = z^I(\omega)$ with $\omega \in \Omega$ and taking expectations we get

(36) $$\mathbf{E}(z^{*k}(g)) = \nu_z^k(g \circ r).$$

If the invariant measures τ_γ are normalized by equation (3) and z is stationary up to the kth order, the formulas (36) and (4) allow us to express the measure κ of the decomposition (33) directly in the form

$$\kappa(g) = \beta^{-1}\,\mathbf{E}(z^{*k}(g)).$$

2.6.8 FACTORIZATION OF RANDOM MEASURES

We now combine the reasonings of Sections 2.6.3 and 2.6.7 by assuming that the space X itself is a product

(37) $$X = P \times S$$

of two locally compact spaces P and S with countable bases. We also assume that we are given a locally compact group $\mathscr{U}$ of transformations of P with a countable base which admits a non-trivial invariant measure λ in P, and only one, up to a factor. Let $\mathscr{G}$ be a group acting in X as before and $\mathscr{V}$ the group of all transformations V of S with the property that the map $(p, \varphi) \to (p, V\varphi)$, where $p \in P$ and $\varphi \in S$, belongs to $\mathscr{G}$. We make the final assumption that any $G \in \mathscr{G}$ has the form

(38) $$G(p,\varphi) = (U_\varphi p, V\varphi)$$

where $V \in \mathscr{V}$ and $U_\varphi \in \mathscr{U}$ for every $\varphi \in S$.

Consider a random measure u in X which can be factored as

(39) $$u = \lambda \otimes y,$$

that is, $u(f \otimes g) = \lambda(f)\, y(g)$ for all $f \in \mathscr{K}(P)$ and $g \in \mathscr{K}(S)$, where y is a random measure in S. Then, by equation (38) and Fubini's theorem, u is strictly stationary under $\mathscr{G}$ if and only if y is strictly stationary under $\mathscr{V}$.

Let k be a positive integer. By equation (37), the space $Y = X^k$ has, upon a permutation of the factors, the form of equation (5) with $Q = P^k$ and $T = S^k$. It follows from equation (39) that

(40) $$\nu_u^k = \lambda^{[k]} \otimes \nu_y^k$$

if u or, equivalently, y is of kth order. Hence u is $\mathscr{G}$-stationary up to the kth order if and only if y is $\mathscr{V}$-stationary up to the same order.

The trivial implication (39)⇒(40) admits the following converse.

Theorem 5 [4]. *Let $k > 1$ and u be a kth-order random measure in X whose kth moment measure can be written as*

(41) $$\nu_u^k = \lambda^{[k]} \otimes \nu'$$

with some measure ν' in T. Then u has the form of equation (39).

Proof. Let Ω_0 be the set of all $\omega \in \Omega$ such that the measure $u^I(\omega)$ vanishes identically. Then $\Omega_0 \in \mathscr{F}$ and

(42) $$u(f \otimes g, \omega) = u((f \circ U) \otimes g, \omega)$$

for all $f \in \mathscr{K}(P)$, $g \in \mathscr{K}(S)$, $U \in \mathscr{U}$ and $\omega \in \Omega_0$. Moreover, since P and S have countable bases, there are sequences $f_n \in \mathscr{K}^+(P)$ and $g_n \in \mathscr{K}^+(S)$ such that

(43) $$\Omega - \Omega_0 = \bigcup_n \{\omega\colon u(f_n \otimes g_n, \omega) > 0\}.$$

Let $f \in \mathscr{K}(P)$, $g \in \mathscr{K}(S)$, $U \in \mathscr{U}$ and $f' = f \circ U$. From equation (41) and $\lambda(f) = \lambda(f')$ it follows that

$$\begin{aligned}\mathbf{E}(u(f \otimes g)\, u(f' \otimes g)\, u(f_n \otimes g_n)^{k-2}) &= \mathbf{E}(u(f \otimes g)^2\, u(f_n \otimes g_n)^{k-2}) \\ &= \mathbf{E}(u(f' \otimes g)^2\, u(f_n \otimes g_n)^{k-2}),\end{aligned}$$

hence

$$\mathbf{E}((u(f \otimes g) - u(f' \otimes g))^2\, u(f_n \otimes g_n)^{k-2}) = 0$$

and therefore $u(f \otimes g) = u(f' \otimes g)$ almost surely on the set

$$\{\omega\colon u(f_n \otimes g_n, \omega) > 0\}.$$

On account of equations (42) and (43) this implies $u(f \otimes g) = u(f' \otimes g)$ almost surely. Making use of the fact that P, S and $\mathscr{U}$ have countable bases we find that, for almost all ω, we have $u(f\otimes g,\omega) = u((f\circ U)\otimes g,\omega)$ for all $f\in\mathscr{K}(P)$, $g\in\mathscr{K}(S)$ and $U\in\mathscr{U}$. By the definition of λ, this entails $u(f\otimes g,\omega) = \lambda(f)\otimes y(g,\omega)$, and it follows now immediately that y is a random measure in S.

Let us look at the case where Y, the diagonal group $\mathscr{H}$ and a certain subset Γ^0 of Γ satisfy the assumption of Section 2.6.3. Clearly, if σ_γ is a $\mathscr{V}$-invariant measure in T carried by T_γ, the measure

(44) $$\tau_\gamma = \lambda^{[k]}\otimes\sigma_\gamma,$$

for $\gamma\in\Gamma^0$, is $\mathscr{G}$-invariant and carried by Y_γ, hence equation (7) holds with $\rho = \lambda^{[k]}$ for every $\mathscr{G}$-invariant measure ν carried by $Q\times T^0$. In this context, we have

Theorem 5, Corollary 1. *Let $k>1$, and suppose that $Y,\mathscr{H}$ and Γ^0 satisfy the assumptions of Section* 2.6.3, *and that τ_γ has the form of equation* (44) *for every $\gamma\in\Gamma^0$. Then every kth-order stationary random measure u with the property $\nu_u^k(Q\times(T-T^0)) = 0$ admits a factorization.*

2.6.9 HYPERPLANE PROCESSES

In the framework of the preceding section we turn to the particular case where $P = R$ is the real line and $S = S_{n-1}$ is the unit hypersphere in the Euclidean space R^n, thus $X = R\times S_{n-1}$ is a hypercylinder. The points $\xi = (p,\varphi)$ of X correspond in a one-to-one fashion to the oriented hyperplanes in R^n, the hyperplane represented by ξ being the set

(45) $$\{\alpha\colon \alpha\in R^n, \langle\varphi,\alpha\rangle + p = 0\},$$

where $\langle\varphi,\alpha\rangle = \sum_{i=1}^n \varphi_i\,\alpha_i$, and the orientation is determined by the direction of φ.

As usual we denote by $\mathscr{O}^+$ the group of all rotations of R^n, that is, all orthonormal transformations with determinant 1. Let $\mathscr{G}$ be the group of all transformations of X induced on the hyperplanes by the Euclidean motions of R^n. By equation (45) the Euclidean motion $\alpha\to V\alpha+\varepsilon$ with $\varepsilon\in R^n$ and $V\in\mathscr{O}^+$ generates in X the map

(46) $$(p,\varphi)\to(p-\langle V\varphi,\varepsilon\rangle, V\varphi).$$

Hence $\mathscr{G}$ is the group of all maps $(p,\varphi)\to(p+t(\varphi), V\varphi)$ where $V\in\mathscr{O}^+$ and t is a linear functional on R^n.

Let $\mathscr{U}$ be the group of all translations of R, and $\mathscr{V}$ the restriction of $\mathscr{O}^+$ to S_{n-1}. By equation (46), $\mathscr{V}$ consists of all transformations V such that $(p,\varphi)\to(p,V\varphi)$ belongs to $\mathscr{G}$, and every $G\in\mathscr{G}$ has the form (32), where $V\in\mathscr{V}$ and $U_\varphi\in\mathscr{U}$ for every $\varphi\in S_{n-1}$.

It is well known [8] that there exists an invariant measure τ on X, and only one, up to a factor, namely $\tau = \lambda \otimes \sigma$ where λ is the one dimensional Lebesgue measure in R and σ the surface measure on S_{n-1}.

Next we are going to study the equivalence classes in a product space $Y = X^k$ with $k \leqslant n$ under the diagonal group defined by $\mathscr{G}$, and the invariant measures carried by them. As noted in Section 2.6.7 it would suffice to consider only the case $k = n$; however, in view of the applications, it appears to be more practical to take up directly the general case.

Rearranging the factors of X^k as we did in Section 2.6.8 we have

$$Y = Q \times T, \quad Q = R^k, \quad T = (S_{n-1})^k.$$

Consider a point $\eta^0 = (p_1^0, \ldots, p_k^0;\ \varphi_1^0, \ldots, \varphi_k^0)$ of Y. The equivalence class Y_γ which contains η^0 consists of all points η of the form

$$(47) \qquad \eta = (p_1, \ldots, p_k;\ V\varphi_1^0, \ldots, V\varphi_k^0), \quad p_i = p_i^0 - \langle V\varphi_i^0, \varepsilon \rangle,$$

where $\varepsilon \in R^n$ and $V \in \mathscr{V}$. We distinguish two cases:

(*a*) Rank $(\varphi_1^0, \ldots, \varphi_k^0) = k$. In this case, for any fixed $V \in \mathscr{V}$, the numbers $p_1, \ldots, p_k$ given by equation (47) take all real values if ε runs through R^n, in particular $(p_1, \ldots, p_k;\ \varphi_1^0, \ldots, \varphi_k^0) \sim (q_1, \ldots, q_k;\ \varphi_1^0, \ldots, \varphi_k^0)$ for all $p_1, \ldots, p_k, q_1, \ldots, q_k$. Hence, denoting by Γ^0 the set of all γs which define equivalence classes of this type, we have the situation outlined in Sections 2.6.3 and 2.6.8. The set T^0 is the set of all $(\varphi_1, \ldots, \varphi_k)$ of rank k, and the equivalence relation induced in T^0 is the equivalence under the diagonal group given by $\mathscr{V}$, that is, $(\varphi_1, \ldots, \varphi_k) \sim (\psi_1, \ldots, \psi_k)$ if and only if there is a transformation V in $\mathscr{V}$ such that $\psi_i = V\varphi_i$, $i = 1, \ldots, k$.

Let $\tilde{\sigma}$ be the Haar measure in $\mathscr{O}^+$ normalized by $\tilde{\sigma}(\mathscr{O}^+) = \sigma(S_{n-1})$, and σ_γ^0 the image of $\tilde{\sigma}$ under the map

$$(48) \qquad V \to (V\varphi_1^0, \ldots, V\varphi_k^0)$$

of $\mathscr{O}^+$ onto T_γ. Denote by $|\det(\varphi_1^0, \ldots, \varphi_k^0)|$ the k-dimensional volume of the parallelepiped spanned by $\varphi_1^0, \ldots, \varphi_k^0$. Then

$$(49) \qquad \tau_\gamma = |\det(\varphi_1^0, \ldots, \varphi_k^0)|^{-1} \lambda^k \otimes \sigma_\gamma^0$$

is an $\mathscr{H}$-invariant measure in Y concentrated on Y_γ, and the only one up to a factor [8]. In the case $k = n$, the map (48) is bijective and τ_γ is called 'Poincaré's kinematic measure'; in the case $k = 1$ we have $\tau_\gamma = \tau$.

(*b*) Rank $(\varphi_1^0, \ldots, \varphi_k^0) = m < k$. To simplify matters let us assume that $\varphi_1^0, \ldots, \varphi_m^0$ are linearly independent, hence

$$\varphi_i^0 = \sum_{j=1}^{m} a_{ij} \varphi_j^0, \quad i = m+1, \ldots, k,$$

with $a_{ij} \in R$. Then, for any fixed $V \in \mathscr{V}$, the numbers $p_i = p_i^0 - \langle V\varphi_i^0, \varepsilon \rangle$ with $i = 1, \ldots, m$ take all real values if ε runs through R^n, whereas

$$p_i = p_i^0 + \sum_{j=1}^{m} a_{ij}(p_j - p_j^0), \quad i = m+1, \ldots, k.$$

Thus every element η of Y_γ can be uniquely represented by

$$\eta \leftrightarrow p_1, \ldots, p_m, \chi, \quad \text{where } \chi = (V\varphi_1^0, \ldots, V\varphi_m^0). \tag{50}$$

As before, let σ_γ^0 stand for the image of $\tilde{\sigma}$ under the map

$$V \to (V\varphi_1^0, \ldots, V\varphi_m^0).$$

Then, the measure

$$|\det(\varphi_1^0, \ldots, \varphi_m^0)|^{-1} \lambda^{[m]} \otimes \sigma_\gamma^0$$

corresponds via the mapping (50) to an $\mathscr{H}$-invariant measure τ_γ in Y carried by Y_γ, and this is the only one, up to a factor. The normalizing factor $|\det(\varphi_1^0, \ldots, \varphi_m^0)|^{-1}$ makes τ_γ independent of the choice of m linearly independent vectors out of the entire sequence $\varphi_1^0, \ldots, \varphi_k^0$.

Given a partition $\mathscr{J}$ of $\{1, \ldots, k\}$ such that $\mathrm{cd}(\mathscr{J}) = m$, we define $\Gamma_{\mathscr{J}}^0$ to be the set of all γs in $\Delta_{\mathscr{J}}$ such that rank $(\varphi_1^0, \ldots, \varphi_k^0) = m$, in particular $\Gamma_{\mathscr{J}_{\max}}^0 = \Gamma^0$ and $\Gamma_{\mathscr{J}_{\min}}^0 = \Delta_{\mathscr{J}_{\min}}^0$. It follows, of course, that $\Gamma_{\mathscr{J}}^0 \subseteq \Gamma_{\mathscr{J}}$. After a suitable permutation of the axes, the equivalence class Y_γ with $\gamma \in \Gamma_{\mathscr{J}}^0$ consists of all points

$$(p_1, \ldots, p_m, p_{m+1}, \ldots, p_k;\ V\varphi_1^0, \ldots, V\varphi_m^0, V\varphi_{m+1}^0, \ldots, V\varphi_k^0),$$

where $\varphi_1^0, \ldots, \varphi_m^0$ are fixed linearly independent vectors in S_{n-1}, V runs through $\mathscr{V}$, $(p_1, \ldots, p_m)$ takes all values in R^m, each φ_i with $i > m$ is equal to some φ_j with $j \leqslant m$, and $\varphi_i = \varphi_j, j \leqslant m < i$ implies $p_i = p_j$. The application $\Pi_{\mathscr{J}}$ will reduce this case to the situation (*a*) with m in the place of k.

Next we consider a few transformations F of Y which, in general, do not belong to $\mathscr{H}$.

(*i*) F has the form

$$(p_1, \ldots, p_k;\ \varphi_1, \ldots, \varphi_k) \to (p_1 + q_1, \ldots, p_k + q_k;\ \varphi_1, \ldots, \varphi_k) \tag{51}$$

with fixed $q_i \in R$. If the q_is are all equal, we obtain the map of Y induced by a 'translation of the hypercylinder X parallel to its generators'.

(*ii*) F_2 is the map

$$(p_1, \ldots, p_k;\ \varphi_1, \ldots, \varphi_k) \to (-p_1, \ldots, -p_k;\ \varphi_1, \ldots, \varphi_k).$$

(*iii*) F_3 is the map of Y induced by a change of the orientation of all hyperplanes, that is,

$$(p_1, \ldots, p_k;\ \varphi_1, \ldots, \varphi_k) \to (-p_1, \ldots, -p_k;\ -\varphi_1, \ldots, -\varphi_k).$$

(*iv*) F is a map of Y induced by a 'reflection' in R^n, that is,

$$(p_1, \ldots, p_k;\ \varphi_1, \ldots, \varphi_k) \to (p_1, \ldots, p_k;\ W\varphi_1, \ldots, W\varphi_k),$$

where W is an orthonormal transformation of R^n with determinant -1.

(*v*) F is a permutation of the axes of Y, that is,

$$(p_1, \ldots, p_k;\ \varphi_1, \ldots, \varphi_k) \to (p_{i_1}, \ldots, p_{i_k};\ \varphi_{i_1}, \ldots, \varphi_{i_k})$$

with some fixed permutation $i_1, \ldots, i_k$ of $1, \ldots, k$.

Clearly, if n is even, $F_3 = F_2 H$ with some $H \in \mathscr{H}$, whereas if n is odd, every F of type (*iv*) can be written as $F = F_2 F_3 H$ with some $H \in \mathscr{H}$.

In all of these examples we have $FHF^{-1} \in \mathscr{H}$ for every $H \in \mathscr{H}$, hence there is a transformation Φ of Γ which satisfies equation (2). We write $\Phi = \Phi_l$ if $F = F_l$, $l = 2, 3$. It also follows from the definition of the τ_γs that $\tau_{\Phi(\gamma)}$ is the image of τ_γ under F. Therefore, given an $\mathscr{H}$-invariant measure ν on Y in its representation (1), we can apply Theorem 2 to it. This involves, of course, a more detailed study of Φ.

The general discussion of Φ is rather tedious; a complete description in the case $n = k = 2$ was given in [5]. The only trivial, and still slightly useful, remarks we can make here are that for n even, we have $\Phi_3 = \Phi_2$, whereas $\Phi = \Phi_2 \Phi_3$ for every F of type (*iv*) if n is odd.

For the applications we have in mind it suffices to consider the case $\gamma \in \Gamma^0_{\mathscr{J}}$ with some $\mathscr{J}$. Then from the shape of the elements of Y_γ exhibited above we derive immediately that $\varphi(\gamma) = \gamma$ if F has the form (51) and $q_i = q_j$ for any i and j in the same component of the partition $\mathscr{J}$, in particular if F is induced by a translation of the hypercylinder X parallel to its generators, or if F is an arbitrary map of type (*i*) and $\mathscr{J} = \mathscr{J}_{\max}$. We also have $\varphi_2(\gamma) = \gamma$. Therefore, $\varphi_3(\gamma) = \gamma$ if n is even, and $\varphi(\gamma) = \varphi_3(\gamma)$ for any F of type (*iv*) if n is odd. Moreover, if $k < n$, we find that $\varphi(\gamma) = \gamma$ for every F of type (*iv*). Finally, in the case $k = 2$, denoting by φ_5 the map φ which corresponds to the permutation

$$F_5 \colon (p_1, p_2;\ \varphi_1, \varphi_2) \to (p_2, p_1;\ \varphi_2, \varphi_1),$$

we have $\Phi(\gamma) = \Phi_5(\gamma)$ for every F of type (*iv*).

On account of these remarks, Theorem 2 has the following corollary.

Theorem 2, Corollary 1. *Let ν be an $\mathscr{H}$-invariant measure on Y given by its representation* (1), *that is, equation* (32), *and suppose that κ is carried by*

$$\bigcup_{\mathscr{I}} \Gamma^0_{\mathscr{I}}.$$

Then ν is invariant under any translation of X parallel to its generators, the map F_2, the map F_3 if n is even, and any map of type (iv) if $k<n$. If n is odd, ν is invariant under F_3 if and only if it is invariant under any, or some, F of type (iv). If $k = 2$, ν is invariant under any, or some, F of type (iv) if and only if it is invariant under F_5. Finally, if κ is carried by Γ^0, ν is invariant under any F of type (i).

The last assertion follows, of course, directly from the fact that by equation (49) the disintegration (32) of an $\mathscr{H}$-invariant measure ν carried by $R^k \times T^0$ takes the factored form

$$\nu = \lambda^{[k]} \otimes \int_{\Gamma^0} \sigma_\gamma \, \kappa(d\gamma), \tag{52}$$

where $\sigma_\gamma = |\det(\varphi_1^0, \ldots, \varphi_k^0)|^{-1} \sigma_\gamma^0$ if $(\varphi_1^0, \ldots, \varphi_k^0) \in T_\gamma$; this is the situation described by equation (7).

By a kth-order non-degenerate stationary point-process z in X we mean a kth-order stationary point-process in X with the property that for almost all ω and every $m \leqslant k$, any m distinct elements $(p_1, \varphi_1), \ldots, (p_m, \varphi_m)$ of the carrier of $z^I(\omega)$ have linearly independent directions $\varphi_1, \ldots, \varphi_m$. This is equivalent to ν_z^k being carried by

$$\bigcup_{\mathscr{I}} r^{-1}(\Gamma^0_{\mathscr{I}}),$$

hence the corollary to Theorem 2 applies in this case to $\nu = \nu_z^k$. In addition, ν_z^k is of course invariant under any map F of type (v).

Next, from Section 2.6.8, in particular Theorem 5, Corollary 1, we get

Theorem 5, Corollary 2. *If $k>1$, any kth-order $\mathscr{G}$-stationary random measure u in X which satisfies*

$$\nu_u^k(R^k \times (T - T^0)) = 0 \tag{53}$$

admits a factorization $u = \lambda \otimes y$ where y is a kth-order $\mathscr{V}$-stationary random measure on S_{n-1}, and

$$\nu_y^k(T - T^0) = 0. \tag{54}$$

Conversely, given any random measure y in S_{n-1}, $u = \lambda \otimes y$ is a random measure in X, and u is strictly or kth-order $\mathscr{G}$-stationary if and only if y has the corresponding property with respect to $\mathscr{V}$. If y is diffuse, u also is. Finally, equations (53) and (54) are equivalent.

Note that, by Theorem 3, Corollary 1, the condition (53) or (54) implies that u or y, respectively, is diffuse.

Let z be the doubly stochastic Poisson process with a diffuse and kth-order stationary mean number of points u, and consider the decomposition (33) of its covariance measure. Then, ν_u^k can be disintegrated according to equation (34), and since $\Gamma^0 \subseteq \Gamma_{\mathcal{J}_{max}}$ it follows that u satisfies equation (53) if and only if

$$\kappa(\Gamma_{\mathcal{J}_{max}} - \Gamma^0) = 0,$$

that is,

(55) $$\nu_z^k(E_{\mathcal{J}_{max}} \cap (R^k \times (T - T^0))) = 0.$$

Assuming this to be true, the moment measure ν_y^k of the random measure y on S_{n-1} obtained from Theorem 5, Corollary 2, is given by

$$\nu_y^k = \int_{\Gamma^0} \sigma_\gamma \, \kappa(d\gamma).$$

By a simple transformation of the coordinates in X we derive from the preceding factorization of u that, given any fixed line L in R^n, the point-process of the intersections of the hyperplanes of z with L is a mixed Poisson process whose mean number of points is a random multiple of the Lebesgue measure on L [3].

Clearly, the condition (55) is, for any kth-order stationary point-process z, necessary for z to be kth-order non-degenerate.

We now return to the case of an arbitrary kth-order non-degenerate stationary point-process z where $k \geqslant 2$. As usual we denote by $||.||_m$ the norm in the space $\mathcal{L}_m(\mathbf{P})$. Given a Borelian subset M of S_{n-1} and an integer i we define

$$M^{(i)} = [i, i+1[\times M$$

and

$$x_i(M) = z(M^{(i)}).$$

The moment measures ν_z^m with $m \leqslant k$ being invariant under any translation of X parallel to its generators by the corollary to Theorem 2, we have

(56) $$||x_i(M)||_m = ||x_0(M)||_m, \quad i = 0, \pm 1, \ldots,$$

and the sequence $(x_i(M))_{i=1,2,\ldots}$ is stationary in $\mathcal{L}_2(\mathbf{P})$, that is,

$$(x_i(M) x_j(M)) = \mathbf{E}(x_{j-i}(M) x_0(M)) \text{ for all } i \text{ and } j.$$

We set

(57) $$y_l(M) = \frac{1}{l} \sum_{i=0}^{l-1} x_i(M) = \frac{1}{l} z([0, l[\times M).$$

By von Neumann's mean ergodic theorem, the limit

$$(58) \qquad y(M) = \lim_{l\to\infty} y_l(M)$$

exists strongly in $\mathscr{L}_2(\mathbf{P})$, and

$$(59) \qquad y(M) = \lim_{l\to\infty} \frac{1}{l} \sum_{i=0}^{l-1} x_{j_i}(M)$$

for every strictly increasing sequence of integers $j_i \geqslant 0$. Of course $y(M)$ is only determined up to a change in a set of probability 0.

The definition (59) of y shows that $0 \leqslant y(M)$, and that

$$y(M_1 \cup M_2) = y(M_1) + y(M_2)$$

almost surely if M_1 and M_2 are disjoint. Let M_j be a sequence of Borel sets in S_{n-1} such that $M_j \searrow \emptyset$. Then $x_0(M_j) \searrow 0$ almost surely, hence strongly in $\mathscr{L}_2(\mathbf{P})$. Therefore, by equations (56), (57) and (58),

$$\lim_{j\to\infty} \| y(M_j) \|_2 = 0,$$

hence $y(M_j) \to 0$ almost surely since $y(M_j)$ decreases. It follows that $y(M)$ can be defined for every M in such a way that y is a random measure in S_{n-1}.

Consider m Borelian subsets $M_1, \ldots, M_m$ of S_{n-1}, where $m \leqslant k$. The sequences $(x_i(M_k))_{i=1,2,\ldots}$ being bounded in $\mathscr{L}_m(\mathbf{P})$ on account of equation (56), there is a sequence of positive integers $(j_i)_{i=1,2,\ldots}$ such that

$$\lim_{l\to\infty} \frac{1}{l} \sum_{i=0}^{l-1} x_{j_i}(M_h), \quad h = 1, \ldots, m,$$

exist strongly in $\mathscr{L}_m(\mathbf{P})$, see, for example [7]. By equation (59), these limits are equal to $y(M_1), \ldots, y(M_m)$, respectively. Therefore, writing

$$N_h^{(l)} = \bigcup_{i=0}^{l-1} M_h^{(j_i)} = \left(\bigcup_{i=0}^{l-1} [j_i, j_i+1[\right) \times M_h, \quad h = 1, \ldots, m,$$

we have

$$y(M_h) = \lim_{l\to\infty} \frac{1}{l} z(N_h^{(l)}), \quad h = 1, \ldots, m,$$

strongly in $\mathscr{L}_m(\mathbf{P})$ which entails

$$(60) \qquad \nu_y^m(M_1 \times \ldots \times M_m) = \lim_{l\to\infty} \frac{1}{l^m} \nu_z^m(N_1^{(l)} \times \ldots \times N_m^{(l)})$$

by the definition of ν_z^m and ν_y^m.

Since z is kth-order stationary and non-degenerate, we can represent ν_z^k in the form

$$\nu_z^k = \sum_{\mathcal{J}} \int_{\Gamma^0_{\mathcal{J}}} \tau_\gamma \,\kappa(d\gamma). \tag{61}$$

As noted in Section 2.6.7, this implies

$$\nu_z^m = \sum_{\mathcal{J}': \mathcal{J}' \prec \mathcal{J}} \int_{\Gamma^0_{\mathcal{J}'}} \Pi_{\mathcal{J}} \tau_\gamma \,\kappa(d\gamma),$$

where $\mathcal{J}$ may be any partition of $\{1, \ldots, k\}$ such that $\mathrm{cd}(\mathcal{J}) = m$. Taking into account the form of the invariant measure τ_γ for $\gamma \in \Gamma^0_{\mathcal{J}'}$ we find that

$$\nu_z^m(N_1^{(l)} \times \ldots \times N_m^{(l)}) = l^m \int_{\Gamma^0_{\mathcal{J}}} (\Pi_{\mathcal{J}} \sigma_\gamma)(M_1 \times \ldots \times M_m)\,\kappa(d\gamma) + \sum_{\substack{\mathcal{J}': \mathcal{J}' \prec \mathcal{J} \\ \mathcal{J}' \neq \mathcal{J}}} l^{\mathrm{cd}(\mathcal{J}')} s_{\mathcal{J}'},$$

where the numbers $s_{\mathcal{J}'}$ do not depend on l and $\Pi_{\mathcal{J}}$ is defined in $(S_{n-1})^k$ in the same way as in X^k. It now follows from equation (60) that

$$\nu_y^m(M_1 \times \ldots \times M_m) = \int_{\Gamma^0_{\mathcal{J}}} (\Pi_{\mathcal{J}} \sigma_\gamma)(M_1 \times \ldots \times M_m)\,\kappa(d\gamma). \tag{62}$$

Let $u = \lambda \otimes y$, hence by equation (62),

$$\nu_u^m = \lambda^{[m]} \otimes \nu_y^m = \lambda^{[m]} \otimes \int_{\Gamma^0_{\mathcal{J}}} \Pi_{\mathcal{J}} \sigma_\gamma \,\kappa(d\gamma) = \int_{\Gamma^0_{\mathcal{J}}} \Pi_{\mathcal{J}} \tau_\gamma \,\kappa(d\gamma). \tag{63}$$

Define $\tilde{z}$ to be the doubly stochastic Poisson process with mean number of points u and write

$$\nu_{\tilde{z}}^k = \sum_{\mathcal{J}} \int_{\Gamma_{\mathcal{J}}} \tau_\gamma \,\tilde{\kappa}(d\gamma) \tag{64}$$

with some measure $\tilde{\kappa}$ in Γ. Then, as shown in Section 2.6.7, we have

$$\nu_u^m = \int_{\Gamma_{\mathcal{J}}} \Pi_{\mathcal{J}} \tau_\gamma \,\tilde{\kappa}(d\gamma) \tag{65}$$

for any $\mathcal{J}$ with $\mathrm{cd}(\mathcal{J}) = m \leqslant k$. Comparing equations (63) and (65) we see that κ and $\tilde{\kappa}$ coincide within $\Gamma_{\mathcal{J}}$ for every $\mathcal{J}$, thus $\kappa = \tilde{\kappa}$. By equations (61) and (64), this amounts to $\nu_z^k = \nu_{\tilde{z}}^k$ which, in turn, implies $\nu_z^m = \nu_{\tilde{z}}^m$ for $m = 1, \ldots, k$. Therefore, we have the following theorem.

Theorem 6. *Given any kth-order non-degenerate $\mathcal{G}$-stationary point-process z in $X = R \times S_{n-1}$ with $1 < k \leqslant m$, there is a doubly stochastic Poisson process $\tilde{z}$ such that $\nu_z^m = \nu_{\tilde{z}}^m$, $m = 1, \ldots, k$.*

The proof of this theorem as well as Theorem 5, Corollary 2 show that the mean number of points of $\tilde{z}$ has the form $u = \lambda \otimes y$ with a random measure y on S_{n-1} which satisfies equation (54).

We note that this theorem, with $k = 2$, does not hold in the case of the space R endowed with the translation group [1, p. 184]. It does not hold either for the space S_{n-1} with the rotation group as shown by the following reasoning.

Let X be a compact space, $\mathscr{G}$ a group acting continuously on it, μ a $\mathscr{G}$-invariant measure in X normalized by $\mu(X) = 1$ and z the point-process defined by a single point distributed in X according to μ. Then z is strictly stationary and of order k for all k, and

$$\mathbf{E}(z(M)^2) = \mathbf{E}(z(M))$$

for every Borelian set $M \subseteq X$. Hence there can be no doubly stochastic Poisson process $\tilde{z}$ such that $\nu_z^2 = \nu_{\tilde{z}}^2$, because if there were such a process with mean number of points u, we would have

$$\mathbf{E}(z(M)^2) = \mathbf{E}(\tilde{z}(M)^2) = \mathbf{E}(u(M)^2) + \mathbf{E}(u(M))$$
$$= \mathbf{E}(u(M)^2) + \mathbf{E}(\tilde{z}(M)) = \mathbf{E}(u(M)^2) + \mathbf{E}(z(M))$$

for any M, thus $u(M) = 0$ almost surely and therefore $\tilde{z} = 0$.

REFERENCES

1. M. S. Bartlett, *An Introduction to Stochastic Processes*, 2nd ed., Cambridge, England (1966).
2. N. Bourbaki, *Topologie générale*, Chap. 3: Groupes topologiques (Actual. scientifiques et industrielles 1143), 3ième éd., Hermann, Paris (1960).
3. R. Davidson, Ph.D. thesis, Cambridge, England (1967). (Largely reprinted here in 2.1.)
4. ——, "Construction of line processes: second-order properties", *Izv. Akad. Nauk Armjan. SSR Ser. Fiz.-Mat. Nauk.* (2.4 of this book.)
5. K. Krickeberg, "Invariance properties of the correlation measure of line processes", *Izv. Akad. Nauk Armjan. SSR Ser. Fiz.-Mat. Nauk.* (2.5 of this book.)
6. ——, "The Cox process", *Sympos. Math.* 9, *Calcolo Probab., Teor. Turbolenza* 1971 (1972), 151–167.
7. F. Riesz and B. Sz. Nagy, *Leçons d'Analyse Fonctionnelle*, 3ième éd., Gauthier-Villars, Paris (1955).
8. L. A. Santaló, *Introduction to Integral Geometry* (Actual. scientifiques et industrielles 1198), Hermann, Paris (1953).
9. W. von Waldenfels, "Zur mathematischen Theorie der Druckverbreiterung von Spektrallinien. II", *Ztschr. Wahrsch'theorie & verw. Geb.* **13** (1969), 39–59.

2.7

On the Palm Probabilities of Processes of Points and Processes of Lines

F. PAPANGELOU†

2.7.1 INTRODUCTION

A useful tool in the study of random systems of points are the Palm probabilities $\mathbf{P}(A|x)$. Briefly $\mathbf{P}(A|x)$ is the conditional probability of an event A given that x is one of the points in the system. The main theorems in the present paper deal with properties which follow from the hypothesis that $\mathbf{P}(\cdot|x)$ is absolutely continuous relative to the absolute probability distribution $\mathbf{P}$ on some σ-field of events describable in terms of what happens in the complement of $\{x\}$. A preliminary announcement of some of the results appeared in [17].

Though we investigate only two cases here (systems of points on the line and systems of lines in the plane) it would obviously be a waste of space to build the machinery needed for these two cases separately. Consequently we discuss in Sections 2.7.2, 2.7.3 and 2.7.4 the general set-up of point systems in a 'homogeneous space'. Stationarity is understood as invariance of the distribution of the process under some group acting on the space. In Section 2.7.2 we describe the sample space Ω used subsequently. We consider the elements ω of Ω as counting measures and topologize Ω with the vague topology to obtain a Polish space. Thus we have a simple setting for the handling of measurability, weak convergence, etc. A similar approach has recently been adopted by others ([7] for instance). We confine ourselves to stating the facts without proofs.

In Section 2.7.3 we review briefly the various ways that have been proposed for dealing with the Palm probabilities and suggest two 'new'

† Supported by NSF grant GP-21339.

ones, which are really variations on known themes. One of them links $\mathbf{P}(\cdot\,|0)$ with an Ambrose type representation of a stationary point-process on the line as a 'flow under a function'. As however we need Palm probabilities for non-stationary processes and for multidimensional ones we prefer an alternative approach: we follow the definition given in [20] but use a different tool for handling it, namely the Fubini theorem. In fact we show in the second half of Section 2.7.4 how practically all the important formulae involving the Palm distribution can be easily deduced from the Fubini theorem.

Theorems 3 and 4 in Section 2.7.5 are stronger versions of results essentially known for the stationary case. (The author was not aware of this when the report [17] was made.) They characterize the Poisson and mixed Poisson processes on the line in terms of the Palm distribution. Both are needed in the form we present them here for the proof of the main results of Sections 2.7.5 and 2.7.6. To describe Theorem 5, given a stationary point-process on the line let $\mathscr{F}_t$ be the σ-field of events happening in $(-\infty, t)$. It is proved that if $\mathbf{P}(\cdot\,|0) \ll \mathbf{P}(\cdot)$ on $\mathscr{F}_0$, equivalently if

$$\mathbf{P}(A\,|\,t) = \int_A X(t,\omega)\,\mathbf{P}(d\omega)$$

whenever $A \in \mathscr{F}_t$ then the past-dependent random transformation

$$t \Rightarrow \int_0^t X(\tau)\,\mathrm{d}\tau$$

of the positive half-line sends the non-negative points of the system onto a Poisson process independent of $\mathscr{F}_0$. With a equal to the intensity of the process, Theorem 7 characterizes $aX(t)$ as the derivative and

$$a\int_0^t X(\tau)\,d\tau$$

as the Burkill integral of the random interval function $\mathbf{E}(N(x,y)\,|\,\mathscr{F}_x) =$ expected number of points of the system in (x,y) given the events that happened to the left of x. Theorem 6 is the resulting limit theorem. Let $0 < \xi_1 < \xi_2 < \ldots$ be a partition of $[0, \infty)$. Trace the interval $(0, \xi_1)$ with uniform velocity so as to cover it in time $\mathbf{E}(N(0, \xi_1)\,|\,\mathscr{F}_0)$. Then trace (ξ_1, ξ_2) uniformly so as to cover it in time $\mathbf{E}(N(\xi_1, \xi_2)\,|\,\mathscr{F}_{\xi_1})$; and so on. The time instants at which one meets the points of the process form a new point-process. When the mesh of the partition $0 < \xi_1 < \xi_2 < \ldots$ goes to zero this new point-process converges in law to the Poisson process with rate one.

All this says roughly that if we move along the half-line with variable speed regulated in such a way that, given up-to-the-moment information on the past, we try to meet expected future points at an instantaneous rate of one per time unit then we meet them at the times of a Poisson process with rate one.

Theorem 8 provides a strong converse to Theorem 5; only the assumption of independence of $\mathscr{F}_0$ is needed for this.

In Section 2.7.6 we consider random sets of lines in the plane which are stationary under Euclidean motions and such that the expected number of lines in the system intersecting a given circle is finite. R. Davidson [5] had asked whether ruling out parallel lines would leave us only with doubly stochastic Poisson processes, i.e. Poisson processes built on a random measure. Under the much stronger hypothesis we make in Theorem 8 the answer is yes and we give a complete description of the processes involved.

The author's interest in the present subject is due to the influence of his late friend R. Davidson. Discussions with him were many and stimulating, especially during the year 1968–69 which the author had the good fortune to spend at the Statistical Laboratory of the University of Cambridge as visiting Fellow of Clare Hall. The present paper actually has its origins in some problems that arose out of Davidson's thesis [5]. The author takes this opportunity to express his indebtedness to D. G. Kendall, to the British Science Research Council and to Clare Hall for making those invaluable exchanges possible.

2.7.2 THE SAMPLE SPACE

Let E be a locally compact second-countability Hausdorff topological space. A subset of E will be called bounded if its closure is compact. Denote by $\mathscr{M}_E$ and $\mathscr{M}_E^+$ the space of all Radon measures and the space of all non-negative Radon measures on E respectively, and let $\mathscr{K}$ be the set of all continuous functions on E with compact support. If $\mu \in \mathscr{M}_E$ and $f \in \mathscr{K}$ we write $\mu(f)$ for $\int f\, d\mu$. The vague topology [3] makes $\mathscr{M}_E$ into a topological vector space. This topology is defined as follows: $\mu_\alpha \to \mu$ if $\mu_\alpha(f) \to \mu(f)$ for every $f \in \mathscr{K}$. The space $\mathscr{M}_E$ is not in general complete relative to the induced uniform structure, but the subset $\mathscr{M}_E^+$ is complete [3, pp. 61–62] and is in fact metrizable and separable, hence a Polish space.

Let Ω_E^* be the subspace of $\mathscr{M}_E^+$ consisting of those $\mu \in \mathscr{M}_E^+$ with non-empty discrete support whose mass at each point of the support is a positive integer. If $\mu \in \Omega_E^*$ then there is a finite or countable discrete closed set $\{x_1, x_2, \ldots\}$ and positive integers $k_1, k_2, \ldots$ such that

$$\mu(f) = \sum_{n=1}^{\infty} k_n f(x_n) \quad (f \in \mathscr{K}).$$

If $\mu \in \Omega_E^*$ and F is the support of μ, then points $x \in F$ with $\mu(\{x\}) = 1$ will be called simple points of μ; points $x \in F$ with $\mu(\{x\}) \geqslant 2$ will be called multiple points.

Ω_E^* is closed in $\mathscr{M}_E^+$. The subset Ω_E of elements of Ω_E^* without multiple points is a dense G_δ in Ω_E^*, hence a Polish space. We shall usually write Ω^*, Ω when it is clear what the underlying space E is and will switch to the symbol ω for the generic element of Ω^*. We shall also write $N(\omega, Q)$ for $\omega(Q)$. If ω has no multiple points it can be identified with its support and in this case $N(\omega, Q)$ is the number of points of ω in Q (see [20]).

Proposition 1. *When E is the real line the following are equivalent:*
(i) $\omega_n \to \omega$ in the vague topology;
(ii) for the every bounded interval I,

$$N(\omega_n, I^0) \geqslant N(\omega, I^0)$$

and $N(\omega_n, \bar{I}) \leqslant N(\omega, \bar{I})$ eventually (I^0 is the interior, $\bar{I}$ the closure of I).

Corollary. *The class $\mathscr{A}$ of sets of the form $\{\omega\colon N(\omega, I^0) \geqslant j, N(\omega, \bar{I}) \leqslant k\}$, where I is a bounded interval, is a sub-basis of the vague topology in Ω^*.*

Returning to an abstract E, denote by $\mathscr{F}^*$ and $\mathscr{F}$ the σ-fields of Borel sets of the Polish spaces Ω^* and Ω respectively (i.e. the σ-fields generated by the open sets). They are the minimal σ-fields which make all $\int f(x)\, N(\omega, dx)\, (f \in \mathscr{K})$ measurable functions of ω in Ω^* and Ω respectively. It is easy to see that they are also the minimal σ-fields that make all $N(\omega, Q)$ (Q a bounded Borel set in E) measurable functions of ω. As is customary in probability theory we shall often suppress the ω and write $N(Q)$ for $N(\omega, Q)$ or $\{N(Q) = j\}$ for $\{\omega\colon N(\omega, Q) = j\}$. If (b, c) is an interval on the real line $N(b, c)$ will stand for $N(\omega, (b, c))$, etc.

A stochastic point-process is a probability measure $\mathbf{P}$ on Ω^*. A stochastic point-process without multiple points is one satisfying $\mathbf{P}(\Omega) = 1$. In the present paper we shall be dealing almost exclusively with processes of the latter kind and we shall often consider the probability measure $\mathbf{P}$ as defined on $(\Omega, \mathscr{F})$. Note that by definition ω is never empty.

As is well known, weak convergence $\mathbf{P}_n \to \mathbf{P}$ for such probabilities is defined as $\int X\, d\mathbf{P}_n \to \int X\, d\mathbf{P}$ for every bounded continuous function X on Ω. For this it is necessary and sufficient that $\mathbf{P}_n(A) \to \mathbf{P}(A)$ for every $\mathbf{P}$-almost boundaryless set $A \in \mathscr{F}$ (i.e. with $\mathbf{P}(\partial A) = 0$ where ∂A denotes the boundary of A); it is sufficient (but not necessary) that $\mathbf{P}_n(A) \to \mathbf{P}(A)$ for every set A

which is a finite intersection of members of the sub-base $\mathscr{A}$. This shows the 'if' part of the following.

Theorem 1. *Let E be the real line. If, for each fixed x,* $\mathbf{P}\{N(\{x\})=0\}=1$, *then* $\mathbf{P}_n \to \mathbf{P}$ *if and only if* $\mathbf{P}_n(B) \to \mathbf{P}(B)$ *for every set B of the form*

$$B=\{N(I_1)=j_1,\dots,N(I_k)=j_k\},$$

where $I_1,\dots,I_k$ *are bounded intervals.*

For the 'only if' part note that for any bounded interval I the set $\{N(I)=j\}$ is $\mathbf{P}$-almost boundaryless since

$$\{N(I^0)\geqslant j, N(\bar{I})\leqslant j\}\subseteq\{N(I)=j\}\subseteq\{N(I^0)\leqslant j, N(\bar{I})\geqslant j\},$$

where the smaller set is open and the bigger one is closed in the vague topology.

In the present paper we shall consider only point-processes for which the measure $\mu(Q)=\mathbf{E}(N(Q))$ is finite whenever Q is bounded. This condition will be assumed throughout, even when not explicitly mentioned. The measure μ will be called the intensity measure of the process.

2.7.3 THE PALM PROBABILITIES

Let E be the real line R and λ the Lebesgue measure on the Borel sets of R. If $Q\subseteq R$, $A\in\mathscr{F}$ and $t\in R$, write $Q+t$ for the set $\{x+t\colon x\in Q\}$ and $A+t$ for $\{\omega+t\colon \omega\in A\}$. A point process in R is called stationary if

$$\mathbf{P}(N(Q_1)=j_1,\dots,N(Q_n)=j_n)=\mathbf{P}(N(Q_1+t)=j_1,\dots,N(Q_n+t)=j_n)$$

for any bounded Borel sets $Q_1,\dots,Q_n$ and any real t. In this case with probability one ω is unbounded on the left and on the right.

Of considerable importance for the study of point-processes are the Palm conditional probabilities: if $A\in\mathscr{F}$ and $t\in R$ we denote by $\mathbf{P}(A\,|\,t)$ the conditional probability of A given that t is in the random set ω. Historically Palm, in his study [16] on telephone calls, introduced the conditional probability of the absence of calls in $(0,t)$ given that a call occurs at 0. Khintchine [9] later proved by analytic methods the existence of this and a whole series of other similar probabilities for stationary processes which he defined as limits of ratios, e.g.

$$\mathbf{P}(N(b,c)=j\,|\,0)=\lim_{\varepsilon\downarrow 0}\frac{\mathbf{P}(N(b,c)=j\text{ and }N(-\varepsilon,0]\geqslant 1)}{\mathbf{P}(N(-\varepsilon,0]\geqslant 1)}.$$

A measure-theoretic basis for these probabilities (for the not necessarily stationary case) was given by Ryll-Nardzewski [20], who defined $\mathbf{P}(A\,|\,t)$

as a Radon–Nikodym derivative:

$$\int_A N(\omega, Q)\,\mathbf{P}(d\omega) = \int_Q \mathbf{P}(A\,|\,t)\,\mu(dt). \tag{1}$$

If the process is stationary then there is a regular version of $\mathbf{P}(A\,|\,t)$ such that $\mathbf{P}(A\,|\,t) = \mathbf{P}(A - t\,|\,0)$. We shall then write $\mathbf{P}_0(A)$ for $\mathbf{P}(A\,|\,0)$. For the stationary case various authors have suggested other ways of defining $\mathbf{P}_0$ and establishing its connection with $\mathbf{P}$. Slivnyak [22, 23] uses Khintchine's analytic arguments (subadditivity) to obtain the conditional probabilities as limits of ratios. Matthes [11] defined $\mathbf{P}_0(A)$ as a quotient of expectations

$$\mathbf{P}_0(A) = \frac{\mathbf{E}Z^*}{\mathbf{E}N^*}, \tag{2}$$

where $Z^*(\omega)$ is the number of points $x \in \omega \cap [0,1]$ such that $\omega - x \in A$, while $N^*(\omega) = N(\omega, [0,1])$.

Neveu [15] has established the correspondence between $\mathbf{P}$ and $\mathbf{P}_0$ through measure-isomorphisms and quotient spaces. This is also the approach of [8]. A conceptually simple way of looking at this was indicated in [17]. We repeat it here. What is involved is a natural representation of the flow of time in Ω $T_t\omega = \omega - t$ as a 'flow under a function' (see [1] or [2] for this concept). It was shown by Ambrose and Kakutani [1, 2] that every flow which 'really moves' is isomorphic with a flow under a function. In the present case we can furnish such a representation as follows. Let $\Omega_0 = \{\omega \in \Omega\colon 0 \in \omega\}$, for each $\omega \in \Omega$ let $-\tau(\omega)$ be the greatest non-positive point in ω and define the mapping

$$\Omega \ni \omega \Rightarrow (\omega + \tau(\omega), \tau(\omega)) \in \Omega_0 \times [0, \infty) \tag{3}$$

This is a one-to-one mapping of Ω onto the space

$$\overline{\Omega} = \{(\omega, t)\colon \omega \in \Omega_0, 0 \leqslant t < \eta_0(\omega)\},$$

where, for $\omega \in \Omega_0$, $\eta_0(\omega)$ denotes the smallest positive point in ω. All the conditions of [1, Theorem 1] for $\overline{\Omega}$ are satisfied and it follows from that theorem (or from the direct treatment of [15]) that there is a finite measure ν on Ω_0 such that the mapping (3) is a measure isomorphism between (Ω, P) and $(\overline{\Omega}, \nu \otimes \lambda)$. The probability $\mathbf{P}_0$ is then the normalization of ν. Notice that the induced transformation on Ω_0 is $T^*\omega = \omega - \eta_0(\omega)$, which is essentially the shift of the discrete time process of interpoint distances $\ldots, \eta_{-1}, \eta_0, \eta_1, \ldots$ defined on Ω_0; the latter is thus a stationary process.

In the present paper we shall adopt Ryll-Nardzewski's definition but proceed by using as our main tool for manipulating the Palm probabilities, the Fubini theorem as it applies to the measure implicitly defined by equation (1) in the space $R \times \Omega$. Let us return to the general case where E is a locally compact second-countability space. Using the notation of Section 2.7.2 let $\mathbf{P}$ be a probability distribution on $(\Omega, \mathscr{F})$ with $\mu(Q) = \mathbf{E}(N(Q)) < \infty$ for bounded Q. If $\mathscr{B}$ is the σ-field of Borel sets in E (i.e. the σ-field generated by the open sets) then the Borel sets of $X \times \Omega$ are the sets which belong to the σ-field $\mathscr{B} \otimes \mathscr{F}$ generated by all rectangles $Q \times A$ ($Q \in \mathscr{B}$, $A \in \mathscr{F}$). For such rectangles set

$$m(Q \times A) = \int_A N(\omega, Q)\, \mathbf{P}(d\omega).$$

Then m can be extended (see [12, p. 15]) to a unique measure m on $\mathscr{B} \otimes \mathscr{F}$ which satisfies the generalized Fubini theorem. We call m the master measure. It is the integral over the space $(\Omega, \mathscr{F}, \mathbf{P})$ of the measures $N(\omega, \cdot)$ each of which lives on a section $E \times \{\omega\}$ of the product space $E \times \Omega$. However m can also be expressed as an integral over $(E, \mathscr{B}, \mu)$ of a family of probability measures $\mathbf{P}(\cdot \mid x)$, each living on a section $\{x\} \times \Omega$. The existence of these probabilities follows from the Radon–Nikodyn theorem as Ryll-Nardzewski has indicated. In fact for fixed $A \in \mathscr{F}$ the measure

$$\sigma(Q) = \int_A N(\omega, Q)\, \mathbf{P}(d\omega) \quad (Q \in \mathscr{B})$$

is absolutely continuous relative to the intensity measure μ hence has a Radon–Nikodym derivative $\mathbf{P}(A \mid x)$ which, as a function of x, is unique to within μ-null sets. Thus for every rectangle $Q \times A$ ($Q \in \mathscr{B}$, $A \in \mathscr{F}$)

$$\int_A N(\omega, Q)\, \mathbf{P}(d\omega) = \int_Q \mathbf{P}(A \mid x)\, \mu(dx)$$

and the two sides have the extension m to $\mathscr{B} \otimes \mathscr{F}$. There is a regular version of $\mathbf{P}(A \mid x)$ (a true probability measure in A for each fixed x and $\mathscr{B}$-measurable in x for each fixed A). By Fubini's theorem if $D \in \mathscr{B} \otimes \mathscr{F}$ then

$$(4) \qquad m(D) = \int_\Omega N(\omega, D^\omega)\, \mathbf{P}(d\omega) = \int_E \mathbf{P}(D_x \mid x)\, \mu(dx),$$

where $D_x = \{\omega \in \Omega \colon (x, \omega) \in D\}$, $D^\omega = \{x \in E \colon (x, \omega) \in D\}$. If $F(x, \omega)$ is a $\mathscr{B} \otimes \mathscr{F}$-measurable and non-negative or m-integrable function, then

$$\int F\, dm = \int_\Omega \int_E F(x, \omega)\, N(\omega, dx)\, \mathbf{P}(d\omega) = \int_E \int_\Omega F(x, \omega)\, \mathbf{P}(d\omega \mid x)\, \mu(dx).$$

(5)

The measures $\mathbf{P}(\cdot\,|\,x)$ of a regular version are called the 'Palm probabilities' of the process and $\mathbf{P}(A\,|\,x)$ can be interpreted as the conditional probability of A given that the process has a point at x.

2.7.4 STATIONARITY UNDER A GROUP

Suppose there is a locally compact second-countability group $\mathscr{G}$ of homeomorphisms of E such that E is a homogeneous space under the action of $\mathscr{G}$. This means: (*i*) the mapping $(T, x) \Rightarrow Tx\,(T \in \mathscr{G}, x \in E)$ is continuous in the pair (T, x); (*ii*) for any $x, y \in E$ there is a $T \in \mathscr{G}$ such that $Tx = y$; (*iii*) if we fix $x_0 \in E$ (to be called below the fixed reference point) and define the mapping π of $\mathscr{G}$ onto E by $\pi(T) = Tx_0$, then π is an open mapping (it is automatically continuous). Then each $x \in E$ can be associated with its inverse image set $H_x = \pi^{-1}(x)$ and this is a natural identification of E with the topological quotient space $\mathscr{G}/\mathscr{G}_0$, where $\mathscr{G}_0$ is the stabilizer of x_0, i.e. $\mathscr{G}_0 = \{T \in \mathscr{G}\colon Tx_0 = x_0\}$. Under the above assumptions this identification is a homeomorphism. Note that $\mathscr{G}_0 = H_{x_0}$ is a closed but not necessarily normal subgroup of $\mathscr{G}$, while for $x \neq x_0$ H_x is a left coset of $\mathscr{G}_0$.

We shall make an additional assumption, namely that E has a non-trivial Radon measure ν invariant under $\mathscr{G}$ ('Haar measure'). The necessary and sufficient condition for this is the equality $\Delta_{\mathscr{G}}(T) = \Delta_{\mathscr{G}_0}(T)$ for every $T \in \mathscr{G}_0$, where $\Delta_{\mathscr{G}}, \Delta_{\mathscr{G}_0}$ are the moduli of the groups $\mathscr{G}$ and $\mathscr{G}_0$ respectively (see [4, p. 59] or [13, p. 140]). This Haar measure ν is unique to within a constant factor and is (roughly) the quotient measure of the Haar measure M of $\mathscr{G}$ and the Haar measure Λ of $\mathscr{G}_0$. To make this precise note that H_x as a coset of $\mathscr{G}_0$ carries a 'translate' Λ_x of the measure Λ; if now S is a Borel set in $\mathscr{G}$ then

$$M(S) = \int_E \Lambda_x(S \cap H_x)\,\nu(dx). \tag{6}$$

Each $T \in \mathscr{G}$ induces a homeomorphism on Ω by $T\omega = \{Tx\colon x \in \omega\}$. A point-process is called stationary under $\mathscr{G}$ if $\mathbf{P} = \mathbf{P}\circ T^{-1}$ for every $T \in \mathscr{G}$; equivalently

$$\mathbf{P}(N(Q_1) = j_1, \ldots, N(Q_n) = j_n) = \mathbf{P}(N(T^{-1}Q_1) = j_1, \ldots, N(T^{-1}Q_n) = j_n)$$

$(Q_1, \ldots, Q_n \in \mathscr{B}; j_1, \ldots, j_n$ non-negative integers). It is called first-order stationary under $\mathscr{G}$ if $\mathbf{E}(N(Q)) = \mathbf{E}(N(T^{-1}Q))$ for every $Q \in \mathscr{B}$. If the process is first-order stationary then its intensity measure μ is the Haar measure of E under $\mathscr{G}$, i.e. $\mu = a.\nu$ for some $a > 0$.

Let us write $\bar{\mathscr{B}}$ for the σ-field of Borel sets of $\mathscr{G}$. Then for fixed $A \in \mathscr{F}$ the function $\mathbf{P}(TA \mid Tx)$ is $\bar{\mathscr{B}} \otimes \mathscr{B}$-measurable in the pair (T, x). In fact if $f(\omega)$ is continuous and bounded on Ω then the function

$$h_f(T, y) = \int_\Omega f(T^{-1}\omega)\, \mathbf{P}(d\omega \mid y)$$

is continuous in T when y is fixed and $\mathscr{B}$-measurable in y when T is fixed, so it is $\bar{\mathscr{B}} \otimes \mathscr{B}$-measurable in (T, y). By a limit argument $h_f(T, y)$ is $\bar{\mathscr{B}} \otimes \mathscr{B}$-measurable for every bounded $\mathscr{F}$-measurable $f(\omega)$ on Ω. Now the function $\mathbf{P}(TA \mid Tx)$ is the composition of $(T, x) \Rightarrow (T, Tx)$ and $(T, y) \Rightarrow h_{\chi_A}(T, y)$.

We shall need the following generalization of [20, Theorem 5]. Let $\mathbf{P}$ be defined on $(\Omega, \mathscr{F})$.

Theorem 2. *If $\mathbf{P}$ is stationary under $\mathscr{G}$ then there is a unique regular version of $\mathbf{P}(A \mid x)$ such that $\mathbf{P}(TA \mid Tx) = \mathbf{P}(A \mid x)$ for all $A \in \mathscr{F}$, all $T \in \mathscr{G}$ and all $x \in E$.*

We prove this in several steps. For the sake of this proof we introduce the following terminology: a measurable subset of a measure space is called full if its complement has measure zero. We shall often suppress mention of the corresponding measure when there is no likelihood of confusion.

Step 1. For fixed $A \in \mathscr{F}$, $T \in \mathscr{G}$ we have $\mathbf{P}(A \mid x) = \mathbf{P}(TA \mid Tx)$ for almost all $x \in E$.

In fact by stationarity, for any $Q \in \mathscr{B}$

$$\int_Q \mathbf{P}(A \mid x)\, \mu(dx) = \int_A N(\omega, Q)\, \mathbf{P}(d\omega) = \int_{TA} N(\omega, TQ)\, \mathbf{P}(d\omega)$$

$$= \int_{TQ} \mathbf{P}(TA \mid y)\, \mu(dy) = \int_Q \mathbf{P}(TA \mid Tx)\, \mu(dx)$$

by the change of variable $y = Tx$.

Step 2. The set $\{(T, x) \in \mathscr{G} \times E \colon \mathbf{P}(TA \mid Tx) = \mathbf{P}(A \mid x) \text{ for all } A \in \mathscr{F}\}$ is full in $\mathscr{G} \times E$ relative to the measure $M \otimes \mu$.

First fix $A \in \mathscr{F}$; the set $\{(T, x) \colon \mathbf{P}(TA \mid Tx) = \mathbf{P}(A \mid x)\}$ is $\bar{\mathscr{B}} \otimes \mathscr{B}$-measurable. By step 1 and Fubini's theorem this set is full in $\mathscr{G} \times E$. Hence the same is true of $\{(T, x) \colon \mathbf{P}(TA \mid Tx) = \mathbf{P}(A \mid x) \text{ for } A \in \mathscr{F}^0\}$ if $\mathscr{F}^0$ is a countable generating subset of $\mathscr{F}$, closed under finite intersections. This implies the assertion.

Step 3. There is a full subset E_0 of E such that if $x_1 \in E_0$, $x_2 \in E_0$ then $\mathbf{P}(TA \mid Tx_1) = \mathbf{P}(A \mid x_1)$ for every $A \in \mathscr{F}$ and every T with $Tx_1 = x_2$.

By step 2 and Fubini's theorem there is $x_0 \in E$ such that the set $S = \{T \in \mathscr{G}: \mathbf{P}(TA \mid Tx_0) = \mathbf{P}(A \mid x_0)$ for all $A \in \mathscr{F}\}$ is full in $\mathscr{G}$. If we take x_0 as our fixed reference point, this implies (by equation (6)) that $S \cap H_x$ is full in H_x for almost all $x \in E$. Write $R_x = S \cap H_x$ and define $E_0 = \{x \in E$: R_x is full in $H_x\}$. To prove the assertion let $A \in \mathscr{F}$, $x_1 \in E_0$, $x_2 \in E_0$ and suppose T is such that $Tx_1 = x_2$. The sets TR_{x_1} and R_{x_2} are both full in H_{x_2} hence so is their intersection $TR_{x_1} \cap R_{x_2}$ .Choose $T_2 \in TR_{x_1} \cap R_{x_2}$; then $T_2 \in R_{x_2}$ and also $T_2 = TT_1$ for some $T_1 \in R_{x_1}$, so that

$$\begin{aligned}\mathbf{P}(TA \mid Tx_1) &= \mathbf{P}(TT_1(T_1^{-1}A) \mid TT_1 x_0) = \mathbf{P}(T_2(T_1^{-1}A) \mid T_2 x_0)\\ &= \mathbf{P}(T_1^{-1}A \mid x_0) = \mathbf{P}(T_1 T_1^{-1} A \mid T_1 x_0) = \mathbf{P}(A \mid x_1)\end{aligned}$$

since $T_1 \in R_{x_1}$, $T_2 \in R_{x_2}$.

Step 4. To complete the existence proof we choose $\bar{x} \in E_0$ and redefine $\mathbf{P}(A \mid x)$ for $x \notin E_0$ as follows: we let T_1 be such that $T_1 \bar{x} = x$ and define $\mathbf{P}(A \mid x) = \mathbf{P}(T_1^{-1} A \mid \bar{x})$; this definition does not depend on the choice of $\bar{x}$ or T_1. We omit the proof that the modified version has the property stated in the theorem.

To prove uniqueness note that if $\mathbf{P}_1(\cdot \mid \cdot)$, $\mathbf{P}_2(\cdot \mid \cdot)$ are two such versions, then for each $A \in \mathscr{F}$ $\mathbf{P}_1(A \mid x) = \mathbf{P}_2(A \mid x)$ for almost all $x \in X$. Using a countable generating subset of $\mathscr{F}$ closed under finite intersections we see that there is an $x_0 \in X$ such that $\mathbf{P}_1(A \mid x_0) = \mathbf{P}_2(A \mid x_0)$ for all $A \in \mathscr{F}$; this implies uniqueness.

Corollary. $\displaystyle\int_\Omega f(\omega)\, \mathbf{P}(d\omega \mid Tx) = \int_\Omega f(T\omega)\, \mathbf{P}(d\omega \mid x).$

Thus in the stationary case we are essentially dealing with one conditional distribution $\mathbf{P}(\cdot \mid x_0)$, as all the others are 'translates' of this. When the homogeneous space is the real line under translations ($E = R$, $\mathscr{G} = R$) we denote, as before, $\mathbf{P}(A \mid 0)$ by $\mathbf{P}_0(A)$.

We now illustrate how the well-known formulae involving $\mathbf{P}_0$ can be derived easily from the Fubini theorem (equations (4) and (5)). We begin with the intuitively obvious statement $\mathbf{P}(\Omega_x \mid x) = 1$, where

$$\Omega_x = \{\omega \in \Omega \colon x \in \omega\},$$

which we prove in the context of an arbitrary homogeneous space E (see [20] for the case of the line). Let Q be a compact set in E with $\mu(Q) > 0$ and apply equation (4) to the set $D = \{(x, \omega) \colon x \in \omega \cap Q\}$ which is closed in

$E \times \Omega$.

$$\int_Q \mathbf{P}(D_x | x)\, \mu(dx) = \int_E \mathbf{P}(D_x | x)\, \mu(dx)$$
$$= \int_\Omega N(\omega, D^\omega)\, \mathbf{P}(d\omega) = \int_\Omega N(\omega, Q)\, \mathbf{P}(d\omega) = \mu(Q)$$

which implies $\mathbf{P}(D_x | x) = 1$ for almost all x and hence $\mathbf{P}(\Omega_x | x) = 1$ for all $x \in E$ by Theorem 2.

For the other formulas we return to the real line and note that by stationarity $\mu(dx) = a.dx$, $\mathbf{P}(A | x) = \mathbf{P}_0(A - x)$ and $\mathbf{P}_0(\Omega_0) = 1$, hence equation (5) takes the form

$$\int_\Omega \int_{-\infty}^{\infty} F(x, \omega)\, N(\omega, dx)\, \mathbf{P}(d\omega) = a \int_{-\infty}^{\infty} \int_\Omega F(x, \omega + x)\, \mathbf{P}_0(d\omega)\, dx$$
$$= a \int_{\Omega_0} \int_{-\infty}^{\infty} F(x, \omega + x)\, dx\, \mathbf{P}_0(d\omega),$$

that is,

$$\int_{R \times \Omega} F\, dm = a \int_{R \times \Omega_0} F^*\, d(\lambda \otimes \mathbf{P}_0),$$

where $F^*(x, \omega) = F(x, \omega + x)\, (x \in R, \omega \in \Omega_0)$. Cf. [20, formula (30)], [8, formula (5), p. 815] and [15, Proposition 4].

Let us enumerate the points of each $\omega \in \Omega$, $\ldots < t_{-1}(\omega) < t_0(\omega) < t_1(\omega) \ldots$, so that $t_0(\omega) \leqslant 0 < t_1(\omega)$. For $\omega \in \Omega_0$ let us follow [20] in setting

$$\eta_k(\omega) = t_{k+1}(\omega) - t_k(\omega).$$

If now $f(\omega)$ is any non-negative random variable on Ω we let $F(x, \omega)$ be equal to $f(\omega)$ if $x = t_0(\omega)$ and to 0 otherwise. Then

$$\int_\Omega f(\omega)\, \mathbf{P}(d\omega) = \int_\Omega f(\omega)\, N(\omega, \{t_0(\omega)\})\, \mathbf{P}(d\omega)$$
$$= \int_{R \times \Omega} F\, dm = a \int_{\Omega_0} \int_{-\infty}^{\infty} F^*(x, \omega)\, dx\, \mathbf{P}_0(d\omega),$$

where it is easy to see that, for $\omega \in \Omega_0$, $F^*(x, \omega)$ is equal to $f(\omega + x)$ if $-\eta_0(\omega) < x \leqslant 0$ and to 0 otherwise, so that

$$\int_\Omega f(\omega)\, \mathbf{P}(d\omega) = a \int_{\Omega_0} \int_{-\eta_0(\omega)}^{0} f(\omega + x)\, dx\, \mathbf{P}_0(d\omega)$$

(7) $$= a \int_{\Omega_0} \int_0^{\eta_0(\omega)} f(\omega - x)\, dx\, \mathbf{P}_0(d\omega)$$

which is formula (36) in [20].

Conversely, in a similar way $\mathbf{P}_0$ can be expressed in terms of $\mathbf{P}$. Given $g(\omega) \geqslant 0$ on Ω define $F^*(x, \omega)$ to be equal to $g(\omega)$ if $0 \leqslant x \leqslant 1$ and to 0 otherwise. Then

$$\int_{\Omega_0} g(\omega)\, \mathbf{P}_0(d\omega) = \int_{\Omega_0} g(\omega) \int_0^1 dx\, \mathbf{P}_0(d\omega)$$

$$= \int_{R \times \Omega_0} F^*\, d(\lambda \otimes \mathbf{P}_0) = \frac{1}{a} \int_{R \times \Omega} F\, dm,$$

where $F(x, \omega) = F^*(x, \omega - x)$; the latter is equal to $g(\omega - x)$ if $0 \leqslant x \leqslant 1$ and to 0 otherwise. Thus

$$\int_{\Omega_0} g(\omega)\, \mathbf{P}_0(d\omega) = \frac{1}{a} \int_{\Omega} \int_0^1 g(\omega - x)\, N(\omega, dx)\, \mathbf{P}(d\omega) \tag{8}$$

which for $g = \chi_A$ reduces to equation (2).

Next we prove that $\mathbf{P}_0$ is stationary under the 'shift' $T^* \omega = \omega - \eta_0(\omega)$ of Ω_0 ([20, formula (43)]). In the same way as equation (7) one can prove

$$\int_{\Omega} f(\omega)\, \mathbf{P}(d\omega) = a \int_{\Omega_0} \int_{\eta_0(\omega)}^{\eta_0(\omega)+\eta_1(\omega)} f(\omega - x)\, dx\, \mathbf{P}_0(d\omega). \tag{9}$$

Now if $g(\omega)$ is a non-negative random variable on Ω_0, we define on Ω:

$$f(\omega) = \frac{g(\omega - t_0(\omega))}{t_1(\omega) - t_0(\omega)}.$$

Noting that if $\omega \in \Omega_0$ then, for $0 \leqslant x < \eta_0(\omega)$, $f(\omega - x)$ is $g(\omega)/\eta_0(\omega)$, while for $\eta_0(\omega) \leqslant x < \eta_0(\omega) + \eta_1(\omega)$, it is $g(\omega - \eta_0(\omega))/\eta_1(\omega)$, we get from equations (7) and (9)

$$\int_{\Omega_0} g(\omega)\, \mathbf{P}_0(d\omega) = \int_{\Omega_0} g(\omega - \eta_0(\omega))\, \mathbf{P}_0(d\omega) = \int_{\Omega} \frac{g(\omega - t_0(\omega))}{t_1(\omega) - t_0(\omega)}\, \mathbf{P}(d\omega).\dagger$$

Finally we prove the Palm–Khintchine formulae [9, p. 40, (10.7)]:

$$\mathbf{P}(N(0, c) > j) = a \int_0^c \mathbf{P}_0(N(0, x) = j)\, dx. \tag{10}$$

Given any reference point ('origin') u on R we can enumerate the points of each $\omega \in \Omega$, $\ldots < t^*_{-1}(\omega, u) < t^*_0(\omega, u) < t^*_1(\omega, u) < \ldots$, in such a way as to have $t^*_{-1}(\omega, u) < u \leqslant t^*_0(\omega, u)$. If now $\chi(\omega)$ is the indicator function of the event $\{N(0, c) > j\}$ and if (cf. the proof of equation (7)) we apply the Fubini theorem to the function $F(x, \omega)$ which is equal to $\chi(\omega)$ if

† The equality of the first and third members can be taken as yet another definition of $\mathbf{P}_0$.

$x = t^*_{-(j+1)}(\omega, c)$ and to 0 otherwise, then we get

$$\mathbf{P}(N(0,c) > j) = a \int_{-\infty}^{\infty} \int_{\Omega_0} F(x, \omega + x)\, \mathbf{P}_0(d\omega)\, dx.$$

Now for $\omega \in \Omega_0$, $F(x, \omega + x)$ is 1 if and only if $N(\omega + x, (0, c)) > j$ and $x = t^*_{-(j+1)}(\omega + x, c)$, i.e. $N(\omega, (-x, c-x)) > j$ and $0 = t^*_{-(j+1)}(\omega, c-x)$ which finally is true if and only if $0 < x < c$ and $N(\omega, (0, c-x)) = j$. Thus

$$\mathbf{P}(N(0,c) > j) = a \int_0^c \mathbf{P}_0(N(0, c-x) = j)\, dx$$

which is the same as equation (10).

We conclude this section by noting that Slivnyak [22, 23] introduced another conditional probability $\mathbf{P}_*$ on Ω_0 as follows

$$\mathbf{P}_*(A) = a \int_A E_{\mathbf{P}_0}(\eta_0 \,|\, \mathscr{I}_0)\, d\mathbf{P}_0,$$

where $\mathscr{I}_0$ is the σ-field of events in $(\Omega_0, \mathbf{P}_0)$ invariant under $T^*\omega = \omega - \eta_0(\omega)$. As proved in [6, Section 11]

$$\mathbf{P}_*(A) = \mathbf{E}_{\mathbf{P}}\left(\frac{\mathbf{E}_{\mathbf{P}}(Z^* \,|\, \mathscr{I})}{\mathbf{E}_{\mathbf{P}}(N^* \,|\, \mathscr{I})}\right),$$

where Z^*, N^* are as in equation (2) and $\mathscr{I}$ is the σ-field of events in $(\Omega, \mathbf{P})$ invariant under the group of translations. Thus to get $\mathbf{P}_*$ we split the process into its ergodic components, form the Palm probability for each component and average.

2.7.5 A PAST-DEPENDENT CHANGE OF TIME

If stationarity is added to the hypotheses of Theorems 3 and 4 below then they reduce to statements which can be deduced from a result of [22, 23]; see also [8] for the stationary version of Theorem 3. (Note that the conditional probability used in [22, Section 8] is $\mathbf{P}_*$ and not $\mathbf{P}_0$.) In applications below we need the stronger versions stated and proved here. The method of proof of Theorem 4 is also needed in Section 6.

An event $A \in \mathscr{F}$ is said to happen in a set $Q_0 \in \mathscr{B}$ if it belongs to the σ-field generated by the random variables $N(\cdot, Q)$ $(Q \subseteq Q_0, Q \in \mathscr{B})$. The σ-field of events happening in Q_0 will be denoted by $\mathscr{F}_{Q_0}$.

In the present section we shall assume that E is the real line R. For each $t \in R$ we let $\mathscr{F}_t = \mathscr{F}_{(-\infty, t)}$.

Theorem 3. *Suppose that a not necessarily stationary point-process* (*without multiple points*) *satisfies the following conditions.*

(*i*) *The measure* μ *is non-atomic and finite on bounded sets.*

(*ii*) *If* $A \in \mathscr{F}_s$ *then* $\mathbf{P}(A \mid t) = \mathbf{P}(A)$ *for* μ*-almost all* $t \geqslant s$. (*Equivalently:* $\mathbf{E}(N[s,t] \mid \mathscr{F}_s) = \mu([s,t])$ *a.s. for any interval* $[s,t]$.†)

Then the process is a (not necessarily homogeneous) Poisson process built on (R, μ).

We proceed by stages.

Lemma 1. *Under the hypotheses of the theorem, if I is a finite interval with end-points* b, c $(b < c)$ *and if* $A \in \mathscr{F}_b$, *then*

$$|\mathbf{P}(A \cap \{N(I) = 0\}) - \mathbf{P}(A)\,\mathbf{P}(N(I) = 0)| \leqslant \mathbf{E}(N(I)) - \mathbf{P}(N(I) \neq 0).$$

Proof. In fact

$$\int_A N(I)\, d\mathbf{P} = \int_I \mathbf{P}(A \mid t)\, \mu(dt) = \mathbf{P}(A)\, \mu(I) = \mathbf{P}(A)\, \mathbf{E}(N(I))$$

hence

$$\mathbf{P}(A \cap \{N(I) \neq 0\}) = \int_A \chi_{\{N(I)\neq 0\}}\, d\mathbf{P} = \int_A N(I)\, d\mathbf{P} - \int_A (N(I) - \chi_{\{N(I)\neq 0\}})\, d\mathbf{P}$$

$$= \mathbf{P}(A)\, \mathbf{E}(N(I)) - \int_A (N(I) - \chi_{\{N(I)\neq 0\}})\, d\mathbf{P}.$$

From this

$$|\mathbf{P}(A \cap \{N(I) \neq 0\}) - \mathbf{P}(A)\, \mathbf{P}(N(I) \neq 0)|$$

$$= \left| \mathbf{P}(A)(\mathbf{E}(N(I)) - \mathbf{P}(N(I) \neq 0)) - \int_A (N(I) - \chi_{\{N(I)\neq 0\}})\, d\mathbf{P} \right|$$

$$\leqslant \mathbf{E}(N(I)) - \mathbf{P}(N(I) \neq 0)$$

which implies the assertion.

Lemma 2. *Let I be a finite interval and for each n let* $I_{n,1}, \ldots, I_{n,k_n}$ *be a partition of I into subintervals. If*

$$\lim_{n\to\infty} \max_{1 \leqslant i \leqslant k_n} \lambda(I_{n,i}) = 0$$

then

$$\lim_{n\to\infty} \sum_{i=1}^{k_n} \mathbf{P}(N(I_{n,i}) \neq 0) = \mathbf{E}(N(I)).$$

† *Note added in proof.* In this equivalent form the theorem is stated in [29, p. 59], as was recently pointed out to me by J. B. Walsh. However, our present method of proof is still needed below.

Proof. This follows from the Lebesgue dominated convergence theorem and the fact that

$$\sum_{i=1}^{k_n} \chi_{\{N(I_{n,i})\neq 0\}} \to N(I) \quad \text{a.s.}$$

(cf. [10]).

Lemma 3. *Under the hypotheses of the theorem if I is a finite interval with endpoints b, c $(b<c)$ and if $A \in \mathscr{F}_b$ then*

$$\mathbf{P}(A \cap \{N(I)=0\}) = \mathbf{P}(A)\,\mathbf{P}(N(I)=0).$$

Proof. For each n subdivide I into n equal subintervals $I_{n,1}, \ldots, I_{n,n}$ (enumerated from left to right). If $B_{n,i} = \{N(I_{n,i})=0\}$ then

$$\begin{aligned}
&|\mathbf{P}(A \cap \{N(I)=0\}) - \mathbf{P}(A)\,\mathbf{P}(N(I)=0)| \\
&= \left|\mathbf{P}\left(A \cap \bigcap_{i=1}^{n} B_{n,i}\right) - P(A)\,\mathbf{P}\left(\bigcap_{i=1}^{n} B_{n,i}\right)\right| \\
&\leqslant \left|\mathbf{P}\left(A \cap \bigcap_{i=1}^{n} B_{n,i}\right) - \mathbf{P}(A)\prod_{i=1}^{n}\mathbf{P}(B_{n,i})\right| \\
&\qquad + \left|\mathbf{P}(A)\prod_{i=1}^{n}\mathbf{P}(B_{n,i}) - \mathbf{P}(A)\,\mathbf{P}\left(\bigcap_{i=1}^{n} B_{n,i}\right)\right| \\
&\leqslant \sum_{k=1}^{n}\left|\mathbf{P}\left(A \cap \bigcap_{i=1}^{k} B_{n,i}\right)\prod_{i=k+1}^{n}\mathbf{P}(B_{n,i}) - \mathbf{P}\left(A \cap \bigcap_{i=1}^{k-1} B_{n,i}\right)\prod_{i=k}^{n}\mathbf{P}(B_{n,i})\right| \\
&\qquad + \mathbf{P}(A)\sum_{k=1}^{n}\left|\mathbf{P}\left(\bigcap_{i=1}^{k} B_{n,i}\right)\prod_{i=k+1}^{n}\mathbf{P}(B_{n,i}) - \mathbf{P}\left(\bigcap_{i=1}^{k-1} B_{n,i}\right)\prod_{i=k}^{n}\mathbf{P}(B_{n,i})\right| \\
&\leqslant \sum_{k=1}^{n}\left|\mathbf{P}\left(A \cap \bigcap_{i=1}^{k} B_{n,i}\right) - \mathbf{P}\left(A \cap \bigcap_{i=1}^{k-1} B_{n,i}\right)\mathbf{P}(B_{n,k})\right| \\
&\qquad + \sum_{k=1}^{n}\left|\mathbf{P}\left(\bigcap_{i=1}^{k} B_{n,i}\right) - \mathbf{P}\left(\bigcap_{i=1}^{k-1} B_{n,i}\right)\mathbf{P}(B_{n,k})\right| \\
&\leqslant 2\sum_{k=1}^{n}[\mathbf{E}(N(I_{n,k})) - \mathbf{P}(N(I_{n,k})\neq 0)] \quad \text{(by Lemma 1)} \\
&= 2[\mathbf{E}(N(I)) - \sum_{k=1}^{n}\mathbf{P}(N(I_{n,k})\neq 0)]
\end{aligned}$$

which goes to 0, by Lemma 2.

Lemma 4. *Under the hypotheses of the theorem, if $I^1, I^2, \ldots, I^k$ are disjoint intervals and if $\alpha_1, \ldots, \alpha_k$ are non-negative integers then*

$$\mathbf{P}\left(\bigcap_{j=1}^{k}\{N(I^j)=\alpha_j\}\right)=\prod_{j=1}^{k}\mathbf{P}(N(I^j)=\alpha_j).$$

Proof. This is shown by a technique of [18]. For every n we subdivide each I^j into n equal subintervals

$$I^j=\bigcup_{i=1}^{n} I^j_{n,i}$$

and let $\chi^j_{n,i}$ be the indicator function of the event $\{N(I^j_{n,1})\neq 0\}$. Then

$$\begin{aligned}\left|\mathbf{P}\left(\bigcap_{j=1}^{k}\{N(I^j)=\alpha_j\}\right)-\prod_{j=1}^{k}\mathbf{P}(N(I^j)=\alpha_j)\right| \\ \leqslant\left|\mathbf{P}\left(\bigcap_{j=1}^{k}\{N(I^j)=\alpha_j\}\right)-\mathbf{P}\left(\bigcap_{j=1}^{k}\left\{\sum_{i=1}^{n}\chi^j_{n,i}=\alpha_j\right\}\right)\right| \\ +\left|\mathbf{P}\left(\bigcap_{j=1}^{k}\left\{\sum_{i=1}^{n}\chi^j_{n,i}=\alpha_j\right\}\right)-\prod_{j=1}^{k}\mathbf{P}\left(\sum_{i=1}^{n}\chi^j_{n,i}=\alpha_j\right)\right| \\ +\left|\prod_{j=1}^{k}\mathbf{P}\left(\sum_{i=1}^{n}\chi^j_{n,i}=\alpha_j\right)-\prod_{j=1}^{k}\mathbf{P}(N(I^j)=\alpha_j)\right|.\end{aligned}$$

The middle term is zero because the $\chi^j_{n,i}$ are independent for each fixed n, by Lemma 3. The other two terms tend to zero as $n\to\infty$, because

$$\begin{aligned}\mathbf{P}\left(N(I^j)\neq\sum_{i=1}^{n}\chi^j_{n,i}\right)&\leqslant\mathbf{E}\left(N(I^j)-\sum_{i=1}^{n}\chi^j_{n,i}\right)\\&=\mathbf{E}(N(I^j))-\sum_{i=1}^{n}\mathbf{P}(N(I^j_{n,i})\neq 0).\end{aligned}$$

Theorem 3 follows from Lemma 4 (cf. for instance [19]).

Recall that a mixed Poisson process is one constructed as follows. A random positive number X is chosen with distribution function $F(x)$ such that

$$\int_0^\infty x\,dF(x)<\infty$$

and a Poisson sample is built with rate X. Equivalently, define the random rate of a stationary process by

$$X(\omega) = \lim_{n\to\infty} \frac{N(\omega, [0, n])}{n} \quad \text{a.s.};$$

then the process is mixed Poisson if and only if its conditional probability distribution, given $X = x$, is that of a Poisson process with rate x. In ergodic terms, its ergodic components are homogeneous Poisson (with different rates).

Theorem 4. *Assume for a point-process (without multiple points) that*

(*i*) $\mathbf{E}(N(Q)) = a\lambda(Q)$.

(*ii*) *If A happens in Q then $\mathbf{P}(A \mid t_1) = \mathbf{P}(A \mid t_2)$ for almost all pairs $(t_1, t_2) \in Q^c \times Q^c$.*

Then the process is stationary and is in fact a mixed Poisson process.

In fact if I_1, I_2 are intervals of equal length and if A happens in $(I_1 \cup I_2)^c$ then

$$\begin{aligned}|\mathbf{P}(A \cap \{N(I_1) = 0\}) - \mathbf{P}(A \cap \{N(I_2) = 0\})| \\ \leqslant (\mathbf{E}(N(I_1)) - \mathbf{P}(N(I_1) \neq 0)) + (\mathbf{E}(N(I_2)) - \mathbf{P}(N(I_2) \neq 0)).\end{aligned}$$

This follows from an argument similar to that of Lemma 1, if we note that the hypotheses of the theorem imply

$$\int_A N(I_1)\, d\mathbf{P} = \int_A N(I_2)\, d\mathbf{P}.$$

As before one can show that we actually have equality

$$\mathbf{P}(A \cap \{N(I_1) = 0\}) = \mathbf{P}(A \cap \{N(I_2) = 0\})$$

and this implies that if $I_1, \ldots, I_k$ is a collection of disjoint intervals and if $I_1', \ldots, I_k'$ is another collection of disjoint intervals such that $\lambda(I_j) = \lambda(I_j')$, $j = 1, 2, \ldots, k$ then

$$\mathbf{P}\left(\bigcap_{j=1}^{k} \{N(I_j) = 0\}\right) = \mathbf{P}\left(\bigcap_{j=1}^{k} \{N(I_j') = 0\}\right)$$

(if there is any overlap between the two collections of intervals we can compare with a third such collection, 'disjoint' from both). From this, as before

$$\mathbf{P}\left(\bigcap_{j=1}^{k} \{N(I_j) = \alpha_j\}\right) = \mathbf{P}\left(\bigcap_{j=1}^{k} \{N(I_j') = \alpha_j\}\right) \tag{11}$$

which shows that the process is stationary. Equation (11) implies exchangeability which is known to characterize the mixed Poisson process (see [5]; cf. also Section 2.7.6 below).

Let now $\dots, \eta_{-1}, \eta_0, \eta_1, \dots$ be a sequence of independent random variables with a common distribution F_0 such that

$$F_0(0, +\infty) = 1 \quad \text{and} \quad \int_0^\infty x \, dF_0(x) < \infty.$$

The process

$$\dots, -\eta_{-1} - \eta_{-2}, -\eta_{-1}, 0, \eta_0, \eta_0 + \eta_1, \dots$$

defines a probability $\mathbf{P}_0$ on Ω_0. The stationary process $\mathbf{P}$ on Ω that has $\mathbf{P}_0$ as its Palm probability (see equation (7)) is called an '(equilibrium) renewal process'. If $F_0(x) = 1 - e^{-ax}$ $(a > 0)$ then $\mathbf{P}$ is a Poisson process with rate a.

Now notice that if F is absolutely continuous relative to λ (denoted $F \ll \lambda$) then the transformation

$$\varphi(y) = \int_0^y \frac{F_0'(t)}{a(1 - F_0(t))} \, dt \tag{12}$$

of $[0, \infty)$ onto itself satisfies the equation $F_0(y) = 1 - e^{-a\varphi(y)}$ and hence carries the distribution F_0 onto the exponential $1 - e^{-ax}$. Suppose then that a renewal process is given, with $F_0 \ll \lambda$. If we run along the line, regulating our velocity in such a way that each time we hit a point of the process we take that point as a new 'origin' and start applying the transformation φ until we hit the next point, then the renewal process will be mapped onto a Poisson process with rate a. Our 'velocity' will depend only on our distance from the last point we left behind. This is an instance of the situation described in Theorem 5 below.

Consider a stationary process with $\mu = a\lambda$. The probabilities $\mathbf{P}$ and $\mathbf{P}_0$ are singular to each other on $\mathscr{F}$, since $\mathbf{P}_0(\Omega_0) = 1$, $\mathbf{P}(\Omega_0) = 0$. However, it is possible that $\mathbf{P}_0$ may be absolutely continuous relative to $\mathbf{P}$ ($\mathbf{P}_0 \ll \mathbf{P}$) on $\mathscr{F}_0$. If this is true let $X(\omega)$ be the Radon–Nikodym derivative

$$\mathbf{P}_0(A) = \int_A X(\omega) \, \mathbf{P}(d\omega) \quad (A \in \mathscr{F}_0). \tag{13}$$

X is $\mathscr{F}_0$-measurable with $\mathbf{E}X = 1$. Then

$$\mathbf{P}(A \mid t) = \int_A X(\omega - t) \, \mathbf{P}(d\omega) \quad (A \in \mathscr{F}_t).$$

Letting $X(t, \omega) = X(\omega - t)$ $(t \in R, \omega \in \Omega)$ we obtain a stationary (under $\mathbf{P}$) and measurable stochastic process adapted to the σ-fields $\mathscr{F}_t$, $-\infty < t < \infty$

(i.e. with $X(t)$ $\mathscr{F}_t$-measurable). As is well known, stationarity, measurability and $\mathbf{E}X = 1$ imply mean continuity

$$\lim_{t \to t_0} \mathbf{E}|X(t) - X(t_0)| = 0.$$

Theorem 5, roughly speaking, states the following. Suppose that starting at 0 say, we trace the positive half-line $[0, \infty)$ in such a way that at the time we are passing position t our speed is $1/X(t)$, which can be ∞. (The value of $X(t)$ is determined by the observation of the past, i.e. of what happened in $(-\infty, t)$.) Then the time instants at which we shall meet all the non-negative points of the process form a homogeneous Poisson process, independent of what happened in $(-\infty, 0)$.

For a more precise formulation define

$$\varphi_\omega(t) = \int_0^t X(t, \omega)\, d\tau \quad (t \geqslant 0).$$

For fixed t we have, by Fubini's theorem,

$$\mathbf{E}\varphi_{.}(t) = \int_0^t \mathbf{E}X(\tau)\, d\tau = t.$$

It follows that a.s. $\varphi_\omega(t) < \infty$ for every t. By the ergodic theorem a.s.

$$\lim_{t \to \infty} \varphi_\omega(t) = \infty.$$

Thus a.s. φ_ω is a non-decreasing (but not necessarily strictly increasing) transformation of $[0, \infty)$ onto itself.

Theorem 5. *For a stationary point-process (without multiple points) assume* $\mathbf{P}_0 \ll \mathbf{P}$ *on* $\mathscr{F}_0$. *If*

$$0 \leqslant t_1(\omega) < t_2(\omega) < t_3(\omega) < \dots$$

are the non-negative points of ω, *we define the points*

$$0 \leqslant \tau_1(\omega) \leqslant \tau_2(\omega) \leqslant \tau_3(\omega) \leqslant \dots \tag{14}$$

by

$$\tau_n(\omega) = \varphi_\omega(t_n(\omega)) = \int_0^{t_n(\omega)} X(u, \omega)\, du.$$

Under the probability $\mathbf{P}$ *the sequence* (14) *is a.s. strictly increasing and, in fact, a homogeneous Poisson process with rate a and independent of the* σ-*field* $\mathscr{F}_0$.

We begin with a lemma.

Lemma 5. *The stochastic processes $X(t,\omega)$, $t \geqslant 0$ and $Y(t,\omega) = \varphi_\omega(t)$, $t \geqslant 0$, are progressively measurable relative to $\mathscr{F}_t$, $t \geqslant 0$, i.e. for each $t_0 > 0$ the mappings*

$$[0, t_0] \times \Omega \ni (t,\omega) \Rightarrow X(t,\omega)$$
$$\Rightarrow Y(t,\omega)$$

are measurable relative to $\mathscr{B}_{[0,t_0]} \otimes \mathscr{F}_{t_0}$.

Proof. For $X(t,\omega)$ the statement follows from the fact that the mapping $(t,\omega) \Rightarrow X(t,\omega)$ is the composition of $(t,\omega) \Rightarrow \omega - t$ and $\omega \Rightarrow X(\omega)$. For $Y(t,\omega)$ note that

$$Y(t,\omega) = \int_0^t X(\tau,\omega)\, d\tau = \lim_{n\to\infty} \sum_{i=0}^{\infty} \frac{i}{2^n} \alpha_i(t,\omega),$$

where $\alpha_i(t,\omega) = \lambda(\{\tau : 0 \leqslant \tau < t \text{ and } (i-1)/2^n \leqslant X(\tau,\omega) < i/2^n\})$ is $\mathscr{F}_t$-measurable for fixed t, by Fubini's theorem. Hence the process $Y(t), t \geqslant 0$ is adapted to $\mathscr{F}_t$, $t \geqslant 0$ and since it has continuous paths it is progressively measurable [12, p. 70].

Proof of Theorem 5. Let $(\Omega', \mathscr{F}')$ be $(\Omega^*_{[0,\infty)}, \mathscr{F}^*_{[0,\infty)})$ (the elements of Ω' can have multiple points). Define the mapping

$$\Omega \ni \omega \Rightarrow S\omega = \{\varphi_\omega(x) : x \in \omega, x \geqslant 0\} \in \Omega'. \tag{15}$$

It will be shown in Lemma 6 that S is a Borel mapping of $(\Omega, \mathscr{F})$ into $(\Omega', \mathscr{F}')$. On Ω' we introduce the following probabilities. For each $B \in \mathscr{F}_0$ with $\mathbf{P}(B) > 0$ we define

$$\mathbf{P}'_B(A) = \mathbf{P}(S^{-1}A \mid B) = \frac{\mathbf{P}(S^{-1}A \cap B)}{\mathbf{P}(B)} \quad (A \in \mathscr{F}');$$

this is the distribution of the process (14) when restricted to B. We shall show that all these $\mathbf{P}'_B$ are Poisson in $[0,\infty)$ with rate a and hence that they coincide. This will imply $\mathbf{P}'_B(A) = \mathbf{P}'_\Omega(A)$, i.e. $\mathbf{P}(S^{-1}A \cap B) = \mathbf{P}(S^{-1}A)\,\mathbf{P}(B)$, which shows the independence of $\mathscr{F}_0$ and the σ-field generated by equation (14).

We first prove however that a.s. there are no multiple points. Let

$$K = \{(t,\omega) : \text{there is } \varepsilon > 0 \text{ such that } \varphi_\omega(t-\varepsilon) = \varphi_\omega(t)\},$$
$$K(n) = \{(t,\omega) : \varphi_\omega(t - (1/n)) = \varphi_\omega(t)\},$$
$$K_t = \{\omega : (t,\omega) \in K\}, \quad K^\omega = \{t : (t,\omega) \in K\},$$
$$K_t(n) = \{\omega : (t,\omega) \in K(n)\}, \quad K^\omega(n) = \{t : (t,\omega) \in K(n)\}.$$

By the above lemma $K_t(n)\in\mathscr{F}_t$. Note also that

$$K=\bigcup_n K(n),\quad K_t=\bigcup_n K_t(n),\quad K^\omega=\bigcup_n K^\omega(n)$$

so that $K\in\mathscr{B}_{[0,\infty)}\otimes\mathscr{F}$. Our assertion that there are no multiple points will follow if we show that a.s. $N(\omega, K^\omega)=0$; in fact this means that if $[a_1(\omega), b_1(\omega)]$, $[a_2(\omega), b_2(\omega)], \ldots$ are the disjoint intervals of constancy of $\varphi_\omega(\cdot)$ then with probability one the sample ω has no points in the set $(a_1(\omega), b_1(\omega)]\cup(a_2(\omega), b_2(\omega)]\cup\ldots$. In particular $\mathbf{P}'_B(0\in\omega')=\mathbf{P}(0\in\omega)=0$.

Fix n and t and note that $K_t(n)=\{\omega\colon X(t,\omega)=0 \text{ for a.a. } \tau\in(t-(1/n),t)\}$. Applying the Fubini theorem to the measure $\lambda\otimes\mathbf{P}$ on $(t-(1/n),t)\times K_t(n)$ we see that

$$\int_{K_t(n)} X(\tau,\omega)\,\mathbf{P}(d\omega)=0$$

for a.a. τ, hence by the mean continuity of $X(t)$

$$\int_{K_t(n)} X(t,\omega)\,\mathbf{P}(d\omega)=0.$$

Since $K_t(n)\in\mathscr{F}_t$ this means $\mathbf{P}(K_t(n)\,|\,t)=0$, hence by equation (4)

$$\int_\Omega N(\omega, K^\omega(n))\,\mathbf{P}(d\omega)=\int_0^\infty \mathbf{P}(K_t(n)\,|\,t)\,\mu(dt)=0.$$

To prove the other assertions of the theorem we shall now show that for $B\in\mathscr{F}_0$ $\mathbf{P}'_B$ satisfies the hypotheses of Theorem 3. Let $Q=[s_0, r_0]$ with $s_0>0$ and let A be an event happening in $[0,s_0)$, i.e. $A\in\mathscr{F}^*_{[0,s_0)}$. Then denoting by S_B the restriction of S to B we have

$$\mathbf{P}(B)\int_A N(\omega', Q)\,\mathbf{P}'_B(d\omega')=\int_{S_B^{-1}A} N(S\omega, Q)\,\mathbf{P}(d\omega)$$

$$=\int_{B\cap S^{-1}A} N(\omega,\varphi_\omega^{-1}Q)\,\mathbf{P}(d\omega)=m(D),$$

where m is the master-measure and $D=\{(t,\omega)\colon \omega\in B\cap S^{-1}A, t\in\varphi_\omega^{-1}Q\}$. Now by equation (4)

$$m(D)=a\int_0^\infty \mathbf{P}(D_t\,|\,t)\,dt,$$

where $D_t=B\cap S^{-1}A\cap\{\omega\colon t\in\varphi_\omega^{-1}Q\}=B\cap S^{-1}A\cap\{\omega\colon \varphi_\omega(t)\in Q\}$. We shall prove in Lemma 6 below that $D_t\in\mathscr{F}_t$ $(t\geqslant 0)$. Thus

$$m(D)=a\int_0^\infty\int_{D_t} X(t,\omega)\,\mathbf{P}(d\omega)\,dt=a\int_{B\cap S^{-1}A}\int_{\varphi_\omega^{-1}Q} X(t,\omega)\,dt\,\mathbf{P}(d\omega)$$

and since

$$\lambda(Q) = \int_{s_0}^{r_0} dx = \int_{\varphi_\omega^{-1}Q} X(t,\omega)\,dt$$

by the change of variable

$$x = \int_0^t X(\tau,\omega)\,d\tau,$$

we see that $m(D) = a\lambda(Q)\,\mathbf{P}(B\cap S^{-1}A)$, i.e.

$$\int_A N(\omega',Q)\,\mathbf{P}'_B(d\omega') = \mu(Q)\,\mathbf{P}'_B(A).$$

Putting $A=\Omega'$ we see that $\mu'_B(Q)=\mu(Q)=a\lambda(Q)$ for all $Q=[s_0,r_0]$ with $s_0>0$, hence for all Q since $\mu'_B(\{0\})=\mu(\{0\})=0$. Further if $A\in\mathscr{F}_s$ then

$$\int_A \mathbf{P}'_B(A\,|\,t)\,\mu'(dt) = \int_A N(\omega',Q)\,\mathbf{P}'_B(d\omega') = \mu'(Q)\,\mathbf{P}'_B(A)$$

for every interval $Q\subseteq[s,\infty)$, hence $\mathbf{P}'_B(A\,|\,t)=\mathbf{P}'_B(A)$ for a.a. $t\geqslant s$ and Theorem 3 (in a version for $E=(0,\infty)$) applies. To complete the proof we turn to the promised

Lemma 6. *(i) The mapping in* (15) *is a Borel mapping of* $(\Omega,\mathscr{F})$ *into* $(\Omega',\mathscr{F}')$.

(ii) $D_t\in\mathscr{F}_t \quad (t\geqslant 0)$.

For each $s\geqslant 0$ and each $\omega\in\Omega$ let $\tau_s(\omega)$ be the smallest t with $\varphi_\omega(t)=s$. Then with $Y(t,\omega)$ as in Lemma 5 we have $\{\tau_s\leqslant t\}=\{s\leqslant Y(t)\}\in\mathscr{F}_t$ so that τ_s is a stopping time relative to $\mathscr{F}_t$, $t\geqslant 0$. We shall show that if $\hat{\mathscr{F}}_s$ denotes the σ-field of events that happen before time τ_s, i.e.

$$\hat{\mathscr{F}}_s=\{A\in\mathscr{F}: A\cap\{\tau_s\leqslant t\}\in\mathscr{F}_t \text{ for every } t\geqslant 0\}$$

then

$$S^{-1}\mathscr{F}^*_{[0,s)}\subseteq\hat{\mathscr{F}}_s. \tag{16}$$

To see this let $C\in\mathscr{F}^*_{[0,s)}$ be of the form $\{\omega'\in\Omega': N(\omega',[0,u))=k\}$, $0<u\leqslant s$. The process $Z_t(\omega)=N(\omega,[0,t))$, $t\geqslant 0$, on Ω has left-continuous paths and is adapted to $\mathscr{F}_t$, $t\geqslant 0$, hence it is progressively measurable relative to $\mathscr{F}_t$, $t\geqslant 0$ and therefore Z_{τ_u} is $\hat{\mathscr{F}}_u$-measurable ([12, p. 70] or [14, p. 96]). Thus

$$S^{-1}C=\{\omega: N(S\omega,[0,u))=k\}=\{\omega: N(\omega,[0,\tau_u(\omega)))=k\}$$
$$=\{Z_{\tau_u}=k\}\in\hat{\mathscr{F}}_u\subseteq\hat{\mathscr{F}}_s.$$

Assertion (*i*) follows immediately from the inclusion (16). As for (*ii*): $A \in \mathscr{F}^*_{[0,s_0)}$ hence from the inclusion (16) again

$$S^{-1}A \in \hat{\mathscr{F}}_{s_0}, \quad S^{-1}A \cap \{\tau_{s_0} \leqslant t\} \in \mathscr{F}_t, \quad B \cap S^{-1}A \cap \{s_0 \leqslant Y(t)\} \in \mathscr{F}_t$$

and finally

$$\begin{aligned} D_t &= B \cap S^{-1}A \cap \{\omega\colon \varphi_\omega(t) \in Q\} \\ &= (B \cap S^{-1}A \cap \{s_0 \leqslant Y(t)\}) \cap \{s_0 \leqslant Y(t) \leqslant r_0\} \in \mathscr{F}_t. \end{aligned}$$

Consider now a renewal process whose increment has distribution F_0. If ω is the set

$$\ldots < t_{-1}(\omega) < t_0(\omega) < t_1(\omega) < \ldots$$

where $t_0(\omega) < 0 \leqslant t_1(\omega)$ then the mapping

$$\omega \Rightarrow (\theta_1(\omega), \theta_2(\omega), \ldots)$$

where $\theta_1(\omega) = -t_0(\omega)$, $\theta_2(\omega) = t_0(\omega) - t_{-1}(\omega)$, ... carries the probability $\mathbf{P}_0 | \mathscr{F}_0$ on Ω onto the product probability $F_0 \otimes F_0 \otimes \ldots$ on $R^+ \times R^+ \times \ldots$ and the probability $\mathbf{P} | \mathscr{F}_0$ onto $F \otimes F_0 \otimes F_0 \otimes \ldots$, where

$$F(x) = a \int_0^x (1 - F_0(u))\, du \quad \text{(by equation (10))}.$$

Now $F_0 \ll F$ if and only if $F_0 \ll \lambda$. In fact if $F_0 \ll \lambda$ and $F(Q) = 0$ then the set $\{u \in Q : 1 - F_0(u) \neq 0\}$ is λ-null, hence F_0-null; also F_0-null, clearly, is the set $\{u \in Q : 1 - F_0(u) = 0\}$. It is easy to see that

$$\frac{dF_0}{dF}(x) = \frac{F_0'(x)}{a(1 - F_0(x))} \quad F\text{-a.s.}$$

which agrees with equation (12). Thus in this case $\mathbf{P}_0 \ll \mathbf{P}$ on $\mathscr{F}_0$ if and only if $F_0 \ll \lambda$; the Radon–Nikodym derivative $X(\omega)$ is

$$X(\omega) = \frac{F_0'(-t_0(\omega))}{a(1 - F_0(-t_0(\omega)))}.$$

The next proposition throws some light on the nature of the condition $\mathbf{P}_0 \ll \mathbf{P}$ on $\mathscr{F}_0$.

***Proposition* 2.** *The following conditions are equivalent for a stationary point-process.*

(*i*) $\mathbf{P}_0 \ll \mathbf{P}$ *on* $\mathscr{F}_0$.

(*ii*) $\lim\limits_{t \downarrow 0} \left(\sup\limits_{A \in \mathscr{F}_0} |\mathbf{P}(A|t) - \mathbf{P}(A|0)| \right) = 0.$

(*iii*) *For each* $A \in \mathscr{F}_0$, $\lim\limits_{t \downarrow 0} \mathbf{P}(A|t) = \mathbf{P}(A|0)$.

Proof. In fact (*i*) implies (*ii*) because

$$\sup_{A \in \mathscr{F}_0} |\mathbf{P}(A|t) - \mathbf{P}(A|0)| \leqslant \mathbf{E}|X(t) - X(0)|$$

and (*ii*) clearly implies (*iii*). Finally if (*iii*) is true and if $\mathbf{P}(A) = 0$, $A \in \mathscr{F}_0$, then for $Q \subseteq [0, \infty)$

$$0 = \int_A N(Q)\, d\mathbf{P} = \int_A \mathbf{P}(A|t)\, \mu(dt)$$

hence $\mathbf{P}(A|t) = 0$ for a.a. $t \geqslant 0$ and by condition (*iii*) $\mathbf{P}(A|0) = 0$.

Note that the mapping $t \Rightarrow \mathbf{P}(\cdot\,|t)$ is always weakly continuous.

To state the next theorem let

$$0 \leqslant t_1(\omega) < t_2(\omega) < \dots \tag{17}$$

be the non-negative points of ω. If $\Delta = \{0 < \xi_1 < \xi_2 < \dots\}$ is any partition of $[0, \infty)$ let

$$0 \leqslant \tau_1^\Delta(\omega) \leqslant \tau_2^\Delta(\omega) \leqslant \tau_3^\Delta(\omega) \leqslant \dots \tag{18}$$

be the time instants at which we arrive at the points (17) if we adopt the following pattern of motion: Starting at 0 we look at the past $\mathscr{F}_0$ to see how many points we should expect in $[0, \xi_1)$, given the events up to 0. This is the expected number $\mathbf{E}(N(0, \xi_1)|\mathscr{F}_0)$. We now trace the interval $[0, \xi_1]$ with uniform velocity so as to arrive at ξ_1 in time $\mathbf{E}(N(0, \xi_1)|\mathscr{F}_0)$ (this way we can anticipate meeting the expected points of the process in $[0, \xi_1)$ at an average rate of one per time unit). On arrival of ξ_1 we look back again to take stock of the pattern of points up to ξ_1 and proceed to ξ_2 so as to cover the interval $[\xi_1, \xi_2]$ with uniform velocity in time

$$\mathbf{E}(N(\xi_1, \xi_2)|\mathscr{F}_{\xi_1}).$$

We continue in this manner. Under $\mathbf{P}$ process (18) is a point-process with a.s. no multiple points (as can be easily proved).

Define the mesh $\|\Delta\|$ of the partition as

$$\max_{i \geqslant 1} (\xi_{i+1} - \xi_i).$$

Theorem 6. *Suppose* $\mathbf{P}_0 \ll \mathbf{P}$ *on* $\mathscr{F}_0$. *If* $\|\Delta\| \to 0$ *then the process* (18) *converges weakly to a homogeneous Poisson process with rate* 1. *In fact if* $\|\Delta_k\| \to 0$ *rapidly enough, then the process* (18) *converges a.s. to*

$$0 \leqslant a\tau_1(\omega) < a\tau_2(\omega) < \dots,$$

where $\tau_1(\omega), \tau_2(\omega), \dots$ *are as in the sequence* (14).

The proof is based on Theorems 5 and 6.

Suppose $Z(\omega, I)$ is a random interval function, defined for $\omega \in \Omega$ and $I \subseteq R$. Z is said to be differentiable in the mean at a point s if there is a random variable V_s (called the derivative of Z at s) such that

$$\lim_{\substack{x \to s- \\ y \to s+}} \frac{Z(\cdot, (x, y))}{y - x} = V_s(\cdot)$$

in the mean. Z is said to be Burkill integrable in the mean over an interval $[b, c)$ if there is a random variable W_b^c (called the integral of Z over $[b, c)$) with the following property. For every $\varepsilon > 0$ there is $\delta > 0$ such that if $b = \xi_0 < \xi_1 < \ldots < \xi_\nu = c$ is a partition of $[b, c)$ with

$$\max_{0 \leqslant i \leqslant \nu - 1} (\xi_{i+1} - \xi_i) < \delta$$

then

$$\mathbf{E}\left|\sum_{i=0}^{\nu-1} Z(\cdot, [\xi_i, \xi_{i+1})) - W_b^c(\cdot)\right| < \varepsilon$$

(all the r.v.s involved are assumed integrable).

Theorem 7. *Suppose* $\mathbf{P}_0 \ll \mathbf{P}$ *on* $\mathscr{F}_0$. *Then the random interval function* $Z(I) = \mathbf{E}(N(I) \mid \mathscr{F}_x)$, *where* x *is the left end-point of* I, *is differentiable in the mean at every* s *with derivative* $aX(s)$ *and also Burkill integrable in the mean over any finite* $[b, c)$, *with integral*

$$W_b^c(\omega) = a \int_b^c X(\tau, \omega)\, d\tau = a(\varphi_\omega(c) - \varphi_\omega(b)).$$

This is a consequence of

Lemma 7. *If* $x \leqslant s \leqslant y$ *then*

$$\|\mathbf{E}(N(x, y) \mid \mathscr{F}_x) - a(y - x)\, X(s)\| \leqslant a \int_x^y \|X(t) - X(x)\|\, dt + a(y - x)\|X(s) - X(x)\|,$$

where $\|X\|$ *denotes* $\mathbf{E}|X|$.

Proof. First note that if $A \in \mathscr{F}_x$ then

$$(19) \quad \int_A \mathbf{E}(N(x, y) \mid \mathscr{F}_x)\, d\mathbf{P} = \int_A N(x, y)\, d\mathbf{P}$$

$$= \int_x^y \mathbf{P}(A \mid t)\, \mu(dt) = a \int_x^y \int_A X(t, \omega)\, \mathbf{P}(d\omega)\, dt$$

$$= \int_A \left(a \int_x^y X(t)\, dt\right) d\mathbf{P}.$$

Now

$$\int_\Omega |\mathbf{E}(N(x,y)|\mathscr{F}_x)-a(y-x)\,X(s)|\,d\mathbf{P}\leqslant \int_\Omega |\mathbf{E}(N(x,y)|\mathscr{F}_x)$$
$$-a(y-x)\,X(x)|\,d\mathbf{P}+a(y-x)\int_\Omega |X(x)-X(s)|\,d\mathbf{P}.$$

The first integral can be split in two

$$\int_{A^+}+\int_{A^-},$$

where

$$A^+=\{\mathbf{E}(N(x,y)|\mathscr{F}_x)\geqslant a(y-x)\,X(x)\}\quad\text{and}\quad A^-=(A^+)^c.$$

Since $A^+, A^-\in\mathscr{F}_x$ we see from equation (19) that

$$\int_\Omega |\mathbf{E}(N(x,y)|\mathscr{F}_x)-a(y-x)\,X(x)|\,dP$$
$$\leqslant a\int_\Omega \left|\int_x^y X(t)\,dt-(y-x)\,X(x)\right|\,d\mathbf{P}$$
$$\leqslant a\int_x^y \|X(t)-X(x)\|\,dt.$$

If we recall that the process $X(t)$ is stationary and mean continuous we see that given $\varepsilon>0$ there is $\delta>0$ such that if $y-x<\delta$ then

$$\|\mathbf{E}(N(x,y)|\mathscr{F}_x)-a\int_x^y X(t)\,dt\|\leqslant a(y-x)\varepsilon$$

from which mean Burkill integrability follows.

Now to show Theorem 6 let $S^\Delta\omega$ be the set of points in the sequence (18). If $\|\Delta_n\|\to 0$, there is a subsequence Δ_k such that $\|\Delta_k\|\to 0$ so rapidly that in Theorem 7 we have a.s. convergence to the integral

$$a\int_0^t X(\tau,\omega)\,d\tau,$$

for any t. This implies that if $S'\omega=\{ay\colon y\in S\omega\}$ ($S\omega$ as in equation (15)), then a.s. $S^{\Delta_k}\omega\to S'\omega$ in the vague topology of $\Omega_{[0,\infty)}$. Then this subsequence of processes converges weakly to the Poisson process and since every subsequence contains such a sub-subsequence Theorem 6 follows.

The intuitive meaning of Theorems 5, 6, 7 was explained in the introduction. Naturally, in the stationary case Theorem 4 (or [22, Theorem 6]) can be proved by the method of Theorem 5 and the random rate can then be identified.

Theorem 4a. *If for a stationary process* $\mathbf{P}(A|t_1) = \mathbf{P}(A|t_2)$ *whenever* $A \in \mathscr{F}_{\min(t_1,t_2)}$, *then* $\mathbf{P}_0 \ll \mathbf{P}$ *on* $\mathscr{F}_0$, *the random variable X in equation* (13) *is invariant under translations and*

$$\ldots < X(\omega)\, t_{-1}(\omega) < X(\omega)\, t_0(\omega) < X(\omega)\, t_1(\omega) < \ldots$$

is a Poisson process in $(-\infty, \infty)$ *with rate a, independent of*

$$\mathscr{F}_{-\infty} = \bigcap_{-\infty < t} \mathscr{F}_t$$

(and of X). The original process is mixed Poisson with rate aX.

In fact by Proposition 2 $\mathbf{P}_0 \ll \mathbf{P}$ on $\mathscr{F}_0$. Also $X(t_1) = \mathbf{E}(X(t_2)|\mathscr{F}_{t_1})$ if $t_1 < t_2$, so that $X(t)$, $-\infty < t < \infty$, is a stationary martingale and hence constant in the sense that $X(t_1) = X(t_2)$ a.s. for any t_1, t_2. Thus X is invariant under translations and we can take $X(\omega) = \liminf X(-n)$, which is $\mathscr{F}_{-\infty}$-measurable. $\mathbf{P}(X = 0) = 0$ since

$$\mathbf{P}(X = 0|t) = \int_{\{X=0\}} X d\mathbf{P} = 0$$

for every t. The rest of the proof imitates that of Theorem 5 (but is easier).

We conclude this section with a kind of converse to Theorem 5.

Theorem 8. *Suppose a stationary point-process in* $(-\infty, \infty)$ *has the following property. There exists a measurable stationary process* $X(t, \omega)$, $t \geqslant 0$, *on* Ω *with* $X(t, \omega) \geqslant 0$ *and such that the process* (14) *as defined in Theorem* 5 *is first-order stationary (under positive translations) and independent of* $\mathscr{F}_0$. *Then* $\mathbf{P}_0 \ll \mathbf{P}$ *on* $\mathscr{F}_0$.

Proof. Define $S\omega$ as in equation (15). If $A \in \mathscr{F}_0$ and $Q \subseteq [0, \infty)$ then by the hypotheses

$$\int_A N(S\omega, Q)\, \mathbf{P}(d\omega) = \mathbf{P}(A) \int_\Omega N(S\omega, Q)\, \mathbf{P}(d\omega) = \mathbf{P}(A)\, a'\lambda(Q)$$

so that the two measures

$$m'(D) = \int_\Omega N(S\omega, D^\omega)\, \mathbf{P}(d\omega)$$

and

$$(\lambda \otimes \mathbf{P})(D) = a' \int_\Omega \lambda(D^\omega)\, \mathbf{P}(d\omega)$$

agree on $([0, \infty) \times \Omega, \mathscr{B}_{[0,\infty)} \otimes \mathscr{F}_0)$.

Let now $A \in \mathscr{F}_0$ and suppose Q is an interval with end-points $b < c$ $(0 \leqslant b)$. Since $N(\omega, Q) \leqslant N(S\omega, [\varphi_\omega b, \varphi_\omega c])$ we have

$$a \int_Q \mathbf{P}(A|t)\, dt = \int_A N(\omega, Q)\, \mathbf{P}(d\omega) \leqslant \int_A N(S\omega, [\varphi_\omega b, \varphi_\omega c])\, \mathbf{P}(d\omega)$$

$$= a' \int_A \lambda([\varphi_\omega b, \varphi_\omega c])\, \mathbf{P}(d\omega) = a' \int_A (\varphi_\omega c - \varphi_\omega b)\, \mathbf{P}(d\omega)$$

$$= a' \int_A \int_b^c X(t, \omega)\, dt\, \mathbf{P}(d\omega) = a' \int_Q \left(\int_A X(t, \omega)\, \mathbf{P}(d\omega) \right) dt.$$

The reverse inequality can be proved using $N(\omega, Q) \geqslant N(S\omega, (\varphi_\omega b, \varphi_\omega c))$. Thus for a.a. $t \geqslant 0$

$$\mathbf{P}(A|t) = \frac{a'}{a} \int_A X(t, \omega)\, \mathbf{P}(d\omega).$$

Using a countable generating subset of $\mathscr{F}_0$, closed under finite intersections, we see that there is a null-set $M \subseteq (0, \infty)$ such that for all $A \in \mathscr{F}_0$ and all $t \notin M$

$$\mathbf{P}(A - t|0) = \frac{a'}{a} \int_A X(t, \omega)\, \mathbf{P}(d\omega) = \frac{a'}{a} \int_{A-t} X(t, \omega + t)\, \mathbf{P}(d\omega)$$

so that $\mathbf{P}_0 \ll \mathbf{P}$ on $\mathscr{F}_{-t}$ and the result follows from the uniform integrability of $X(t)$, $t \geqslant 0$.

2.7.6 RANDOM SETS OF LINES

We turn now to random sets of lines in the plane which are stationary under the group $\mathscr{G}$ of Euclidean motions. Here the underlying space E is the space L of all oriented lines in the plane. We choose for a natural reference system a fixed line $l_0 \in L$ and a fixed point O on l_0. The normal coordinates of an oriented line l are (θ, p), where θ is the angle it makes with l_0 and p its signed distance from O $(0 \leqslant \theta < 2\pi,\ -\infty < p < \infty)$. If we identify l with (θ, p) then we see that we are really considering point-processes on the cylinder $C \times R$ $(C = \{\theta\colon 0 \leqslant \theta < 2\pi\})$ whose distributions are stationary under a three-parameter group acting on $C \times R$. (On L we introduce the product topology of $C \times R$ and assume as before that $\mu(Q) = \mathbf{E}(N(Q))$ is finite for bounded Q.) It is well-known [21] that $d\nu = d\theta\, dp$ is the (unique) measure invariant under $\mathscr{G}$. Hence by stationarity there is $a > 0$ such that $\mu(Q) = a.\nu(Q)$ for every Borel subset Q of L.

R. Davidson [5] has asked whether such a stationary random set with a.s. no parallel lines is doubly stochastic Poisson (i.e. it is a 'Poisson set' built on a random measure with stationary distribution). We shall give a partial answer by determining the structure of such random sets under a condition similar to that of Theorems 5, 6, 7.

Write $l_1 \| l_2$ if l_1, l_2 are parallel with the same orientation.

Lemma 8. *If $\mathbf{P}(A \mid l_1) = \mathbf{P}(A \mid l_2)$ whenever $l_1 \| l_2$ and A happens in $L - \{l_1, l_2\}$, then the process is doubly stochastic Poisson. More precisely it is a Poisson process built on a random measure of the form $a(\bar{m} \otimes \lambda)$, where λ is the Lebesgue measure on R and $\bar{m}$ a finite random measure on the circle C, stationary under the rotations of C.*

Outline of proof. A rectangle in $L = C \times R$ is a set of the form $I \times J$, where I, J are finite intervals in C and R respectively. If Q is a Borel set in L, A an event in Ω_L and $-\infty < q < \infty$, we write

$$Q + q = \{(\theta, p+q) \colon (\theta, p) \in Q\}, \quad A + q = \{\omega + q \colon \omega \in A\}.$$

The fact that $\mathbf{P}(A \mid l_1) = \mathbf{P}(A \mid l_2)$ whenever $l_1 \| l_2$ and A happens in $L - \{l_1, l_2\}$ can be seen, as in Theorem 4, to imply the following. If $Q_1, \ldots, Q_k$ are disjoint rectangles in L and $q_1, \ldots, q_k$ real numbers such that $Q_1 + q_1, \ldots, Q_k + q_k$ are also disjoint then

$$\begin{aligned} (20) \qquad & \mathbf{P}(N(Q_1) = \alpha_1, \ldots, N(Q_k) = \alpha_k) \\ & = \mathbf{P}(N(Q_1 + q_1) = \alpha_1, \ldots, N(Q_k + q_k) = \alpha_k). \end{aligned}$$

An important consequence of this is that the point-process is stationary under the group $\mathscr{H}$ of 'translations' of the cylinder $C \times R$

$$Sq \colon (\theta, p) \Rightarrow (\theta, p+q)$$

(which are not in the Euclidean group). By equation (20) if Q_1, Q_2 are two rectangles in $C \times R$ such that $Q_1 \cap (Q_2 + q) = \emptyset$ for every $q \geqslant 0$ and if $A \in \mathscr{F}_{Q_1}$, $B \in \mathscr{F}_{Q_2}$ then $\mathbf{P}(A \cap B) = \mathbf{P}(A \cap (B + q))$ for every $q \geqslant 0$. If $\mathscr{I}_{\mathscr{H}}$ is the σ-field of events invariant under the group $\mathscr{H}$ and if $D \in \mathscr{I}_{\mathscr{H}}$ then

$$D \in \bigcap_p \mathscr{F}_{C \times [p, \infty)}$$

(to within a $\mathbf{P}$-null set), hence with A, B as above

$$\mathbf{P}(A \cap B \cap D) = \mathbf{P}(A \cap (B+q) \cap D), \quad q \geqslant 0,$$

i.e.

$$\int_D \chi_A(\omega)\, \chi_B(\omega)\, \mathbf{P}(d\omega) = \int_D \chi_A(\omega)\, \chi_B(\omega - q)\, \mathbf{P}(d\omega).$$

Applying the ergodic theorem to χ_B we obtain

$$\int_D \chi_A \chi_B \, d\mathbf{P} = \int_D \chi_A \, \mathbf{E}(\chi_B | \mathscr{I}_{\mathscr{H}}) \, d\mathbf{P},$$

hence

$$\mathbf{P}(A \cap B | \mathscr{I}_{\mathscr{H}}) = \mathbf{P}(A | \mathscr{I}_{\mathscr{H}}) \, \mathbf{P}(B | \mathscr{I}_{\mathscr{H}}) \quad \text{a.s.}$$

Returning to equation (20) and working with the 'right-most' rectangle Q_i and then proceeding inductively we get

$$\mathbf{P}\left(\bigcap_{j=1}^{k} \{N(Q_j) = \alpha_j\} \middle| \mathscr{I}_{\mathscr{H}}\right) = \prod_{j=1}^{k} \mathbf{P}(N(Q_j) = \alpha_j | \mathscr{I}_{\mathscr{H}})$$

so that the ergodic components of the process relative to the group $\mathscr{H}$ are Poisson. If $D \in \mathscr{I}_{\mathscr{H}}$ then

$$\int_D N(Q) \, d\mathbf{P} = \int_D N(Q+q) \, d\mathbf{P},$$

so that $\mathbf{E}(N(Q) | \mathscr{I}_{\mathscr{H}}) = \mathbf{E}(N(Q+q) | \mathscr{I}_{\mathscr{H}})$ and hence (using a regular decomposition) the random measure $\mathbf{E}(N(\cdot) | \mathscr{I}_{\mathscr{H}})$ has the form $a(\bar{m} \otimes \lambda)$, where $\bar{m}(\omega)$ is a finite random measure on C. It is easy to see that $\bar{m}$ is stationary under the rotations of C.

Lemma 9. (R. Davidson.) *Suppose $X(l)$, $l \in L$, is a stochastic process parametrized by the elements of L, stationary under $\mathscr{G}$ and continuous in the mean (we assume $\mathbf{E}|X(l)| < \infty$). If $l_1 \| l_2$ then $X(l_1) = X(l_2)$ a.s.*

In fact fix a point γ on l_1 and let $\bar{l}$ be the line passing through γ and forming an angle ε with l_1. Then

$$\mathbf{E}|X(l_1) - X(l_2)| \leqslant \mathbf{E}|X(l_1) - X(\bar{l})| + \mathbf{E}|X(\bar{l}) - X(l_2)| = 2\mathbf{E}|X(l_1) - X(\bar{l})|$$

by stationarity. If we let $\varepsilon \to 0$ we get $\mathbf{E}|X(l_1) - X(l_2)| = 0$.

Lemma 10. *If a process $X(l, \omega)$ is measurable in (l, ω), stationary under $\mathscr{G}$ and satisfies $\mathbf{E}|X(l)| < \infty$, then it is mean continuous.*

Only an adaptation of the standard argument is needed. For the sake of completeness we show 'rotation continuity', as this was all we needed in Lemma 9; i.e. if γ is a point on the line l and T_θ denotes the rotation around γ by angle θ then

$$\lim_{\varepsilon \to 0} \mathbf{E}|X(T_\varepsilon l) - X(l)| = 0.$$

If $\varphi(\varepsilon) = \mathbf{E}|X(T_\varepsilon l) - X(l)|$ then by stationarity, Fubini's theorem and the generalized Lebesgue convergence theorem (uniform integrability)

$$\varphi(\varepsilon) = \frac{1}{\theta_0}\int_0^{\theta_0}\int_\Omega |X(T_{\theta+\varepsilon}l, \omega) - X(T_\theta l, \omega)|\, d\mathbf{P}\, d\theta$$

$$= \frac{1}{\theta_0}\int_\Omega \int_0^{\theta_0} |X(T_{\theta+\varepsilon}l, \omega) - X(T_\theta l, \omega)|\, d\theta\, d\mathbf{P} \to 0$$

as $\varepsilon \to 0$.

Suppose now $\mathbf{P}(\cdot\,|\,l) \ll \mathbf{P}(\cdot)$ on the σ-field of events happening in $L - \{l\}$, with Radon–Nikodym derivative $X(l, \omega)$. Here $X(l, \omega)$ is a stationary process with $\mathbf{E}X(l) = 1$ and there is a measurable version (for instance we can fix $l_0 \in L$, choose a continuous mapping $L \ni l \Rightarrow T_l \in \mathscr{G}$ such that $T_l l = l_0$ and take $X(l, \omega) = X(l_0, T_l\omega)$). By Lemmata 9 and 10 $X(l_1) = X(l_2)$ a.s. when $l_1 \| l_2$ and this implies that $\mathbf{P}(A\,|\,l_1) = \mathbf{P}(A\,|\,l_2)$, whenever $l_1 \| l_2$ and A happens in $L - \{l_1, l_2\}$, and that the process $X(l, \omega)$ is essentially parametrized by the first coordinate θ of $l = (\theta, p)$, i.e. it is a stochastic process $X(\theta, \omega)$ on the circle $0 \leqslant \theta < 2\pi$, stationary under rotations. To be exact, for almost every pair (θ, ω) there is a number $X(\theta, \omega)$ such that $X((\theta, p), \omega) = X(\theta, \omega)$ for a.a. p.

Theorem 9. *Suppose that for some (and then for all) $l \in L$ $\mathbf{P}(\cdot\,|\,l) \ll \mathbf{P}(\cdot)$ on the σ-field of events happening in L-$\{l\}$. Then the process has the following structure. On the circle C we put the (stationary under rotations) random measure $\bar{m}(\omega)$ with density $X(\theta, \omega)$, $0 \leqslant \theta < 2\pi$, then form the product measure $\bar{m} \otimes \lambda$ where λ is the Lebesgue measure and finally pick a random 'Poisson set' from the measure space $(L, a(\bar{m} \otimes \lambda))$.*

This follows from Lemmata 8, 9 and 10. For the identification of $\bar{m}$ note that if Q is any rectangle and if $D \in \mathscr{I}_{\mathscr{H}}$ then there is an event D' happening in Q^c such that $\mathbf{P}(D \triangle D') = 0$. Then

$$\int_D N(\omega, Q)\, \mathbf{P}(d\omega) = \int_{D'} N(\omega, Q)\, \mathbf{P}(d\omega) = \int_Q \mathbf{P}(D'\,|\,l)\, \mu(dl)$$

$$= a\int_Q \int_{D'} X(l, \omega)\, \mathbf{P}(d\omega)\, \nu(dl)$$

$$= \int_D \int_Q a X(l, \omega)\, \nu(dl)\, \mathbf{P}(d\omega),$$

that is

$$\mathbf{E}(N(Q)\,|\,\mathscr{I}_{\mathscr{H}}) = \int_Q aX(l,\cdot)\,\nu(dl)$$

which completes the proof.

Theorem 10. *Another condition under which Lemma* 8 *holds is this: for any* l_1, l_2 *the probabilities* $\mathbf{P}(\cdot\,|\,l_1), \mathbf{P}(\cdot\,|\,l_2)$ *are mutually absolutely continuous on the σ-field of events happening in* $L-\{l_1, l_2\}$.

We omit the proof.

Note (added January 1972). The results of this paper which was submitted in the autumn of 1971 have now been improved and an extensive account can be found in [27]. (See also [28] for a summary.) For instance, without the assumption '$\mathbf{P}_0 \ll \mathbf{P}$ on $\mathscr{F}_0$', if for each partition

$$\Delta = \{b = \xi_0 < \xi_1 < \ldots < \xi_{n+1} = c\}$$

of $[b, c]$ we set

$$S_\Delta(\omega) = \sum_{\nu=0}^{n} \mathbf{E}(N[\xi_\nu, \xi_{\nu+1})\,|\,\mathscr{F}_{\xi_\nu})$$

then

$$\lim_\Delta S_\Delta(\omega) = W(\omega, [b, c))$$

exists a.s. and in the mean when

$$\max_{0\leqslant\nu\leqslant n} (\xi_{\nu+1} - \xi_\nu) \to 0.$$

If now the transformation $t \Rightarrow W(\omega, [0, t))$ of $[0, \infty)$ onto itself is a.s. continuous, then it transforms the non-negative points of the process into a Poisson process with rate 1 and independent of $\mathscr{F}_0$. The condition '$\mathbf{P}_0 \ll \mathbf{P}$ on $\mathscr{F}_0$' of the present paper is necessary and sufficient for the almost sure absolute continuity of $t \Rightarrow W(\omega, [0, t))$.

Shortly after the submission of this paper, the work of G. Kummer and K. Matthes [24] came to my attention. In this, the master measure m is introduced under the name 'Campbell measure' but used for a different purpose; namely the characterization of infinitely divisible point-processes in a Polish space (see also [25]).

If we weaken Theorem 3 by making, instead of (*ii*), the two-sided assumption 'if $A \in \mathscr{F}_{[a,b]^c}$ then $\mathbf{P}(A\,|\,t) = \mathbf{P}(A)$ for μ-almost all $t \in [a, b]$', then we obtain in effect Satz 3.1 of [26] for the real line.

Finally, the editors have renumbered theorems in the present paper, so that Theorem k referred to in [27] or [28], for instance, is in reality Theorem $(k+1)$ here.

REFERENCES

1. W. Ambrose, "Representation of ergodic flows", *Ann. Math.* **42** (1941), 723–739.
2. W. Ambrose and S. Kakutani, "Structure and continuity of measurable flows", *Duke Math. J.* **9** (1942), 25–42.
3. N. Bourbaki, *Intégration*, 1re édition, Chaps. 1–4, Hermann, Paris (1952).
4. ——, *Intégration*, Chaps. 7–8, Hermann, Paris (1963).
5. R. Davidson, Thesis (Chapter 6, "Stochastic processes of flats"; Chapter 7, "Exchangeable stochastic point processes"), University of Cambridge (1968). (Reprinted in 2.1 of this book.)
6. P. Franken, A. Liemant and K. Matthes, "Stationäre zufällige Punktfolgen III", *Jahresbericht D.M.V.* **67** (1965), 183–202.
7. T. E. Harris, "Random measures and motions of point processes", *Ztschr. Wahrsch'theorie & verw. Geb.* **8** (1971), 85–115.
8. J. Kerstan und K. Matthes, "Verallgemeinerung eines Satzes von Sliwnjak", *Rev. Roum. Math. Pures et Appl.* **9** (1964), 811–829.
9. A. Y. Khintchine, *Mathematical Methods in the Theory of Queueing*, Griffin, London (1960).
10. M. R. Leadbetter, "On three basic results in the theory of stationary point processes", *Proc. A.M.S.* **19** (1968), 115–117.
11. K. Matthes, "Stationäre zufällige Punktfolgen I", *Jahresbericht D.M.V.* **66** (1963), 66–79.
12. P. A. Meyer, *Probability and Potentials*, Blaisdell, Waltham, Mass. (1966).
13. L. Nachbin, *The Haar Integral*, D. van Nostrand, Princeton (1965).
14. J. Neveu, *Bases Mathématiques du Calcul des Probabilités*, Masson, Paris (1964).
15. ——, "Sur la structure des processus ponctuels stationnaires", *C. R. Acad. Sci. Paris* **267** Série A (1968), 561–564.
16. C. Palm, "Intensitätsschwankungen im Fernsprechverkehr", *Ericsson Technics* **44** (1943), 1–189.
17. F. Papangelou, "The Ambrose–Kakutani theorem and the Poisson process", *Contributions to Ergodic Theory and Probability*, pp. 234–240, Springer-Verlag, Lecture notes in Math. **160** (1970).
18. A. Rényi, "Remarks on the Poisson process", *Proc. of Symposium on Probability Methods in Analysis*, pp. 280–286, Springer-Verlag, Lecture Notes in Math. **31** (1967).
19. C. Ryll-Nardzewski, "On the non-homogeneous Poisson process (I)", *Studia Math.* **14** (1954), 124–128.
20. ——, "Remarks on processes of calls", *Proc. of Fourth Berkeley Symposium*, vol. II, pp. 455–465 (1961).
21. L. A. Santaló, *Introduction to Integral Geometry*, Actual. Sci. et Ind. 1198, Hermann, Paris (1953).
22. I. M. Slivnyak, "Some properties of stationary flows of homogeneous random events" (in Russian), *Teor. Verojatnost. i ee Primenen.* **7** (1962), 347–352. (Transl.: *Theory of Prob. and its Appl.* **7**, 336–341).
23. ——, "Stationary streams of homogeneous random events", *Vestnik Harkov. Gos. Univ., Ser. Mech.-Math.* **32** (1966), 73–116. (English translation by

D. J. Daley and R. K. Milne, available from the Statistics Department, Australian National University.)

24. G. Kummer und K. Matthes, "Verallgemeinerung eines Satzes von Sliwnjak II", *Rev. Roum. Math. Pures et Appl.* **15** (1970), 845–870.
25. —— ——, "Verallgemeinerung eines Satzes von Sliwnjak III", *Rev. Roum. Math. Pures et Appl.* **15** (1970), 1631–1642.
26. J. Mecke, "Stationäre zufällige Masse auf lokalkompakten Abelschen Gruppen", *Ztschr. Wahrsch'theorie & verw. Geb.* **9** (1967), 36–58.
27. F. Papangelou, "Integrability of expected increments of point processes and a related random change of scale", *Trans. Am. math. Soc.* **165** (1972), 483–506.
28. ——, "Summary of some results on point and line processes" *Stochastic Point Processes* (Ed. P. A. W. Lewis), pp. 522–532, Wiley, New York (1972).
29. S. Watanabe, "On discontinuous additive functionals and Lévy measures of a Markov process', *Japan J. Math.* **34** (1964), 53–70.

2.8

Various Concepts of Orderliness for Point-processes

DARYL J. DALEY

2.8.1 SUMMARY

Orderliness of a point-process is loosely speaking the property that points are distinct or that probabilistically they are not infinitesimally close. Four essentially different definitions of orderliness are proposed, and their interrelationships are explored for point-processes that may be stationary or non-stationary, and have a σ-finite or infinite parametric measure. Counter-examples abound.

2.8.2 INTRODUCTION

The motivation behind the various definitions that have been proposed concerning 'orderliness' of a point-process lies embedded in the apparently similar ideas:

(*a*) no two points of the process coincide;

(*b*) the chance of two or more points lying in a small interval is negligible compared with the size of the interval;

(*c*) the chance of exactly one point lying in a small interval is negligibly different from the chance of one or more points lying in the interval.

In this note we propose what seems to us a more systematic terminology for these different concepts; that they are different will be shown by various examples. What we call *almost sure orderliness* (a.s. orderliness) stems from (*a*), and generally speaking it is the weakest concept. I recall Rollo stressing this point in the Cambridge University Statistical Laboratory; it may well be that he had an intuitive understanding of some of the results below stemming from counter-examples which he was generally adept at devising.

2.8.3 DEFINITIONS

We confine our attention to not necessarily stationary point-processes on the real line R. That is, our sample space Ω has as generic element ω any countable set $\{x_i\} \equiv \{x_i(\omega)\}$ ($i \in$ finite or infinite subset of $Z = \{0, 1, \ldots\}$) of not necessarily distinct points of R, and we study the set-function†

$$N(A) \equiv N(A, \omega) = \text{card}\{i: x_i(\omega) \in A\}$$

for each $A \in \mathscr{B} \equiv \mathscr{B}(R)$, the family of Borel sets in R. The σ-algebra $\mathscr{F}$ of our probability triple $(\Omega, \mathscr{F}, \text{pr})$ is such that all sets in the ring generated by

$$\{\omega: N(A) = k\} \quad (A \in \mathscr{B},\ k \in Z)$$

are pr-measurable. It is always assumed that

(1) $$\text{pr}\{N(A) < \infty \text{ for bounded } A \in \mathscr{B}\} = 1,$$

i.e. $\text{pr}\{\omega: \omega \text{ has a finite cluster point}\} = 0$.

We turn now to our definitions arising from (*a*), (*b*) and (*c*).

Definition 1. A point-process is *almost surely orderly* (*a.s. orderly*) if

(2) $$\text{pr}\{\omega: N(\{x\}, \omega) = 0 \text{ or } 1 \text{ for every } x \in R\} = 1.$$

It has been more common in the literature (e.g. Leadbetter [8]) to speak of a.s. orderliness as 'without multiple points with probability one' or more briefly as 'without multiple points'. Belyaev [1] uses the word 'regular', but a less frequently occurring terminology seems preferable. Observe that equation (2) is equivalent to

(2*i*) $$\text{pr}\{x_i(\omega) \neq x_j(\omega) \text{ for } i \neq j\} = 1.$$

From equations (2*i*) and (1) it follows that with probability one there exist unique permutations of the suffixes in

$$\{i: x_i(\omega) < 0\} \quad \text{and} \quad \{i: x_i(\omega) \geqslant 0\},$$

yielding $\{y_i(\omega)\}$ and $\{z_i(\omega)\}$, say, such that (if the sets are non-empty) $0 > y_0(\omega) > y_1(\omega) > \ldots$ and $0 \leqslant z_0(\omega) < z_1(\omega) < \ldots$, with $y_i(\omega) \to -\infty$ and $z_i(\omega) \to \infty$ ($i \to \infty$) when the sets are countably infinite. In other words, the condition at equation (2) implies that almost surely there exists an essentially unique ordering of the points in ω; for this reason we propose the terminology in Definition 1. Equation (2*i*) can be rephrased as

(2*ii*) $$\text{pr}\{\omega: |x_i(\omega) - x_j(\omega)| > 0 \text{ for every } i \neq j\} = 1,$$

† When A is an interval such as $(a, b]$, we write $N(a, b]$ for $N((a, b])$.

i.e. the 'distance between points of ω is always positive'; this terminology is used in Vasil'ev [11]. Slivnyak [9, 10] calls an a.s. orderly point-process a 'calling time process'.

The reason for considering (*b*) or (*c*) is that we may wish that for intervals I_h of small length h whether the intervals are open or closed or half-open, the approximations

$$\mathsf{E}(N(I_h)) \simeq \mathrm{pr}\{N(I_h) \geqslant 1\} \simeq \mathrm{pr}\{N(I_h) = 1\} \tag{3}$$

should hold and be valid in certain algebraic manipulations. In order that the first approximation hold the expectation must be finite, while the second is equivalent to the convergence to 0 with h of

$$\mathrm{pr}\{N(I_h) \geqslant 2\}/\mathrm{pr}\{N(I_h) \geqslant 1\}.$$

In *stationary* point-processes (where stationarity means the invariance with respect to t of the joint distributions of $\{N(A_i+t), i = 1, \ldots, r\}$ for every finite positive integer r), this ratio will depend only on the length h of the interval I_h. This need not be the case in non-stationary processes, so we arrive at the next two definitions.

Definition 2. A point-process is *Khinchin orderly* if to each $x \in R$ and $\varepsilon > 0$ there exists $\delta \equiv \delta(x, \varepsilon)$ such that

$$\mathrm{pr}\{N(x-h_1, x+h_2] \geqslant 2\} < \varepsilon\, \mathrm{pr}\{N(x-h_1, x+h_2] \geqslant 1\} \quad \text{for} \quad 0 < h_1 < \delta,\ 0 < h_2 < \delta. \tag{4}$$

Definition 3. A point-process is *uniformly Khinchin orderly* (*unif. Khinchin orderly*) if to each $\varepsilon > 0$ there exists $\delta \equiv \delta(\varepsilon)$ such that

$$\mathrm{pr}\{N(x, x+h] \geqslant 2\} < \varepsilon\, \mathrm{pr}\{N(x, x+h] \geqslant 1\} \quad \text{for } 0 < h < \delta \text{ and all } x \in R. \tag{5}$$

The distinction made here between Khinchin orderly and unif. Khinchin orderly is believed to be new. The same property as in (5) is described in Khinchin [6] and Fieger [3] as 'ordinary', and Belyaev [1] in his generalization of this property to point-processes on more general spaces than R uses the same word; Leadbetter [8] uses 'regular'.

It should be noted that the same word in the Russian literature, as for example Khinchin [5], Vasil'ev [11] and Slivnyak [9] has been variously translated into English as 'orderly' and 'ordinary'. Hitherto the term 'orderly' or 'ordinary' has mostly been applied to stationary processes satisfying the condition

$$\limsup_{h \to 0} \mathrm{pr}\{N(0, h] \geqslant 2\}/h = 0. \tag{6}$$

It will be shown that Khinchin's generalization of orderliness to include the non-stationary case is not entirely satisfactory, for it does not always imply nor is it implied by a.s. orderliness. What seems to be more satisfactory is the following definition in which for any interval I, $\mathscr{D}(I)$ denotes the family of all finite partitions of I into mutually disjoint subintervals I_i (i.e. $\{I_i\} \in \mathscr{D}(I)$ if $I_i \cap I_j = \emptyset$ $(i \neq j)$ and $\bigcup_i I_i = I$).

Definition 4. A point-process is *ordinary* if for every bounded interval I

(7) $$\inf_{\{I_i\} \in \mathscr{D}(I)} \sum_i \operatorname{pr}\{N(I_i) \geq 2\} = 0.$$

As another plausible generalization of property (6) and stemming from (*b*) above we propose (cf. Definitions 2 and 3)

Definition 5. A point-process is *analytically orderly* (*analyt. orderly*) if to each x in R

(8) $$\lim_{h_1, h_2 \to 0} \frac{\operatorname{pr}\{N(x-h_1, x+h_2] \geq 2\}}{h_1 + h_2} = 0.$$

Definition 6. A point-process is *uniformly analytically orderly* (*unif. analyt. orderly*) if

(9) $$\lim_{h \to 0} \sup_{x \in R} \frac{\operatorname{pr}\{N(x, x+h] \geq 2\}}{h} = 0.$$

The major contrast in the relations between these various properties lies in the finiteness or otherwise of the *parametric measure* of the process. The basic idea behind this measure appears to be due independently to Fieger [3] and Belyaev [1] (see also Leadbetter [8]); these last two, unlike Fieger, consider point-processes on spaces more general than R. For the present purposes the following suffices.

Definition 7. *The parametric measure* $\lambda(A)$ $(A \in \mathscr{B})$ is the possibly infinite-valued measure induced by the set function defined on intervals I by

(10) $$\lambda(I) = \sup_{\{I_i\} \in \mathscr{D}(I)} \sum_i \operatorname{pr}\{N(I_i) > 0\}.$$

A point-process is called *finite* when

$$\lambda(I) < \infty \quad \text{for every bounded interval } I.$$

In other words, the parametric measure of a finite point-process is a σ-finite measure.

An area of some interest in point-process theory is the relation between the parametric measure $\lambda(\cdot)$ and the first moment measure $M(\cdot)$ defined below.

Definition 8. *The first moment measure* $M(A)$ $(A \in \mathscr{B})$ *of a point-process* is

(11) $$M(A) = \mathbf{E}(N(A)) = \int_{\Omega} N(A, \omega)\,\mathrm{pr}\,(d\omega) \quad (A \in \mathscr{B}).$$

Korolyuk's theorem as originally given by Khinchin [5] states conditions for equality of $\lambda(\cdot)$ and $M(\cdot)$, and we follow Belyaev [1] and Leadbetter [8] in our statement of the

Generalized Korolyuk Theorem. *For an a.s. orderly point-process*

(12) $$\lambda(A) = M(A) \quad (A \in \mathscr{B}).$$

Conversely, if the point-process is finite and (12) *holds, then it is a.s. orderly.*

There have been other generalizations of Korolyuk's theorem, as for example Fieger [4], and they are similar to Lemma 1 below. For this, the process need not be a.s. orderly, for indeed in that case all the *generalized parametric measures* are the same.

Definition 9. *The generalized parametric measures*

$$\lambda_k(A)\ (k = 1, 2, \ldots;\ A \in \mathscr{B})$$

of a point-process are determined by the possibly infinite valued set functions defined on intervals I by

(13) $$\lambda_k(I) = \sup_{\{I_i\} \in \mathscr{D}(I)} \sum_i \mathrm{pr}\,\{0 < N(I_i) \leqslant k\}.$$

To obtain $\lambda(A)$ $(A \in \mathscr{B})$ from $\lambda(I)$ ($I \in$ class $\mathscr{I}$ of all bounded intervals), extend $\lambda(\cdot)$ additively from all $I \in \mathscr{I}$ to all subsets of R in the ring $\mathscr{R}$ generated by $\mathscr{I}$. The subadditivity of the set function $\varphi(A) \equiv \mathrm{pr}\,\{N(A) > 0\}$ ensures that equation (10) continues to hold with I replaced by sets $C \in \mathscr{R}$, and $\lambda(\cdot)$ can be shown to be σ-additive on $\mathscr{R}$. Then the standard extension theorem for measures on rings yields uniquely the measure $\lambda(\cdot)$ on elements of the σ-ring $\sigma(\mathscr{R}) = \mathscr{B}$. Similarly we obtain $\lambda_k(A)$ $(A \in \mathscr{B})$ from $\lambda_k(I)$ $(I \in \mathscr{I})$.

Lemma 1.

(14) $$M(A) = \sum_{k=1}^{\infty} k[\lambda_k(A) - \lambda_{k-1}(A)] = \sum_{k=0}^{\infty} [\lambda(A) - \lambda_k(A)].$$

Proof. It suffices to prove the result for intervals I. It is easily shown that $\lambda_k(I) \uparrow \lambda(I)$ $(k \to \infty)$, and that $M(I) \geqslant \lambda(I)$, so equation (14) follows immediately for $\lambda(I) = \infty$. Assume then that $\lambda(I) < \infty$, and let $\{I_{ni}\} \in \mathscr{D}(I)$ be a nested family of subintervals partitioning I such that

(15) $$\sum_i \mathrm{pr}\,\{0 < N(I_{ni}) \leqslant k\} \to \lambda_k(I) \quad (n \to \infty)$$

for each integer k; the existence of such a family can be verified by the construction

$$\{I_{ni}\} = \bigcap_{k=1}^{n} \{I_{ni}^{(k)}\}$$

starting from nested families of subintervals $\{I_{ni}^{(k)}\}$ $(n \geqslant k)$ yielding approximations to $\lambda_k(I)$, using the sub-additivity property of non-negative random variables X and Y that

$$\text{pr}\{0 < X+Y \leqslant u\} \leqslant \text{pr}\{0 < X \leqslant u\} + \text{pr}\{0 < Y \leqslant u\}.$$

Then

$$M(I) = \sum_i M(I_{ni}) = \sum_i \sum_{j=1}^{\infty} j\,\text{pr}\{N(I_{ni}) = j\}$$

$$= \sum_i \sum_{k=1}^{\infty} \text{pr}\{N(I_{ni}) \geqslant k\} = \sum_{k=1}^{\infty} \sum_i \text{pr}\{N(I_{ni}) \geqslant k\}.$$

$$\geqslant \sum_{k=1}^{\infty} \liminf_{n\to\infty} \sum_i \text{pr}\{N(I_{ni}) \geqslant k\} = \sum_{k=1}^{\infty} [\lambda(I) - \lambda_{k-1}(I)];$$

while, on the other hand,

$$M(I) = \lim_{r\to\infty} \mathbf{E}(\min(N(I), r)) = \lim_{r\to\infty} \lim_{n\to\infty} \mathbf{E}\left(\min\left(\sum_i N(I_{ni}), r\right)\right)$$

$$\leqslant \lim_{r\to\infty} \lim_{n\to\infty} \sum_i \mathbf{E}(\min(N(I_{ni}), r))$$

$$= \lim_{r\to\infty} \lim_{n\to\infty} \sum_i \sum_{j=1}^{\infty} \min(j, r)\,\text{pr}\{N(I_{ni}) = j\}$$

$$= \lim_{r\to\infty} \lim_{n\to\infty} \sum_i \sum_{k=1}^{r} \text{pr}\{N(I_{ni}) \geqslant k\} = \lim_{r\to\infty} \sum_{k=1}^{r} \lim_{n\to\infty} \sum_i \text{pr}\{N(I_{ni}) \geqslant k\}$$

$$= \lim_{r\to\infty} \sum_{k=1}^{r} [\lambda(I) - \lambda_{k-1}(I)] = \sum_{k=1}^{\infty} [\lambda(I) - \lambda_{k-1}(I)].$$

2.8.4 RELATIONS BETWEEN VARIOUS TYPES OF ORDERLINESS

We can now state some relations that exist between the various types of orderliness defined earlier. These are summarized best by the four diagrams below in which '$\rightarrow$' and '$\nrightarrow$' denote 'implies' and 'does not imply' respectively, these statements being justified by the numbered assertions (for $\rightarrow$) and examples (for $\not\rightarrow$); strictly these latter are all counter-examples to any possible implication ($\rightarrow$).

Note in particular under ($\mathscr{D}$) that a stationary point-process with σ-finite parametric measure—which is then proportional to Lebesgue measure—is orderly in one sense if and only if it is orderly in any other sense, in which case it can be described unambiguously as 'orderly' as has been customary.

In a very real sense the only relevant stationary point-processes are finite, for Slivnyak [10] has proved that a metrically transitive stationary point-process is necessarily finite. Little wonder then that our counter-examples with infinite $\lambda(\cdot)$ rely on forming probabilistic mixtures of simpler point-processes.

(*a*) $\lambda(\cdot)$ *arbitrary*

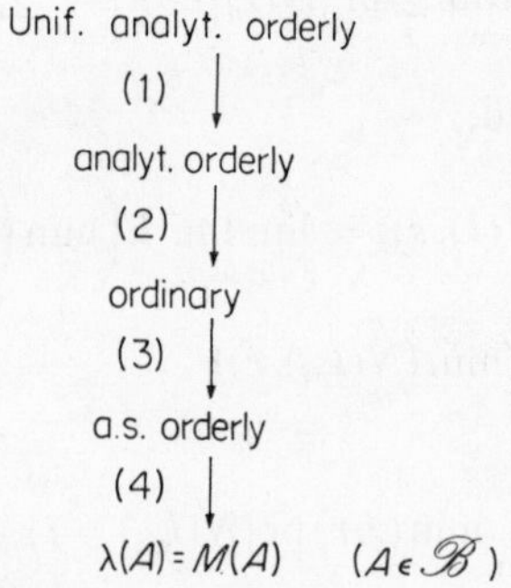

(*b*) *Stationary with infinite* $\lambda(\cdot)$

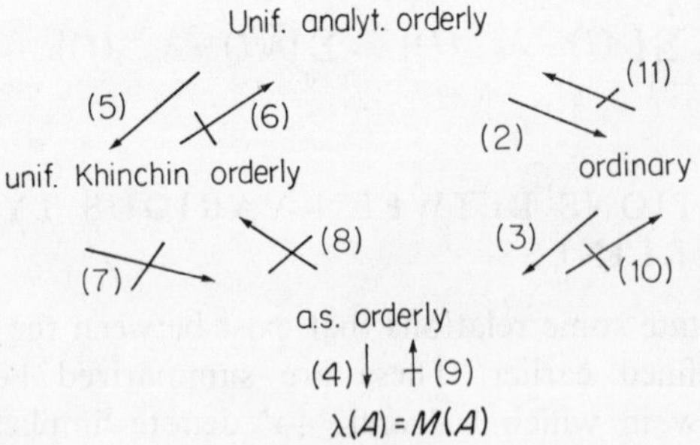

(*c*) $\lambda(\cdot)$ *σ-finite*

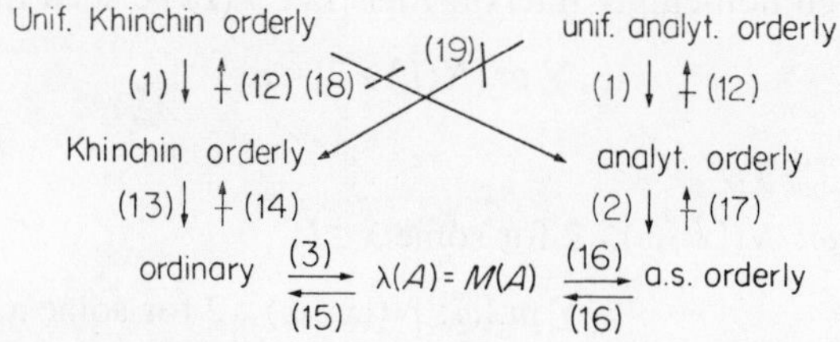

(*d*) *Stationary with σ-finite* $\lambda(\cdot)$

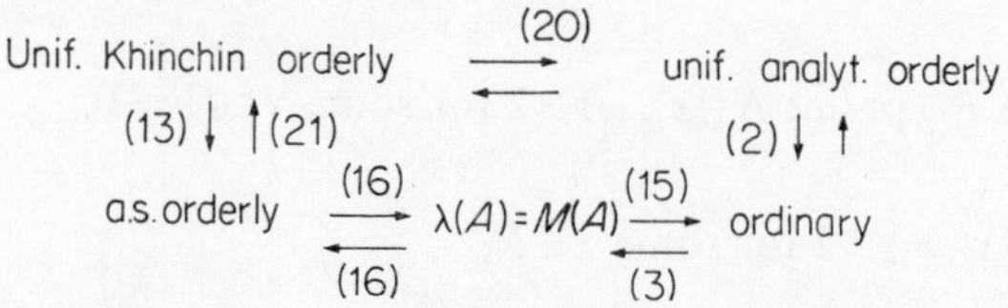

2.8.5 PROOFS AND COUNTER-EXAMPLES

Assertion 1. Trivially, uniformity of a property implies that property.

Assertion 2. Suppose without loss of generality that $I = [0, 1]$, and for $\varepsilon > 0$ let

$$\text{pr}\{N[x-h_1(x), x+h_2(x)] \geqslant 2\} \leqslant \varepsilon(h_1(x)+h_2(x)) \quad \text{for}$$
$$0 < h_1(x) < \delta(x, \varepsilon),\ 0 < h_2(x) < \delta(x, \varepsilon),$$

where for definiteness we may as well take $\delta(x, \varepsilon) \leqslant 1$ and

$$h_1(x) = h_2(x) = h(x) = \tfrac{1}{2}\delta(x, \varepsilon),$$

say. Then the intervals $(x-h(x), x+\frac{1}{2}h(x))$ $(0 \leqslant x \leqslant 1)$ form an open covering of the bounded closed interval $[0, 1]$, so by the Heine–Borel theorem a finite collection of open intervals

$$(x_j - h(x_j),\ x_j + \tfrac{1}{2}h(x_j)) \quad (j = 1, \ldots, n)$$

[covers $[0, 1]$. Without loss of generality we can suppose that $x_1 < \ldots < x_n$, and we may then find $\xi_1, \ldots, \xi_{n+1}$ such that $\xi_1 < x_1 < \xi_2 < x_2 < \ldots < x_n < \xi_{n+1}$ with $-\frac{1}{2} \leqslant \xi_1 < 0$ and $1 \leqslant \xi_{n+1} < 1\frac{1}{2}$ and $(\xi_j, \xi_{j+1}] \subset (x_j - h(x_j), x_j + \frac{1}{2}h(x_j))$. Then

$$\sum_{j=1}^{n} \text{pr}\{N(\xi_j, \xi_{j+1}] \geqslant 2\} \leqslant \varepsilon \sum_{j=1}^{n} (\xi_{j+1} - \xi_j) \leqslant 2\varepsilon,$$

and ordinariness of the process follows.

Assertion 3. (Cf. Leadbetter's [8] proof of his Theorem 2.2.) For arbitrary $\varepsilon > 0$ and given non-empty interval I let $\{I_i\} \in \mathscr{D}(I)$ be such that

$$\sum_i \text{pr}\{N(I_i) \geqslant 2\} < \varepsilon.$$

Then

$$\begin{aligned}\text{pr}\{\omega: N(\{x\}, \omega) \geqslant 2 \text{ for some } x \in I\} \\ &\leqslant \sum_i \text{pr}\{\omega: N(\{x\}, \omega) \geqslant 2 \text{ for some } x \in I_i\} \\ &\leqslant \sum_i \text{pr}\{\omega: N(I_i) \geqslant 2\} < \varepsilon,\end{aligned}$$

so

$$\text{pr}\{\omega: N(\{x\}, \omega) \geqslant 2 \text{ for some } x \in I\} = 0.$$

Hence

$$\begin{aligned}\text{pr}\{\omega: N(\{x\}, \omega) \geqslant 2 \text{ for some } x \in R\} \\ &\leqslant \sum_{j=-\infty}^{\infty} \text{pr}\{\omega: N(\{x\}, \omega) \geqslant 2 \text{ for some } x \in (j, j+1]\} = 0.\end{aligned}$$

Assertion 4 is part of the generalized Korolyuk theorem.

Assertion 5. For any stationary point-process

$$\lim_{h \downarrow 0} \text{pr}\{N(0, h] \geqslant 1\}/h$$

exists, finite or infinite (and here, it is infinite) by Khinchin's existence theorem (e.g. Leadbetter [7, 8]). Then

$$\frac{\text{pr}\{N(0,h] \geqslant 2\}}{\text{pr}\{N(0,h] \geqslant 1\}} = \frac{\text{pr}\{N(0,h] \geqslant 2\}}{h} \Big/ \frac{\text{pr}\{N(0,h] \geqslant 1\}}{h} \to \frac{0}{\infty} = 0 \quad (h \to 0).$$

Example 6. *We shall make repeated use of*

Example A. *Mixtures of stationary deterministic point-processes* (Slivnyak [10]). *A sample realization consists of the points* $\{nX + Y: n = 0, \pm 1, \pm 2, \ldots\}$ *where the random variable* X *is distributed on* $(0, \infty)$ *with distribution function* $F(x), F(0+) = 0$, *and conditional on* X, Y *is distributed uniformly on* $(0, X)$. *Such a process is a.s. orderly by construction; for the rest we shall need*

$$\text{pr}\{N(0,h] = 0\} = \int_h^{\infty} \left(1 - \frac{h}{x}\right) dF(x) = 1 - h \int_h^{\infty} \frac{F(x)\, dx}{x^2}, \tag{16}$$

$$\text{pr}\{N(0,h] \geqslant 2\} = h \int_{\frac{1}{2}h}^{h} \frac{F(x)\, dx}{x^2}. \tag{17}$$

For the present counter-example, let $F(x) = x\ (0<x<1)$, *so that for* $0<h<1$,

$$\frac{\mathrm{pr}\{N(0,h]\geqslant 2\}}{\mathrm{pr}\{N(0,h]\geqslant 1\}} = \frac{h\log 2}{-h\log(h/\mathrm{e})} \to 0 \quad (h\to 0),$$

and the process is thus unif. Khinchin orderly. As Slivnyak asserts however,

$$\mathrm{pr}\{N(0,h]\geqslant 2\}/h = \tfrac{1}{2}[1+\log 2] \not\to 0 \quad (h\to 0).$$

Vasil'ev [11] *also gives an example of a stationary, a.s. orderly, but not unif. analyt. orderly point-process. His example is a mixture of Poisson processes, and it is not difficult to verify that it is unif. Khinchin orderly.*

Example 7. *Form a mixture of the process of Example* 6 *with probability* $\tfrac{1}{2}$, *and with probability* $\tfrac{1}{2}$ *of a stationary deterministic point-process with two points at each of the points* $\{n+Y;\ n=0, \pm 1, \ldots\}$, *where* Y *is uniformly distributed on* $(0,1)$. *Then the resulting point-process is not a.s. orderly, yet since for* $0<h<1$,

$$\mathrm{pr}\{N(0,h]\geqslant 2\} = \tfrac{1}{2}h(1+\log 2),$$

$$\mathrm{pr}\{N(0,h]\geqslant 1\} = \tfrac{1}{2}h - \tfrac{1}{2}h\log(h/\mathrm{e})$$

it is unif. Khinchin orderly.

Example 8. *In Example A let* $F(x) = x^{\frac{1}{2}}\ (0<x<1)$. *Then it follows from equations* (16) *and* (17) *that*

$$\mathrm{pr}\{N(0,h]\geqslant 2\} = \tfrac{1}{2}h^{\frac{1}{2}}(2^{\frac{1}{2}}-1),$$

$$\mathrm{pr}\{N(0,h]\geqslant 1\} = \tfrac{1}{2}h^{\frac{1}{2}}(1-h^{\frac{1}{2}}),$$

so this stationary a.s. orderly process is not unif. Khinchin orderly.

Example 9. (Zitek [12]). *Take a stationary point-process with infinite* $\lambda(\cdot)$ (*e.g. Example* 6) *and replace its sample realizations* $N(\cdot)$ *by* $N'(\cdot) = 2N(\cdot)$, *so that the new process is stationary but no longer a.s. orderly, and has first moment measure* $M'(A) = 0$ *or* ∞ *according as its parametric measure* $\lambda'(A) = \lambda(A) = 0$ or ∞.

Example 10. *A stationary ordinary process has the property that to arbitrary* $\varepsilon>0$ *there exists* $\{u_i\}$ *with* $0<u_i<1$, $\sum u_i = 1$, and

$$\varepsilon > \sum_i \mathrm{pr}\{N(0,u_i]\geqslant 2\} = \sum_i \mathrm{pr}\{N(0,u_i]\geqslant 2\}/\sum_i u_i.$$

Hence,

$$\liminf_{u\to 0} \mathrm{pr}\{N(0,u]\geqslant 2\}/u = 0,$$

which is not the case in Example 8, where

$$\text{pr}\{N(0,h]\geqslant 2\}/h \geqslant h^{\frac{1}{2}}/4h \to \infty \quad (h \downarrow 0).$$

Example 11. *We use Example A again, with a mixing distribution $F(\cdot)$ specified for $0<x<1$ by*

$$F(x) = \xi_r \quad (\xi_r \leqslant x < \xi_{r-1};\ r = 1, 2, \ldots),$$

where

$$\xi_r = t_r\, \xi_{r-1}, \quad \xi_0 = 1, \quad 0 < t_r < 1, \quad t_r \to 0 \quad (r \to \infty).$$

Then for any ε in $(0,1)$, there exist for all sufficiently large r integers n_r for which with $h_r = n_r^{-1}$,

$$\xi_r < (1-\varepsilon)\,\xi_{r-1} < h_r < \xi_{r-1},$$

and

$$\begin{aligned}\inf_{\{I_i\}\in\mathscr{D}((0,1])} \sum_i \text{pr}\{N(I_i)\geqslant 2\} &\leqslant n_r\,\text{pr}\{N(0,h_r]\geqslant 2\}\\ &= n_r h_r \int_{\frac{1}{2}h_r}^{h_r} \frac{F(x)\,dx}{x^2}\\ &\leqslant n_r\,\xi_r < \xi_r/[(1-\varepsilon)\,\xi_{r-1}] = t_r/(1-\varepsilon) \to 0 \quad (r\to\infty),\end{aligned}$$

so the process is ordinary. On the other hand, to any $\varepsilon > 0$ there exist for all sufficiently large r integers $n'_r = 1/h'_r$ such that $\xi_r < \frac{1}{2}h'_r < (1+\varepsilon)\,\xi_r$ and $h'_r < \xi_{r-1}$. Then

$$\begin{aligned}\text{pr}\{N(0,h'_r]\geqslant 2\}/h'_r &= \xi_r/h'_r\\ &> 1/[2(1+\varepsilon)]\end{aligned}$$

so the process is not unif. analyt. orderly.

Example 12. *Let $N(0,x]$ be the number of births during $(0,x]$ in a pure linear birth process $Z(\cdot)$ on $(0,\infty)$ starting from $Z(0)=1$ and with birth rate α. As is easily shown (e.g. Feller [2], Example XVII 3(b)),*

$$\text{pr}\{N(0,x] = n\} = e^{-\alpha x}(1-e^{-\alpha x})^{n-1},$$

$$\text{pr}\{Z(x+h) = n+r \mid Z(x) = n\} = \binom{n-1+r}{r} e^{-n\alpha h}(1-e^{-\alpha h})^r$$

so

$$\begin{aligned}\text{pr}\{N(x,x+h]\geqslant 1\} &= \sum_{n=1}^{\infty} e^{-\alpha x}(1-e^{-\alpha x})^{n-1} \sum_{r=1}^{\infty} \binom{n-1+r}{r} e^{-n\alpha h}(1-e^{-\alpha h})^r\\ &= \sum_{n=1}^{\infty} e^{-\alpha x}(1-e^{-\alpha x})^{n-1}(1-e^{-n\alpha h})\\ &= \frac{1-e^{-\alpha h}}{1-e^{-\alpha h}(1-e^{-\alpha x})},\end{aligned}$$

and similarly

$$\text{pr}\{N(x, x+h] \geqslant 2\} = \left(\frac{1-e^{-\alpha h}}{1-e^{-\alpha h}(1-e^{-\alpha x})}\right)^2,$$

so the process is Khinchin orderly but not uniformly so. It also follows that it is analyt. orderly but not uniformly so.

Assertion 13. Our proof adapts Leadbetter's [8] proof that unif. Khinchin orderliness implies a.s. orderliness. By the generalized Korolyuk theorem and Lemma 1 it is enough to show that $\lambda(I) = \lambda_1(I)$ for every bounded interval I. Suppose then that $\{I_{ni}\} \in \mathscr{D}(I)$ is a nested family of subintervals such that

$$\sum_i \text{pr}\{N(I_i) > 0\} \to \lambda(I) \quad (n \to \infty)$$

and

$$\sum_i \text{pr}\{0 < N(I_i) \leqslant 1\} \to \lambda_1(I) \quad (n \to \infty),$$

assuming without loss of generality that each subinterval is of length $\leqslant 2^{-n}$. Then

$$\lambda(I) - \lambda_1(I) = \lim_{n\to\infty} \sum_i \text{pr}\{N(I_{ni}) \geqslant 2\}$$

$$= \lim_{n\to\infty} \sum_i \frac{\text{pr}\{N(I_{ni}) \geqslant 2\}}{\text{pr}\{N(I_{ni}) \geqslant 1\}} \text{pr}\{N(I_{ni}) \geqslant 1\}$$

$$\leqslant \lim_{n\to\infty} \int_I f_n(x)\,\lambda(dx), \tag{18}$$

where for $x \in I_{ni}$, $f_n(x) = \text{pr}\{N(I_{ni}) \geqslant 2\}/\text{pr}\{N(I_{ni}) \geqslant 1\}$ and this ratio is defined to be zero when the denominator is zero. Now $x \in I_{n+1, j_{n+1}} \subset I_{n, j_n}$ for some j_n and all n, and for arbitrary $h_1, h_2 \geqslant 0$ with $h_1 + h_2 > 0$,

$$I_{nj_n} \subset (x - h_1, x + h_2]$$

for all sufficiently large n, so by the assumed Khinchin orderliness, $f_n(x) \to 0$ $(n \to \infty)$ for each x. Since $\lambda(I) < \infty$ and $f_n(x) \leqslant 1$ we can now apply the dominated convergence theorem to equation (18) to conclude that $\lambda(I) - \lambda_1(I) = 0$.

Example 14. *Consider a point-process whose sample realizations are with probability one sequences of the form $\omega = \{n^2 X: n = 1, 2, \ldots\}$, where the random variable X is uniformly distributed on $(0, 1)$. By construction the process is a.s. orderly, and for $0 < h < 1$*

$$\text{pr}\{N(0, h] \geqslant k\} = h/k^2 = \text{pr}\{N(0, h] \geqslant 1\}/k^2.$$

Thus the process is not Khinchin orderly.

Assertion 15 has been proved in the proof of Assertion 13.

Assertion 16 is the generalized Korolyuk theorem.

Example 17. *Example 14 will do.*

Example 18. *Consider a non-stationary Poisson process on* (0, 1) *with instantaneous rate parameter* $\frac{1}{2}x^{-\frac{1}{2}}$. *Then* $M(0,x] = x^{\frac{1}{2}}$, *and since*

$$\text{pr}\{N(0,h]\geqslant 2\}/h = [1-e^{-h^{\frac{1}{2}}}(1+h^{\frac{1}{2}})]/h \not\to 0 \quad (h\to 0),$$

the process is not analyt. orderly. It is not difficult to check that

$$\frac{\text{pr}\{N(x,x+h]\geqslant 2\}}{\text{pr}\{N(x,x+h]\geqslant 1\}} = \frac{1-\exp[-(x+h)^{\frac{1}{2}}+x^{\frac{1}{2}}][1+(x+h)^{\frac{1}{2}}-x^{\frac{1}{2}}]}{1-\exp[-(x+h)^{\frac{1}{2}}+x^{\frac{1}{2}}]} = O(h^{\frac{1}{2}}),$$

uniformly in x, as $h\to 0$, *so the process is unif. Khinchin orderly.*

Example 19. *Consider again Example* 14 *with the distribution of X replaced by* $F(x)\equiv\text{pr}\{X\leqslant x\} = x^2\ (0<x<1)$ *and all points lying outside* (0, 1) *deleted. Then for* $0<h<1$,

$$\text{pr}\{N(0,h]\geqslant k\} = F(h/k^2) = h^2/k^4 = \text{pr}\{N(0,h]\geqslant 1\}/k^4,$$

so the process is not Khinchin orderly. We now show that it is unif. analyt. orderly noting immediately that $\text{pr}\{N(0,h]\geqslant 2\}/h\to 0$. *For* $0<x<x+h<1$,

$$\text{pr}\{N(x,x+h]\geqslant 2\} = \sum_{n=n_1}^{n_2-1}\left[F\left(\frac{x+h}{(n+1)^2}\right)-F\left(\frac{x}{n^2}\right)\right]+F\left(\frac{x+h}{(n_2+1)^2}\right),$$

where the integers n_1 *and* n_2 *are defined by*

$$n_1 = \inf\{n\geqslant 1: x/n^2<(x+h)/(n+1)^2\},$$
$$n_2 = \inf\{n\geqslant 1: x/n^2<(x+h)/(n+2)^2\}.$$

Then

$$\text{pr}\{N(x,x+h]\geqslant 2\}\leqslant\frac{(x+h)^2}{(n_2+1)^4}+\int_{n_1}^{n_2}\frac{[(x+h)^2-x^2]\,du}{u^4}$$
$$= (x+h)^2/(n_2+1)^4+h(2x+h)(n_1^{-3}-n_2^{-3})/3.$$

For fixed x, h can be taken so small that $n_1\simeq 2x/h$, $n_2\simeq 4x/h$, *and the analyt. orderliness follows. The uniformity can be verified by tedious algebra.*

Assertion 20. Using Khinchin's existence theorem as in Assertion 5 together with the fact that

$$\lim_{h\to 0}\text{pr}\{N(0,h]\geqslant 1\}/h$$

is now finite,

$$\frac{\text{pr}\{N(0,h]\geqslant 2\}}{h}=\frac{\text{pr}\{N(0,h]\geqslant 2\}}{\text{pr}\{N(0,h]\geqslant 1\}}\frac{\text{pr}\{N(0,h]\geqslant 1\}}{h}\to 0 \quad (h\to 0).$$

Assertion 21 is (in view of Assertion 20) Dobrushin's lemma, and is proved for example in Leadbetter [7].

REFERENCES

1. Yu. K. Belyaev, "Elements of the general theory of random streams", Appendix 2 to Russian edition of *Stationary and Related Stochastic Processes* by H. Cramér and M. R. Leadbetter, MIR, Moscow (1969) (English translation: Department of Statistics, University of North Carolina Mimeo Series No. 703.)
2. W. Feller, *An Introduction to Probability Theory and its Applications*, vol. I, Wiley, New York (1957).
3. W. Fieger, "Zwei Verallgemeinerungen der Palmschen Formeln", *Trans. Third Prague Conf. on Information Theory*, pp. 107–122, Czech. Acad. Sci. Publishing House, Prague (1964).
4. W. Fieger, "Eine für beliebige Call-Prozesse geltende Verallgemeinerung der Palmschen Formeln", *Math. Scand.* **16** (1965), 121–147.
5. A. Ya. Khinchin, *Mathematical Methods in the Theory of Queueing*, Trudy Mat. Inst. Steklov. **49**, Moscow (1955). (English translation: ed. M. H. Quenouille; Griffin, London, 1960.)
6. ——, "On Poisson sequences of chance events", *Teor. Veroyatnost. i Primenen* **1** (1956), 320–327. (English translation: *Theory Prob. Applns.* **1**, 291–297.)
7. M. R. Leadbetter, "On three basic results in the theory of stationary point processes", *Proc. Am. math. Soc.* **19** (1968), 115–117.
8. ——, "On basic results of point process theory", *Proc. Sixth Berkeley Symp. Math. Stat. Prob.* (1971), to appear.
9. I. M. Slivnyak, "Stationary streams of homogeneous random events", *Teor. Veroyatnost. i Primenen* **7** (1962), 347–352. (English translation: *Theory Prob. Applns.* **7**, 336–341.)
10. ——, "Stationary flows of homogeneous random events", *Vestnik Harkov. Gros. Univ.* **32** (1966), 73–116. (English translation: Statistics Department. (IAS), Australian National University.)
11. P. I. Vasil'ev, "On the question of ordinariness of a stationary stream" (in Russian). *Kisinev. Gos. Univ. Ucen. Zap.* **82** (1965), 44–48.
12. F. Zitek, "On a theorem of Korolyuk", *Czech. math. J.* **7** (82) (1957), 318–319.

3

SPECIAL PROBLEMS

3.1

Mean Values and Curvatures

L. A. SANTALÓ

[Reprinted from *Izv. Akad. Nauk Armiansk. S.S.S.R.*, **5** (1970), 286–295.]

We divide this exposition into two parts. Section 3.1.1 refers to the mean value of the Euler–Poincaré characteristic of the intersection of two convex hypersurfaces in E_4. Section 3.1.2 deals with the definition of qth total absolute curvatures of a compact n-dimensional variety imbedded in Euclidean space of $n+N$ dimensions, extending some results given in [10].

3.1.1 ON CONVEX BODIES IN E_4

Introduction

Let K be a convex body in 4-dimensional Euclidean space E_4 and let W_i $(i = 0, 1, 2, 3, 4)$ be its Minkowski *Quermass integral* (see for instance Bonnesen–Fenchel [1]). Recall that

$$\left\{\begin{aligned} W_0 &= V = \text{volume of } K,\\ 4W_1 &= F = \text{area of } \partial K,\\ W_4 &= \pi^2/2 \end{aligned}\right. \tag{1}$$

and, if K has sufficiently smooth boundary, we have also

$$\left\{\begin{aligned} 4W_2 &= M_1 = \text{first mean curvature} = \frac{1}{3}\int_{\partial K}\left(\frac{1}{R_1}+\frac{1}{R_2}+\frac{1}{R_3}\right)d\sigma,\\ 4W_3 &= M_2 = \text{second mean curvature} = \frac{1}{3}\int_{\partial K}\left(\frac{1}{R_1R_2}+\frac{1}{R_1R_3}+\frac{1}{R_2R_3}\right)d\sigma, \end{aligned}\right.$$

(2)

where R_i are the principal radii of curvature and $d\sigma$ is the element of area of ∂K.

For instance, if $K =$ sphere of radius r, we have

$$(3) \qquad V = \tfrac{1}{2}\pi^2 r^4, \quad F = 2\pi^2 r^3, \quad M_1 = 2\pi^2 r^2, \quad M_2 = 2\pi^2 r.$$

We will use throughout the invariants V, F, M_1, M_2 because they have a more geometrical meaning; however, we do not assume smoothness of ∂K, so that as definition of M_1, M_2 we take $M_1 = 4W_2$, $M_2 = 4W_3$.

The invariants V, F, M_1, M_2 are not independent. They are related by certain inequalities which may be written in the following symmetrical form (following Hadwiger [6]).

$$(4) \qquad W_\alpha^{\beta-\gamma} W_\beta^{\gamma-\alpha} W_\gamma^{\alpha-\beta} \geqslant 1, \quad 0 \leqslant \alpha \leqslant \beta \leqslant \gamma \leqslant 4.$$

In explicit form and using the invariants V, F, M_1, M_2 the inequalities (4) give the following non-independent inequalities

$$(5) \qquad \begin{cases} F^1 \geqslant 4VM_1, & F^3 \geqslant 16V^2 M_2, \quad F^4 \geqslant 128\pi^2 V^3, \\ M_1^3 \geqslant 4VM_2, & M_1^2 \geqslant 2\pi^2 V, \quad M_2^4 \geqslant 32\pi^6 V, \\ M_1^2 \geqslant FM_2, & M_1^3 \geqslant 2\pi^2 F^2, \quad M_2^3 \geqslant 4\pi^4 F, \\ M_2^2 \geqslant 2\pi^2 M_1. \end{cases}$$

We will represent throughout the paper by O_i the volume of the i-dimensional unit sphere, that is

$$(6) \qquad O_i = \frac{2\pi^{(i+1)/2}}{\Gamma(\frac{1}{2}[i+1])};$$

for instance,

$$(7) \quad O_0 = 2, \quad O_1 = 2\pi, \quad O_2 = 4\pi, \quad O_3 = 2\pi^2, \quad O_4 = \tfrac{8}{3}\pi^2, \quad O_5 = \pi^3.$$

Mean value of $\chi(\partial K \cap g\,\partial K)$

Let G be the group of isometries of E_4. For any $g \in G$ we represent by $g\,\partial K$ the image of ∂K under the isometry g. Let dg denote the invariant volume element of G (= kinematic density for E_4). Assume the convex body K fixed and consider the intersections $\partial K \cap g\,\partial K$, $g \in G$. Then, Federer [5] and Chern [2] have proved the following integral formula

$$(8) \qquad \int_G \chi(\partial K \cap g\,\partial K)\,dg = 64\pi^2 FM_2,$$

where $\chi(\partial K \cap g\,\partial K)$ denotes the Euler–Poincaré characteristic of the surface $\partial K \cap g\,\partial K$.

On the other hand, the so-called fundamental kinematic formula of integral geometry gives

$$(9) \qquad \int_{K \cap gK \neq \emptyset} dg = 8\pi^2(4\pi^2 V + 2FM_2 + \tfrac{3}{2}M_1^2).$$

Therefore the expected value of $\chi(\partial K \cap g\,\partial K)$ is

$$(10) \qquad \mathbf{E}(\chi(\partial K \cap g\,\partial K)) = \frac{8FM_2}{4\pi^2 V + 2FM_2 + \tfrac{3}{2}M_1^2}.$$

Notice that, K being convex, the intersections $\partial K \cap g\,\partial K$ are closed orientable surfaces. Thus the possible values of χ are either $\chi = 2, 4, 6, \ldots$ or $\chi = 0, -2, -4, -5, \ldots$. If K is an Euclidean sphere, obviously we have $\mathbf{E}(\chi) = 2$.

Conjecture. *For all convex sets K of E_4 the inequality*

$$(11) \qquad \mathbf{E}(\chi(\partial K \cap g\,\partial K)) \leqslant 2$$

holds good, with equality for the Euclidean sphere.†

Putting

$$(12) \qquad \Delta = 8\pi^2 V + 3M_1^2 - 4FM_2$$

the conjecture is equivalent to $\Delta \geqslant 0$. For the Euclidean sphere, according to equations (3) we have $\Delta = 0$.

In support of this conjecture we will prove it for rectangular parallelepipeds. Let a, b, c, d be the sides of a rectangular parallelepiped in E_4 and assume

$$(13) \qquad a \leqslant b \leqslant c \leqslant d.$$

It is known that (Hadwiger [6])

$$V = abcd, \quad F = 2(abc + abd + acd + bcd),$$
$$M_1 = \tfrac{2}{3}\pi(ab + ac + ad + bc + bd + cd), \quad M_2 = \tfrac{4}{3}\pi(a + b + c + d).$$

† H. Hadwiger (personal communication to the author) has shown that the conjecture is not true. The counter-example is a 4-dimensional right cylinder with a 3-dimensional solid unit sphere as section and altitude equal to 1

$$(V = 4\pi/3,\ F = 20\pi/3,\ M_1 = (4/3)\pi\,(\pi + 2),\ M_2 = 20\pi/3).$$

Another counter-example is the 3-dimensional solid sphere considered as a flattened convex body of E_4 ($V = 0$, $F = 8\pi/3$, $M_1 = 4\pi^2/3$, $M_2 = 16\pi/3$, assuming the radius $r = 1$).

With these values we verify the identity

$$\frac{3}{4\pi}\Delta = (4-\pi)\,[a^2c^2+a^2(c-b)^2+b^2(c-a)^2$$
$$+a^2(d-b)^2+c^2(d-a)^2+b^2(d-c)^2+c^2(d-b)^2]$$
$$+(18\pi-56)\,abcd+(4\pi-12)\,(a^2b^2+a^2c^2+b^2c^2)$$
$$+(8-2\pi)\,[(b-a)\,acd+(c-b)\,abd+(d-c)\,acb]$$
$$+(4-\pi)\,d^2[(2A^2-B^2)\,(a^2+b^2)+(Ac-Ba)^2+(Ac-Bb)^2],$$

where $A^2=(3\pi-8)/(8-2\pi)$, $B^2=(8-2\pi)/(3\pi-8)$.

Since all terms are positive, we have $\Delta>0$.

For an ellipsoid of revolution whose semiaxes are $a, a, a, \lambda a$ we have (Hadwiger [6])

$$(14)\qquad \begin{cases} V=(\tfrac{1}{2}\pi)\,\lambda a^4, \quad F=2\pi^2\lambda^2a^3F(\tfrac{5}{2},\tfrac{1}{2},2;\,1-\lambda^2), \\ M_1=2\pi^2\lambda^3a^2F(\tfrac{5}{2},1,2;\,1-\lambda^2), \\ M_2=2\pi^2\lambda^4aF(\tfrac{5}{2},\tfrac{3}{2},2;\,1-\lambda^2), \end{cases}$$

where F denotes the hypergeometric function. In this case the conjecture becomes

$$(15)\qquad 1+3\lambda^5F_1^2-4\lambda^5F_{\frac{1}{2}}F_{\frac{3}{2}}\geqslant 0,$$

where

$$F_{\frac{1}{2}}=F(\tfrac{5}{2},\tfrac{1}{2},2;\,1-\lambda^2),$$
$$F_1=F(\tfrac{5}{2},1,2;1-\lambda^2),$$
$$F_{\frac{3}{2}}=F(\tfrac{5}{2},\tfrac{3}{2},2;1-\lambda^2).$$

I do not know if the inequality (15) holds for all values of λ.

3.1.2 TOTAL ABSOLUTE CURVATURES OF COMPACT MANIFOLDS IMMERSED IN EUCLIDEAN SPACE

Introduction

In this section we extend and complete the contents of [10]. We shall first state some known formulae which will be used in the sequel.

Let L_h be an h-dimensional linear subspace in the $(n+N)$-dimensional Euclidean space E_{n+N}. We will call it, simply, an h-space. Let $L_h(O)$ be an h-space in E_{n+N} through a fixed point O. The set of all oriented $L_h(O)$ constitute the Grassman manifold $G_{h,n+N-h}$. We shall represent by $dL_h(O)$ the element of volume of $G_{h,n+N-h}$, which is the same thing as the density for oriented h-spaces through O. The expression of $dL_h(O)$ is well known, but we will recall it briefly for completeness (see [9], [2]).

Let $(O; e_1, e_2, \ldots, e_{n+N})$ be an orthonormal frame in E_{n+N} of origin O. In the space of all orthonormal frames of origin O we define the differential forms

$$\omega_{im} = -\omega_{mi} = e_m \, de_i. \tag{16}$$

Assuming $L_h(O)$ spanned by the unit vectors $e_1, e_2, \ldots, e_h$, then

$$dL_h(O) = \Lambda \omega_{im}, \tag{17}$$

where the right-hand side is the exterior product of the forms ω_{im} over the range of indices

$$i = 1, 2, \ldots, h; \quad m = h+1, h+2, \ldots, n+N.$$

The $(n+N-h)$-space $L_{n+N-h}(O)$ orthogonal to $L_h(O)$ is spanned by $e_{h+1}, \ldots, e_{n+N}$ and equations (2) give the duality

$$dL_h(O) = dL_{n+N-h}(O). \tag{18}$$

The measure of the set of all oriented $L_h(O)$ (= volume of the Grassman manifold $G_{h,n+N-h}$) may be computed directly from equations (2) (see [9]), or applying the result that it is the quotient space

$$SO(n+N)/SO(h) \times SO(n+N-h)$$

(see [2]). The result is

$$\int_{G_{h,n+N-h}} dL_h(O) = \frac{O_{n+N-1}\, O_{n+N-2} \cdots O_{n+N-h}}{O_1\, O_2 \cdots O_{h-1}} \tag{19}$$

$$= \frac{O_h\, O_{h+1} \cdots O_{n+N-1}}{O_1\, O_2 \cdots O_{n+N-h-1}},$$

where O_i is the area of the i-dimensional unit sphere (equation (6)).

Another known integral formula which we will use is the following.

Consider the unit sphere Σ_{n+N-1} of dimension $n+N-1$ of centre O. Let V^s be an s-dimensional variety in Σ_{n+N-1}. Let $\mu_{s+h-n-N}(V^s \cap L_h)$ be the $(s+h-n-N)$-dimensional measure of the variety $V^s \cap L_h(O)$ of dimension $s+h-(n+N)$ and let $\mu_s(V^s)$ be the s-dimensional measure of V^s (all these measures considered as measures of subvarieties of the Euclidean space E_{n+N}). Then

$$\int_{G_{h,n+N-h}} \mu_{s+h-n-N}(V^s \cap L_h(O))\, dL_h(O)$$

$$= \frac{O_{n+N-h}\, O_{n+N-h+1} \cdots O_{n+N-1}\, O_{n+s-n-N}}{O_1\, O_2 \cdots O_{h-1}\, O_s} \mu_s(V^s). \tag{20}$$

Note that this formula assumes the h-spaces L_h oriented (see [8]). In particular, if $s = 1$ and $h = n+N-1$, that is, for a curve V^1 of length U, we have

$$\int_{G_{n+N-1,1}} \nu dL_{n+N-1}(O) = \frac{2O_{n+N-1}}{O_1} U, \tag{21}$$

where ν is the number of points of the intersection $V^1 \cap L_{n+N-1}(O)$.

Definitions

Let X^n be a compact n-dimensional differentiable manifold (without boundary) of class C^∞ in E_{n+N}. To each point $p \in X^n$ we attach the p-space $T^{(q)}(p)$ spanned by the vectors

$$\frac{\partial}{\partial x_1}, \ldots, \frac{\partial}{\partial x_n}; \quad \frac{\partial^2}{\partial x_1^2}, \ldots, \frac{\partial^2}{\partial x_n^2}; \quad \ldots; \quad \frac{\partial^q}{\partial x_1^q}, \ldots, \frac{\partial^q}{\partial x_n^q}, \tag{22}$$

which we will call the qth tangent fibre over p. Its dimension is

$$\rho(n,q) = \sum_{i=1}^{q} \binom{n+i-1}{i}. \tag{23}$$

Assuming

$$1 \leqslant r \leqslant n+N-1, \quad \rho \leqslant n+N-1, \tag{24}$$

we define the rth total absolute curvature of order q of X^n as follows.

(*a*) *Case* $1 \leqslant r \leqslant \rho$. Let O be a fixed point of E_{n+N} and consider an $(n+N-r)$-space $L_{n+N-r}(O)$. Let Γ_r be the set of all r-spaces L_r of E_{n+N} which are contained in some of the fibres $T^{(q)}(p)$, $p \in X^n$, pass through p, and are orthogonal to $L_{n+N-r}(O)$. The intersection $\Gamma_r \cap L_{n+N-r}(O)$ will be a compact variety in $L_{n+N-r}(O)$ whose dimension δ we shall compute in the next section. Let $\mu(\Gamma_r \cap L_{n+N-r}(O))$ be the measure of this variety as subvariety of the Euclidean space $L_{n+N-r}(O)$; if $\delta = 0$, then μ means the number of intersection points of Γ_r and $L_{n+N-r}(O)$.

Then we define the rth total absolute curvature of order q of $X^n \subset E_{n+N}$ as the mean value of the measures μ for all $L_{n+N-r}(O)$, that is, according to equality (19)

$$K_{r,N}^{(q)}(X^n) = \frac{O_1 O_2 \ldots O_{n+N-r-1}}{O_r O_{r+1} \ldots O_{n+N-1}} \int_{G_{n+N-r,r}} \mu(\Gamma_r \cap L_{n+N-r}(O))\, dL_{n+N-r}(O). \tag{25}$$

The coefficient of the right-hand side may be replaced by

$$O_1 O_2 \ldots O_{r-1} / O_{n+N-r} \ldots O_{n+N-1}.$$

(*b*) *Case* $\rho \leqslant r \leqslant n+N-1$. Instead of the set of L_r which *are contained* in some $T^{(q)}(p)$ we consider now the set of L_r which *contain* some $T^{(q)}(p)$, $p \in X^n$, and are orthogonal to $L_{n+N-r}(O)$. As before we represent this set by Γ_r and the rth total absolute curvature of order q of $X^n \subset E_{n+N}$ is defined by the same mean value (25).

Properties

We proceed now to compute the dimension of $\Gamma_r \cap L_{n+N-r}(O)$.

(*a*) *Case* $1 \leqslant r \leqslant \rho$. The set of all $L_r \subset E_{n+N}$ is the Grassman manifold $G_{r+1,n+N-r}$ whose dimension is $(r+1)(n+N-r)$. The set of all L_r which are contained in $T^{(q)}(p)$ and pass through p is the Grassman manifold $G_{r,\rho-r}$ of dimension $r(\rho-r)$; therefore the set of all L_r which are contained in some $T^{(q)}(p)$, $p \in X^n$, has dimension $r(\rho-r)+n$. On the other hand, the set of all $L_r \subset E_{n+N}$ which are orthogonal to $L_{n+N-r}(O)$ has dimension $n+N-r$. Consequently, the intersection of both sets, as sets of points of $G_{r+1,n+N-r}$, has dimension

$$r(\rho-r)+n+n+N-r-(r+1)(n+N-r) = r\rho+n-r(n+N).$$

Since to each L_r orthogonal to $L_{n+N-r}(O)$ corresponds one and only one intersection point with this linear space, the preceding dimension coincides with the dimension δ of $\Gamma_r \cap L_{n+N-r}$, that is,

$$\delta = \dim(\Gamma_r \cap L_{n+N-r}(O)) = r\rho+n-r(n+N).$$

Hence, in order that $K_{r,N}^{(q)}(X^n) \neq 0$, it is necessary and sufficient that

$$r\rho+n \geqslant r(n+N). \tag{26}$$

(*b*) *Case* $\rho \leqslant r \leqslant n+N-1$. The set of all $L_r \subset E_{n+N}$ which contain a fixed L_ρ, constitute the Grassman manifold $G_{r+\rho,n+N-r}$ and therefore the dimension of the set of all L_r which contain some $T^{(q)}(p)$, $p \in X^n$, is $(r-\rho)(n+N-r)+n$. The remaining dimensions are the same as in the case (*a*), so that the dimension of the set of all L_r which contain some $T^{(q)}(p)$, $p \in X^n$, and are orthogonal to $L_{n+N-r}(O)$, is

$$(r-\rho)(n+N-r)+n+n+N-r-(r+1)(n+N-r) = r\rho+n-\rho(n+N),$$

that is,

$$\delta = \dim(\Gamma_r \cap L_{n+N-r}(O)) = \rho r+n-\rho(n+N).$$

In order that $K_{r,N}^{(q)}(X^n) \neq 0$, it is necessary and sufficient that

$$\rho r+n \geqslant \rho(n+N). \tag{27}$$

Of course, to inequalities (26) and (27) we must add the relations (24).

The most interesting cases correspond to $\delta = 0$, for which the measure μ in equation (25) is a positive integer and the total absolute curvature is invariant under similitudes. In this case the set of points $p \in X^n$ for which L_r contains or is contained in $T^{(q)}(p)$ can be divided according to the index of p, and we get different curvatures in the style of those defined by Kuiper for the case $q = 1$, $r = n+N-1$ [7]. We will not go into details here.

Examples

(1) Curves, $n = 1$. For $n = 1$ the condition (26) is

$$1 \geqslant r + r(N-\rho)$$

and since $\rho \leqslant N$ the only possibility is $\rho = N$, $r = 1$, which gives $\delta = 0$. The corresponding curvature $K_{1,N}^{(N)}(X^1)$ is

$$K_{1,N}^{(N)}(X^1) = \frac{1}{O_N} \int_{G_{N,1}} \nu_1 \, dL_N(O), \tag{28}$$

where ν_1 is the number of lines in E_{n+N} orthogonal to $L_N(O)$ which are contained in some Nth tangent fibre of the curve X^1. Notice that $G_{N,1}$ is the unit sphere Σ_N and $dL_N(O)$ is the element of area of this sphere in consequence of the duality (18). If $e_1, e_2, \ldots, e_{N+1}$ are the principal normals of X^1 then the formula (21) says that the right-hand side of equation (28) is equal to the length of the spherical curve $e_{N+1}(s)$ (s = arc length of X^1) up to the factor $1/\pi$. That is, if κ_N is the Nth curvature of X^1 (see, for instance, Eisenhart [4], p. 107) we have

$$K_{1,N}^{(N)}(X^1) = \frac{1}{\pi} \int_{X_1} |\kappa_N| \, ds. \tag{29}$$

For the case of curves in E_3, $N = 2$, κ_N is the torsion of the curve and $K_{1,N}^{(2)}$ is up to the factor π^{-1}, the *absolute total torsion* of X^1.

The condition (27) gives $1 \geqslant \rho + \rho(N-r)$ and since $r \leqslant N$, this condition implies $\rho = 1$, $r = N$. We have the curvature

$$K_{N,N}^{(1)}(X^1) = \frac{1}{O_N} \int_{G_{1,N}} \nu_N \, dL_1(O), \tag{30}$$

where ν_N is the number of hyperplanes L_N of E_{N+1} orthogonal to $L_1(O)$ which contain some tangent line of X^1. The same formula (21) gives now that the right-hand side of equation (30) is equal to the length of the curve $e_1(s)$ (= spherical tangential image of X^1), up to the factor $1/\pi$. Therefore, if κ_1 is the first curvature of X^1, equation (30) becomes

$$K_{N,N}^{(1)}(X^1) = \frac{1}{\pi} \int_{X^1} |\kappa_1| \, ds. \tag{31}$$

Notice that for each direction $L_1(O)$ there are at least two hyperplanes orthogonal to $L_1(O)$ which contain a tangent line of X^1 (the hyperplanes which separate the hyperplanes which have a common point with X^1 from those which do not). Therefore the mean value $K_{N,N}^{(1)}$ is $\geqslant 2$ and equation (31) gives the classical Fenchel inequality

$$\int_{X^1} |\kappa_1|\, ds \geqslant 2\pi. \tag{32}$$

If the curve X^1 has at least four hyperplanes orthogonal to an arbitrary direction $L_1(O)$ which contain a tangent line of X^1 (as it happens for instance for knotted curves in E_3), the mean value $K_{N,N}^{(1)}(X)$ will be $\geqslant 4$, and we have the Fary inequality

$$\int_{X^1} |\kappa_1|\, ds \geqslant 4\pi. \tag{33}$$

(2) Surfaces, $n = 2$.

(*i*) *Total absolute curvatures of order* 1. We have $n = 2$, $\rho = 2$ and condition (26) becomes $2 \geqslant rN$. Therefore the possible cases are $r = 1$, $N = 1$; $r = 2$, $N = 1$ and $r = 1$, $N = 2$. For $2 \leqslant r \leqslant N+1$, condition (27) gives $r \geqslant N+1$ and therefore the only possible case is $r = N+1$.

(*a*) *Case* $r = 1$, $N = 1$. Surfaces in E_3. Taking into account that $G_{2,1}$ is the unit sphere Σ_2, the curvature (25) is

$$K_{1,1}^{(1)}(X^2) = \frac{1}{4\pi} \int_{\Sigma_2} \lambda\, dL_2(O), \tag{34}$$

where λ is the length of the curve in the plane $L_2(O)$ generated by the intersections of $L_2(O)$ with the lines of E_3 which are tangent to X^2 and are orthogonal to $L_2(O)$. If H denotes the mean curvature of X^2 and $d\sigma$ denotes the element of area of X^2, it is known that (34) is equivalent to the *total absolute mean curvature*

$$K_{1,1}^{(1)}(X^2) = \frac{1}{2} \int_{X^2} |H|\, d\sigma. \tag{35}$$

(*b*) *Case* $r = 2$, $N = 1$. Surface $X^2 \subset E_3$. The Grassman manifold $G_{1,2}$ is the unit sphere Σ_2 and equation (25) can be written

$$K_{2,1}^{(1)}(X^2) = \frac{1}{4\pi} \int_{\Sigma_2} \nu_3\, dL_1(O), \tag{36}$$

where ν_3 is the number of planes in E_3 which are tangent to X^2 and are orthogonal to the line $L_1(O)$. If K denotes the Gaussian curvature of X^2, since $dL_1(O)$ is the element of area on Σ_2, it is easy to see that equation (36) is equivalent, up to a constant factor, to the *total absolute Gaussian curvature* of X^2, that is,

$$K_{2,1}^{(1)}(X^2) = \frac{1}{2\pi} \int_{X^2} |K|\, d\sigma. \tag{37}$$

(*c*) *Case* $r = 1$, $N = 2$. Surfaces $X^2 \subset E_4$. In this case, writing Σ_3 = unit 3-dimensional sphere, instead of $G_{3,1}$, we have

$$K_{1,2}^{(1)}(X^2) = \frac{1}{2\pi^2} \int_{\Sigma_3} \nu_1\, dL_3(O), \tag{38}$$

where ν_1 is the number of tangent lines to X^2 which are orthogonal to the hyperplane $L_3(O)$. The properties of this total absolute curvature seem not to be known. A geometrical interpretation was given in [10].

(*d*) *Case* $r = N+1$. Surfaces $X^2 \subset E_{N+2}$. According to (25) we have the following curvature

$$K_{N+1,N}^{(1)}(X^2) = \frac{1}{O_{N+1}} \int_{\Sigma_N} \nu_{N+1}\, dL_1(O), \tag{39}$$

where ν_{N+1} is the number of hyperplanes of E_{N+2} which are tangent to X^2 and are orthogonal to the line $L_1(O)$ and Σ_N denotes the N-dimensional unit sphere. Up to a constant factor this curvature coincides with the *curvature of Chern–Lashof* [3]. Since obviously $\nu_{N+1} \geqslant 2$ we have the inequality $K_{N+1,N}^{(1)} \geqslant 2$, with the equality sign only if X^2 is a convex surface contained in a linear subspace L_3 of E_4.

For $N = 2$, X^2 is a surface imbedded in E_4 and the curvature (39) is a kind of dual of the curvature (38) (see [10]).

(*ii*) *Total absolute curvatures of order* $q = 2$. We have $n = 2$, $\rho = 5$ and the inequalities (26) and (27) say that the only possible cases are: (*a*) $r = 1$, $N = 4$; (*b*) $r = 2$, $N = 4$; (*c*) $r = 1$, $N = 5$.

(*a*) *Case* $r = 1$, $N = 4$. Surface X^2 in E_6. The Grassman manifold $G_{5,1}$ is the unit sphere Σ_5 and equation (25) can be written

$$K_{1,4}^{(2)}(X^2) = \frac{1}{O_5} \int_{\Sigma_5} \lambda\, dL_5(O), \tag{40}$$

where λ is the length of the curve in $L_5(O)$ generated by the intersections of $L_5(O)$ with the lines of E_6 which are orthogonal to $L_5(O)$ and belong to some of the second tangent fibres of X^2.

(*b*) *Case* $r = 2$, $N = 4$. Surface X^2 in E_6. We have

$$K_{2,4}^{(2)}(X^2) = \frac{O_1}{O_4\, O_5} \int_{G_{4,2}} \nu_2\, dL_4(O), \tag{41}$$

where ν_2 is the number of 2-spaces of E_6 which are orthogonal to $L_4(O)$ and are contained in some second tangent fibre of X^2.

(*c*) *Case* $r = 1$, $N = 5$. Surfaces X^2 in E_7.

We have

$$K_{1,5}^{(2)}(X^2) = \frac{1}{O_6} \int_{\Sigma_6} \nu_1 \, dL_6(O), \tag{42}$$

where ν_1 is the number of lines of E_7 which are contained in some second tangent fibre of X^2 and are orthogonal to $L_6(O)$.

The expression of these absolute total curvatures of order 2 by means of differential invariants of X^2 is not known.

REFERENCES

1. T. Bonnesen and W. Fenchel, *Theorie der konvexen Körper*, *Erg. der Mathematik*, Berlin (1934).
2. S. S. Chern, "On the kinematic formula in integral geometry, *J. Math. Mechanics*, **16** (1966), 101–118.
3. S. S. Chern and R. K. Lashof, "On the total curvature of immersed manifolds", *Am. J. Math.* **79** (1957), 306–318.
4. L. P. Eisenhart, *Riemannian Geometry*, Princeton University Press, Princeton (1949).
5. H. Federer, "Curvature measures", *Trans. Am. Math. Soc.* **93** (1959), 418–491.
6. H. Hadwiger, *Vorlesungen über Inhalt, Oberfläche und Isoperimetrie*, Springer, Berlin (1957).
7. N. Kuiper, "Der Satz von Gauss–Bonnet für Abbildungen in E^N und damit verwandte Probleme", *Jahr. Deutsch. Math. Ver.* **69** (1967), 77–88.
8. L. A. Santaló, "Geometria integral en espacios de curvatura constante", *Publicaciones Com. Energia Atómica, Serie Matem. Buenos Aires* (1952).
9. ——, "Sur la mesure des espaces linéaires qui coupent un corps convexe et problèmes que s'y rattachent", *Colloque sur les Questions de Réalité en Géométrie, Liege*, (1955), 177–190.
10. ——, "Curvaturas absolutas totales de variedades contenidas en un espacio euclidiano", *Coloquio de Geometria Diferencial, Santiago de Compostela (España)* (1967), pp. 29–38.

3.2

Convex Polygons and Random Tessellations

R. V. AMBARTZUMIAN

3.2.1 INTRODUCTION

Random tessellation is defined to be random subdivision of the plane into non-overlapping convex polygons. We are concerned with the *marked point-process* $\{P_i, \psi_i\}$ which arises on the (arbitrary) fixed line L intersecting the homogeneous and isotropic (h.i.) random tessellation $\mathscr{T}$, $\{P_i\}$ being a sequence of points (random point-process) of intersections of L with the edges of $\mathscr{T}$, ψ_i being the angle of intersection at P_i.

The $\{P_i, \psi_i\}$ process was first introduced by Rollo Davidson, for the case of the random tessellations generated by line-processes on the plane, in [1]. It was there proved under quite general conditions that any homogeneous and isotropic line-process on the plane, for which $\{P_i\}$ and $\{\psi_i\}$ are independent and $\{\psi_i\}$ is a sequence of independent angles, is a mixture of Poisson line-processes.

It has been shown in [2] that this statement remains true in the much more general framework of *regular* h.i. random tessellations (not necessarily generated by infinite straight lines) that possess no T-junctions (Figure 1).

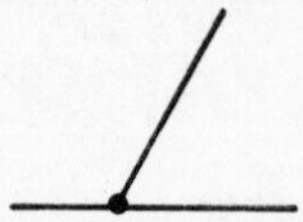

Figure 1

This suggests that $\{P_i, \psi_i\}$ carries rather rich information about h.i. random tessellations and the present article contains further investigations on the subject. Attention is confined to a very particular feature of the

$\{P_i, \psi_i\}$-process: namely, to averages with respect to the joint distribution of I_i, ψ_i and ψ_{i+1} (where I_i is the distance between an arbitrary pair of consecutive points in $\{P_i\}$, and ψ_i, ψ_{i+1} are the corresponding angles). This distribution, of course, does not depend on i.

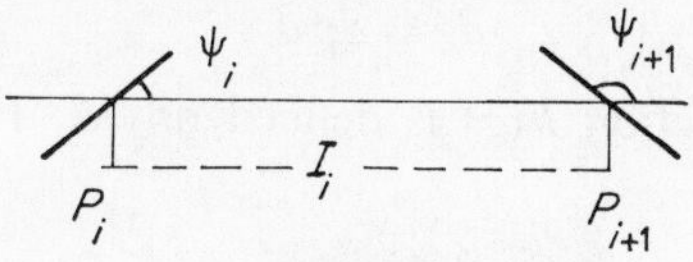

Figure 2

In dealing with $\mathscr{T}$ we pursue that general scheme of reasoning, under which facts about random objects stochastically invariant with respect to a group of transformations are deduced from corresponding facts of integral geometry (see also [2], [3]). Thus we begin with the derivation of an inregral geometric identity for polygons (Section 3.2.2), and then apply it in the theory of h.i. random tessellations.† As a result, the Laplace transform of the distribution of the length of an 'arbitrary edge' of a h.i. random tessellation is obtained.

The results of the present paper were to be discussed with Rollo Davidson on his expected arrival in Armenia in September 1970 on a Royal Society–Soviet Academy Exchange Programme, an event which, so tragically, never happened. This paper is therefore dedicated to his memory.

3.2.2 CONVEX POLYGONS

Denote by

0—the origin on the plane,
$C(r, Q)$—the circle of radius r centred at Q,
g—a directed line on the plane,
$\mathscr{M}$—the set of all directed lines intersecting $C(1, 0)$,
K_r—the planar domain between $C(r+1, 0)$ and $C(r-1, 0)$.

Define the one-to-one mapping $g(\cdot)$: $K_r \to \mathscr{M}$ in the following way; $g(Q)$ is the directed line tangent to $C(r, Q)$ at its point of intersection with the line segment $0Q$, $g(Q)$ being directed so that $C(r, Q)$ lies in the right-hand half-plane with respect to $g(Q)$.

The mapping $g(\cdot)$ induces on $\mathscr{M}$ a measure μ_r which corresponds to Lebesgue measure on K_r. Obviously

(1) $$d\mu_r = (p+r)\, d\varphi\, dp,$$

† Some other corollaries of the identity derived in Section 3.2.2 are presented in [4].

where (φ, p) are polar coordinates of the base of the perpendicular from 0 to g ($p<0$ if 0 is in the left-hand half-plane with respect to g).

Recall that the element dg of the measure on $\mathscr{M}$ invariant with respect to all Euclidean motions of the plane may be written in the form

$$dg = d\varphi\, dp.$$

Assume that a function $F(g)$ is defined on $\mathscr{M}$. From equation (1) it follows that

$$\text{(2)} \qquad \iint_{\mathscr{M}} F(g)\, d\mu_r = r \iint_{\mathscr{M}} F(g)\, dg + \iint_{\mathscr{M}} pF(g)\, dg.$$

For any function $F(r, Q)$, $Q \in K_r$, satisfying the condition

$$\text{(3)} \qquad \iint_{\mathscr{M}} F(g)\, d\mu_r - \iint_{K_r} F(r, Q)\, dS = o(r) \quad (r \to \infty),$$

where dS is an element of plane Lebesgue measure on K_r, we find from equation (2) that

$$\text{(4)} \qquad \iint_{K_r} F(r, Q)\, dS = r \iint_{\mathscr{M}} F(g)\, dg + o(r).$$

Thus it follows, if the limit exists, that

$$\text{(5)} \qquad \lim_{r\to\infty} \frac{d}{dr} \iint_{K_r} F(r, Q)\, dS = \iint_{\mathscr{M}} F(g)\, dg.$$

If $F(g)$ does not depend on the direction of g, then equation (5) may be rewritten in the form

$$\text{(6)} \qquad \lim_{r\to\infty} \frac{d}{dr} \iint_{K_r} F(r, Q)\, dS = 2 \iint_{\mathscr{M}'} F(g)\, dg,$$

where $\mathscr{M}'$ is the set of non-directed lines intersecting $C(1, 0)$.

We apply equation (5) to the functions $F(g)$ and $F(r, Q)$ now to be defined.

Let a_1 and a_2 be two closed line segments inside $C(1, 0)$.

Assumption $\mathscr{A}$. The line containing the segment a_1 does not intersect a_2, and that containing a_2 does not intersect a_1.

Introduce the sets

$$\mathscr{M}^* = \{g\colon g \cap a_1 \neq \varnothing,\ g \cap a_2 \neq \varnothing\},$$
$$K_r^* = \{Q\colon C(r, Q) \cap a_1 \neq \varnothing,\ C(r, Q) \cap a_2 \neq \varnothing\}$$

($\varnothing$ denotes the empty set).

Denote by $\chi(g)$ the length of the interval on g with $a_1 \cap g$ and $a_2 \cap g$ as end-points.

Similarly, denote by $\chi(r, Q)$ the length of the (shorter, if r is large enough) interval on $C(r, Q)$ with $C(r, Q) \cap a_1$ and $C(r, Q) \cap a_2$ as end-points.

Further, introduce the pairs of angles ψ', ψ'' and φ', φ'' associated with each $g \in \mathscr{M}^*$ and with each $Q \in K_r^*$ respectively, as shown in Figure 3.

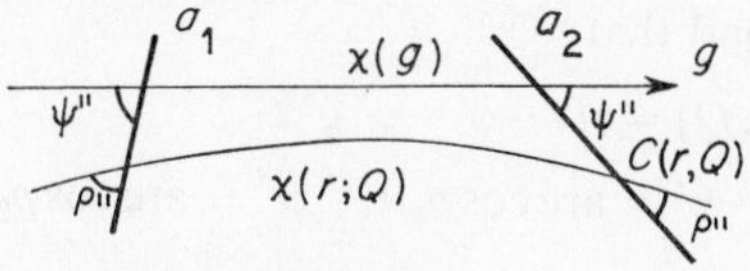

Figure 3

Note that $\mathscr{A}$ assures the correctness of these definitions (at least for the larger values of r that refer to $C(r, Q)$). Put

$$F(g) = \begin{cases} f(\chi(g)) & (g \in \mathscr{M}^*), \\ 0 & (g \notin \mathscr{M}^*), \end{cases}$$

$$F(r, Q) = \begin{cases} f(\chi(r, Q)) & (Q \in K_r^*), \\ 0 & (Q \notin K_r^*), \end{cases}$$

with $f(x) \in C_1(0, \infty)$. By virtue of $\mathscr{A}$, for every such f, equation (3) is fulfilled. Another benefit of $\mathscr{A}$ is that when it holds the angles ψ', ψ'' and (for the larger values of r) the angles φ', φ'' do not approach 0 or π, so that the operations of passing to the limit under integral signs, which are encountered below, may be justified.

Take

$$I(r) = \iint_{Kr} f(\chi(r, Q))\, dS;$$

to calculate $dI(r)/dr$, choose $h > 0$ and put

$$G_1 = K_r^* - G_3,$$

$$G_2 = K_{r+h}^* - G_3,$$

$$G_3 = K_r^* \cap K_{r+h}^*,$$

so that

$$I(r) = \left(\iint_{G_3} + \iint_{G_1}\right) f(\chi(r, Q))\, dS,$$

and

$$(7) \qquad I(r+h) = \left(\iint_{G_3} + \iint_{G_2}\right) f(\chi(r+h, Q))\, dS.$$

First consider

$$\Delta_1(r, h) = \iint_{G_3} \{f(\chi(r+h, Q)) - f(\chi(r, Q))\}\, dS.$$

From Figure 4 we find that

$$\chi(r, Q) = r(\varepsilon - \varphi' - \varphi''),$$

$$\varphi' = \operatorname{arc\,cos} p_1/r, \quad \varphi'' = \operatorname{arc\,cos} p_2/r.$$

Accordingly

$$\frac{d}{dr}\chi(r, Q) = \frac{\chi(r, Q)}{r} - \cot\varphi' - \cot\varphi''.$$

Since the area of G_2 is $O(h)$, it follows that

$$(8) \qquad \lim_{h\to 0} \frac{\Delta_1(r, h)}{h} = \iint_{K_r^*} f'(\chi) \left\{\frac{\chi(r, Q)}{r} - \cot\varphi' - \cot\varphi''\right\} dS.$$

To find the limit as $r \to \infty$ of this expression, note that (Figure 4)

$$\varphi' = \psi_1' - \frac{\chi(r, Q)}{2r}, \quad \varphi'' = \psi_1'' - \frac{\chi(r, Q)}{2r}$$

(unlike ψ' and ψ'', ψ_1' and ψ_1'' are functions defined on K_r^*).

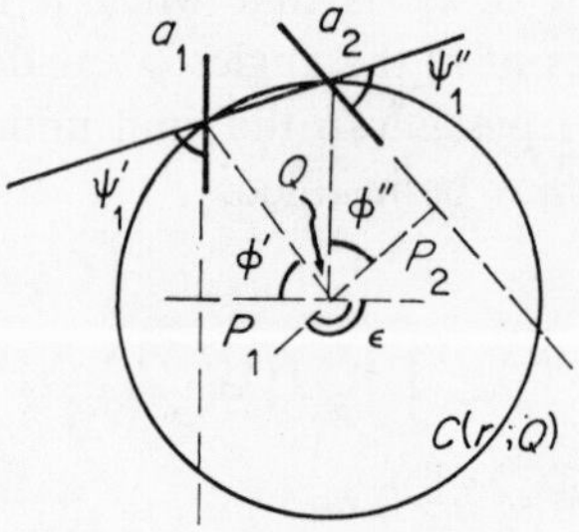

Figure 4

This leads to the expansion

$$(9) \quad \lim_{h\to 0} \frac{\Delta_1(r, h)}{h} = \iint_{K_r^*} f'(\chi(r, Q))$$

$$\times \left\{\frac{\chi(r, Q)}{r} - \cot\psi_1' - \frac{\chi(r, Q)}{2r\sin^2\psi_1'} - \cot\psi_1'' - \frac{\chi(r, Q)}{2r\sin^2\psi_1''}\right\} dS + o\left(\frac{1}{r}\right).$$

By applying equation (4) to the functions,

(10)
$$\begin{cases} F(g) = \begin{cases} f'(\chi(g))\,\chi(g)\left\{1-\dfrac{1}{2\sin^2\psi'}-\dfrac{1}{2\sin^2\psi''}\right\} & (g\in\mathscr{M}^*), \\ 0 \quad (g\notin\mathscr{M}^*), \end{cases} \\ F(r,Q) = \begin{cases} f'(\chi(r,Q))\,\chi(r,Q)\left\{1-\dfrac{1}{2\sin^2\psi_1'}-\dfrac{1}{2\sin^2\psi_1''}\right\} & (Q\in K_r^*), \\ 0 \quad (Q\notin K_r^*), \end{cases} \end{cases}$$

we obtain

(11)
$$\lim_{r\to\infty}\frac{1}{r}\iint_{K_r*} F(Q)\,dS = \iint_{\mathscr{M}*} F(g)\,dg.$$

To find

(12)
$$-\lim_{r\to\infty}\iint_{K_r*} f'(\chi(r,Q))\{\cot\psi_1'+\cot\psi_1''\}\,dS$$

we have to apply a special lemma, whose complete proof may be found in [4].

Lemma. *Let P_1Q_1 and P_2Q_2 be two line segments on the plane satisfying* $\mathscr{A}$ (Figure 5), *and let δ_i be the length of P_iQ_i. Denote by $|S_{P_1P_2}(r)|$ and $|S_{Q_1Q_2}(r)|$ the areas, respectively, of the two sets*

$$S_{P_1P_2}(r)\equiv\{Q\colon P_1P_2\in\operatorname{int}C(r,Q),\quad Q_1Q_2\notin\operatorname{int}C(r,Q)\},$$

$$S_{Q_1Q_2}(r)\equiv\{Q\colon Q_1Q_2\in\operatorname{int}C(r,Q),\quad P_1P_2\notin\operatorname{int}C(r,Q)\}.$$

The difference $|S_{P_1P_2}|-|S_{Q_1Q_2}|$ remains constant for all sufficiently large values of r. Asymptotically, as $\delta_i\to 0$ for $i=1,2$,

$$|S_{P_1P_2}(r)|-|S_{Q_1Q_2}(r)| = -\sin(P_1+P_2)\,\delta_1\,\delta_2+o(\delta_1\,\delta_2).$$

(The angles P_1 and P_2 are shown on Figure 5.)

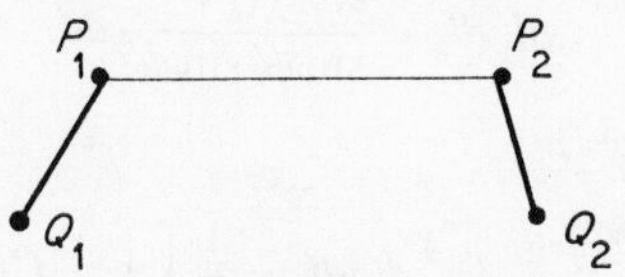

Figure 5

Choose pairs of points $P_1, Q_1 \in a_1$, $P_2, Q_2 \in a_2$ (Figure 6), so that

$$\delta_1 = dl_1, \quad \delta_2 = dl_2.$$

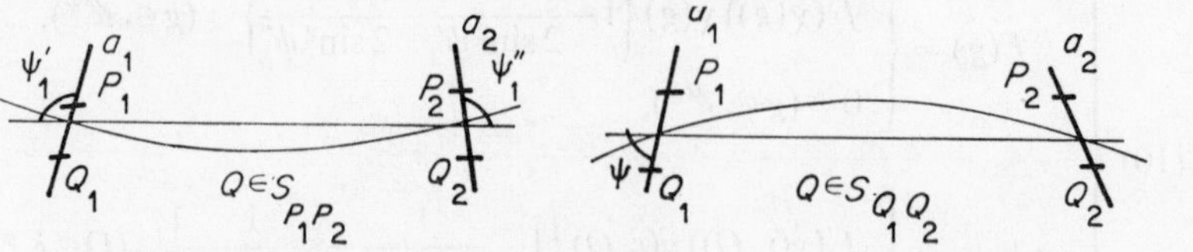

Figure 6

In the notation of the lemma, we may write

$$\psi_1' = P_1, \quad \psi_1'' = P_2 \quad \text{if } Q \in S_{P_1P_2},$$
$$\psi_1' = \pi - P_1, \quad \psi_1'' = \pi - P_2 \quad \text{if } Q \in S_{Q_1Q_2}.$$

Hence the contribution to the integral in (12) from the elements $S_{P_1P_2}$ and $S_{Q_1Q_2}$ is

$$\begin{aligned} f'(\chi(r,Q))(\cot P_1 + \cot P_2)(|S_{Q_1Q_2}| - |S_{P_1P_2}|) \\ = f'(\chi(r,Q))(\cot P_1 + \cot P_2)\sin(P_1+P_2)\,dl_1\,dl_2 \\ = f'(\chi(r,Q))\frac{\sin^2(P_1+P_2)}{\sin P_1 \sin P_2}\,dl_1\,dl_2. \end{aligned}$$

In other words,

$$(13)\qquad -\iint_{K_r^*} f'(\cot\psi_1' + \cot\psi_1'')\,dS = \iint_{a_1\times a_2} f'\frac{\sin^2(\psi'+\psi'')}{\sin\psi'\sin\psi''}\,dl_1\,dl_2.$$

Note that in equation (13), $a_1 \times a_2$ corresponds to the set of all non-directed lines intersecting both a_1 and a_2, while the integrand depends on r only through χ. Obviously

$$-\lim_{r\to\infty}\iint_{K_r^*} f'(\cot\psi_1' + \cot\psi_1'')\,dS$$

$$(14)\qquad = \iint_{a_1\times a_2} f'(\chi(g))\frac{\sin^2(\psi'+\psi'')}{\sin\psi'\sin\psi''}\,dl_1\,dl_2.$$

Since

$$dl_1\,dl_2 = \frac{\chi(g)}{\sin\psi'\sin\psi''}\,dg,$$

we have

$$\iint_{a_1\times a_2} f'(\chi(g))\frac{\sin^2(\psi'+\psi'')}{\sin\psi'\sin\psi''}\,dl_1\,dl_2 = \frac{1}{2}\iint_{\mathscr{M}^*} f'(\chi)\,\chi\frac{\sin^2(\psi'+\psi'')}{\sin^2\psi'\sin^2\psi''}\,dg.$$

Together with equations (14), (11), (10) and (9) this leads to the result

$$\text{(15)}\qquad \lim_{r\to\infty}\lim_{h\to 0}\frac{\Delta_1(r,h)}{h}=\iint_{\mathscr{M}^*} f'(\chi(g))\,\chi(g)\cot\psi'\cot\psi''\,dg.$$

Now consider the difference,

$$\Delta_2(r,h)=\int_{G_2} f(\chi(r+h,Q))\,dS-\int_{G_1} f(\chi(r,Q))\,dS.$$

Introduce the sets,

$\gamma_1(r,A_{ij})=C(r,A_{ij})\cap\{Q\colon C(r,Q)$ encloses a_i and meets $a_s, s\neq i\}$,

$\gamma_2(r,A_{ij})=C(r,A_{ij})\cap\{Q\colon C(r,Q)$ meets $a_s, s\neq i$, and a_i falls outside $C(r,Q)\}$,

where A_{ij} $(j=1,2)$ are the two end-points of a_i $(i=1,2)$.
Obviously

$$\text{(16)}\qquad \lim_{h\to 0}\frac{\Delta_2(r,h)}{h}=\sum_{i,j}\left(\int_{\gamma_2(r,A_{ij})}-\int_{\gamma_1(r,A_{ij})}\right)f(\chi(r,Q))\,dl,$$

where dl is an element of arc-length. To find the limit of this expression as $r\to\infty$, change the variable of integration to φ, defined as the direction of the oriented tangent to

$$C(r,Q)\equiv C(r,\varphi)\quad (Q\in C(r,A_{ij}))$$

at A_{ij} which leaves $C(r,Q)$ in the right half-plane.

Let $\Phi_k(r,A_{ij})$ be the sets of directions corresponding to $\gamma_k(r,A_{ij})$, so that

$$\int_{\gamma_k(r,A_{ij})} f(\chi(r,Q))\,dl=r\int_{\Phi_k(r,A_{ij})} f(\chi(r,\varphi))\,d\varphi.$$

Introduce the sets

$\Lambda_k(r,A_{ij})=\Phi_k(r,A_{ij})\cap\{\varphi\colon$ both $C(r,\varphi)$ and $C(r,\varphi+\pi)$ intersect $a_s, s\neq i\}$

(addition of angles is modulo 2π). Define $\varepsilon_k(r,A_{ij})$ by the 'set-equations'

$$\Phi_k(r,A_{ij})=\Lambda_k(r,A_{ij})+\varepsilon_k(r,A_{ij}).$$

(+ denotes disjoint union). The measure of each $\varepsilon_k(r,A_{ij})$ is equal to $Z_k(A_{ij})/r+o(r)$ (see Figure 7), so that

$$\text{(17)}\qquad \lim_{r\to\infty}\int_{\varepsilon_k(r,A_{ij})} f(\chi(r,\varphi))\,r\,d\varphi=Z_k(A_{ij})f(Z_k(A_{ij})).$$

At the same time

$$\left(\int_{\Lambda_2(r,A_{ij})} - \int_{\Lambda_1(r,A_{ij})}\right) f(\chi(r,\varphi))\, r\, d\varphi = \int_{\Lambda_2(r,A_{ij})} \{f(\chi(r,\varphi)) - f(\chi(r,\varphi+\pi))\}\, r\, d\varphi.$$

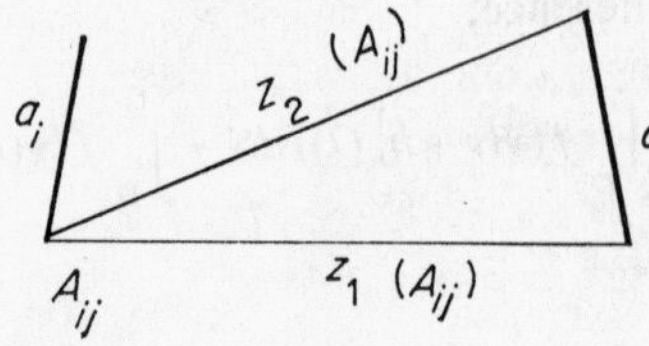

Figure 7

Since

$$f(\chi(r,\varphi)) - f(\chi(r,\varphi+\pi)) = -f'(\rho)\frac{\rho^2 \cot \nu}{r} + o\left(\frac{1}{r}\right)$$

($\rho = \rho(\varphi)$ and $\nu = \nu(\varphi)$ are shown on Figure 8),

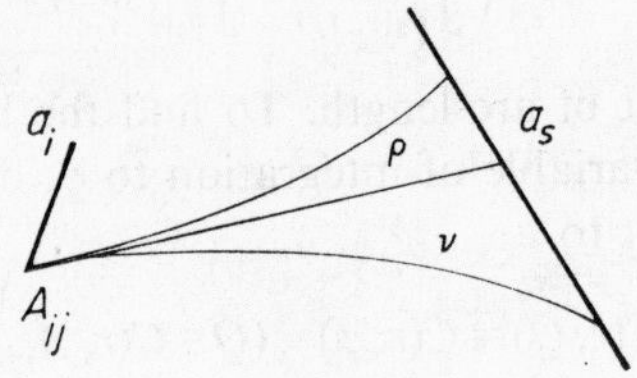

Figure 8

we find that

$$\text{(18)} \qquad \lim_{r\to\infty}\left(\int_{\Lambda_2} - \int_{\Lambda_1}\right) f(\chi(r,\varphi))\, r\, d\varphi = -\int_{\Phi_2(A_{ij})} f'(\rho)\,\rho^2 \cot \nu\, d\varphi,$$

where $\Phi_2(A_{ij}) = \Phi_2(\infty, A_{ij})$. Combining equations (16)–(18), we arrive at the result

$$\text{(19)} \qquad \lim_{r\to\infty}\lim_{h\to 0} \frac{\Delta_2(r,h)}{h} = \sum_{i,j} R_{ij},$$

where

$$R_{ij} = Z_2(A_{ij}) f(Z_2(A_{ij})) - Z_1(A_{ij}) f(Z_1(A_{ij})) - \int_{\Phi_2(A_{ij})} f'(\rho)\,\rho^2 \cot \nu\, d\varphi$$

$$\text{(20)} \qquad = \int_{\Phi_2(A_{ij})} f(\rho)\,\rho \cot \nu\, d\varphi.$$

In obtaining the last equality integration by parts is used and $d\rho = \rho \cot \nu \, d\varphi$. Since, by equation (7),

$$\lim_{r\to\infty} \frac{dI(r)}{dr} = \lim_{r\to\infty} \lim_{h\to 0} \left\{ \frac{\Delta_1(r,h)}{h} + \frac{\Delta_2(r,h)}{h} \right\},$$

by equations (15) and (20) we conclude that, for a pair of segments a_1, a_2 satisfying $\mathscr{A}$,

$$\text{(21)} \qquad \iint_{\mathscr{M}^*} f(\chi)\, dg = \iint_{\mathscr{M}^*} f'(\chi)\, \chi \cot \psi' \cot \psi''\, dg + \sum_{i,j} R_{ij}.$$

To dispense with the condition $\mathscr{A}$, rewrite R_{ij} in the form

$$R_{ij} = \int_{a_i} f(\rho) \cos \nu \, dl,$$

where dl is an element of length given by

$$dl = \rho\, d\varphi / \sin \nu.$$

Fix the position of a_1, and alter that of a_2 so that it tends to a limiting position a_2^* (Figure 9). Assume that a_1, and a_2 in all its positions, satisfy

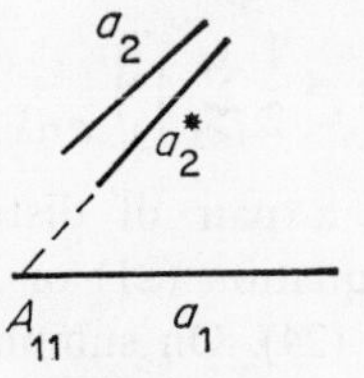

Figure 9

$\mathscr{A}$, but that the end-point A_{11} of a_1 lies on the continuation of a_2^*. Then the limit of R_{11} is obviously

$$\int_{x_1}^{x_2} f(u)\, du.$$

Hence for the case of two segments a_1, a_2 'making an angle' (Figure 10), the result (21) becomes

$$\text{(22)} \qquad \iint_{\mathscr{M}^*} f(\chi)\, dg = \iint_{\mathscr{M}^*} f'(\chi)\, \chi \cot \psi' \cot \psi''\, dg + R_1 + R_2$$

$$+ \int_0^{s_1} f(u)\, du + \int_0^{s_2} f(u)\, du,$$

where R_1 and R_2 are obtained from equation (20) by replacing A_{ij} by A_1 and A_2 respectively, and s_i is the length of a_i.

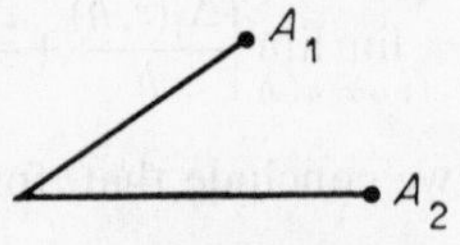

Figure 10

We are now in a position to obtain the main result of this paragraph.
Let D be a convex polygon, $\chi(g)$ the length of $g \cap D$ for $g \in \mathscr{M}'$, ψ' and ψ'' the angles between g and ∂D where they meet (lying in the same half-plane, inside D), and s_i $(i = 1, \ldots, n)$ the lengths of the sides of D.

Theorem 1. *For every* $f \in C_1(0, \infty)$ *and any convex polygon* $D(s_1, s_2, \ldots, s_n)$,

$$\iint f(\chi)\, dg = \iint f'(\chi)\, \chi \cot \psi' \cot \psi''\, dg + \sum_{i=1}^{n} \int_0^{s_i} f(u)\, du, \tag{23}$$

where the double integrals are taken over the set of those non-directed lines that intersect D.

The proof follows from equations (21) and (22). Observe that

$$\iint F(g)\, dg = \frac{1}{2} \sum_{i<j} \iint_{\mathscr{M}^*(a_i, a_j)} F(g)\, dg, \tag{24}$$

where $\mathscr{M}^*(a_i, a_j)$ refers to a pair of distinct sides of D. Now take $F(g) = f(\chi(g))$ and apply equation (21) or (22) to each integral on the right-hand side of equation (24). On summation the R-terms disappear, and the theorem follows finally from equation (24) with, now,

$$F(g) = f'(\chi)\, \chi \cot \psi' \cot \psi''.$$

3.2.3 RANDOM TESSELLATIONS

It is easily verified that the general form of equation (23) for an arbitrary convex domain D is

$$\iint f(\chi)\, dg = \iint f'(\chi)\, \chi \cot \psi' \cot \psi''\, dg + f(0)\left(H - \sum s_k\right) + \sum \int_0^{s_k} f(u)\, du, \tag{25}$$

where H is the length of the (perhaps not polygonal) perimeter of D and the s_k are the lengths of the linear segments thereon. Let the *fixed* tessellation $\mathscr{T}_0$ of the plane into convex polygons be given. Denote by

D_i the convex domains into which $\mathscr{T}_0$ divides the interior K_r of $C(r, 0)$ (see Figure 11). Give suffix i to the variables in equation (25) when D is D_i and sum over all $D_i \subset K_r$; we thus obtain

$$\text{(26)} \qquad \iint \sum_{D_i \cap g \neq \emptyset} f(\chi_i)\, dg = \iint \sum_{D_i \cap g \neq \emptyset} f'(\chi_i)\, \chi_i \cot \psi_i' \cot \psi_i''\, dg + \sum_{D_i \subset K_r} \sum_k F(s_{ki}) + 2\pi r f(0),$$

where

$$F(x) = \int_0^x f(u)\, du$$

and the double integrals are taken over the set of all lines intersecting K_r.

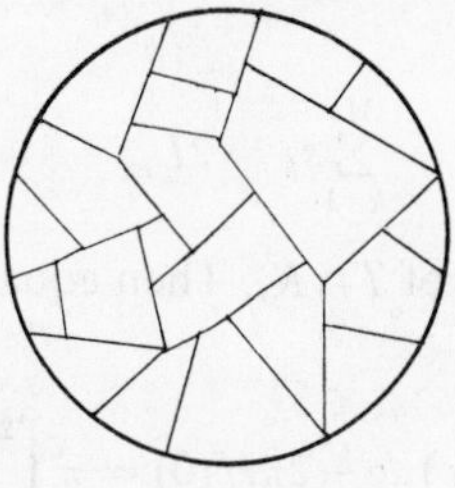

Figure 11

Now suppose equation (26) averaged over the ensemble of realizations of a *random* tessellation $\mathscr{T}$. Writing the result with integration and averaging (shown by a bar) reversed in order, we obtain

$$\text{(27)} \qquad \iint \overline{\sum_{D_i \cap g \neq \emptyset} \{f(\chi_i) - f'(\chi_i)\, \chi_i \cot \psi_i' \cot \psi_i''\}}\, dg = \overline{\sum_{D_i \subset K_r} \sum_k F(s_{ki})} + 2\pi r f(0).$$

If we assume that $\mathscr{T}$ is homogeneous and isotropic, then the integrands in equation (27) depend on g only through Z (the length of $K_r \cap g$) and r, and can obviously be expressed in terms of the distribution of $\{P_i, \psi_i\}$ (see Section 3.2.1). Indeed, for any fixed line g,

$$\text{(28)} \qquad \alpha(Z, r) = \overline{\sum_{D_i \cap g \neq \emptyset} \{f(\chi_i) - f'(\chi_i)\, \chi_i \cot \psi_i' \cot \psi_i''\}}$$
$$= \overline{\sum_{k=1}^{N} \{f(I_k) + f'(I_k)\, I_k \cot \psi_k \cot \psi_{k+1}\}},$$

where $I_1, \ldots, I_N$ are the random intervals (N also random) into which the segment $g \cap K_r$ is divided by the points of $\{P_i\}$, and the angles ψ_k are as shown on Figure 12. Note that

$$\cot\psi_1 = Z^{-1}\sqrt{(4r^2-Z^2)}, \quad \cot\psi_{N+1} = -Z^{-1}\sqrt{(4r^2-Z^2)},$$

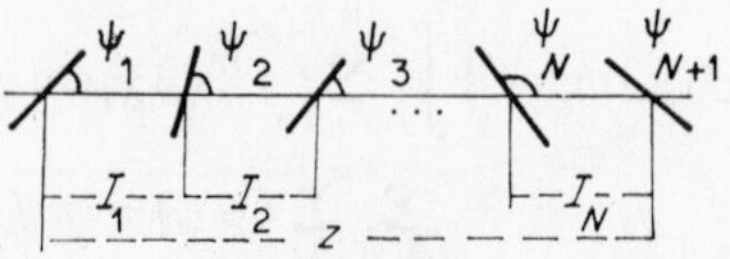

Figure 12

and that in fact only the angles $\psi_2, \ldots, \psi_N$ belong to $\{\psi_i\}$. Now put

(29)
$$\sum_{D_i \subset K_r} \sum_k F(s_{ki}) = \sum_{j=1}^{M_r} F(s_j),$$

in which it is understood that an edge of $\mathscr{T}$ common to two polygons is counted twice, so that

$$\sum_{k=1}^{M_r} s_k = 2L_r,$$

where L_r is the total length of $T \cap K_r$. Then equation (27) may be rewritten as follows:

$$\overline{\sum_{j=1}^{M_r} F(s_j)} = \iint_{g \cap K_r \neq \emptyset} \alpha(Z,r)\,dg - 2\pi r f(0) = \pi \int_0^{2r} \frac{\alpha(Z,r)Z}{\sqrt{(4r^2-Z^2)}}\,dZ - 2\pi r f(0).$$
(30)

Now equation (30) is simplified in the limiting case as $r \to \infty$, if we assume the existence of the limit

(31)
$$\lim_{Z\to\infty} \frac{\alpha(Z,r)}{\bar{N}} = \alpha$$

($\cot\psi_1$ and $\cot\psi_{N+1}$ being kept bounded).

Assume that $\{P_i\}$ is a stationary point-process of finite intensity λ. In this case

(32)
$$\bar{N} - 1 = \lambda Z.$$

By Crofton's theorem,

$$\iint_{K_r \cap g \neq \emptyset} (N-1)\,dg = 2L_r,$$

so that

(33)
$$\iint_{K_r \cap g \neq \emptyset} (\bar{N}-1)\,dg = 2\overline{L_r}.$$

From equations (32) and (33), and the fact that

$$\iint_{K_r \cap g \neq \emptyset} Z\, dg = \pi^2 r^2,$$

we find that

$$2\overline{L_r} = \pi^2 r^2 \lambda.$$

It follows that

$$\iint_{K_r \cap g \neq \emptyset} \alpha(Z, r)\, dg = \iint_{K_r \cap g \neq \emptyset} \frac{\alpha(Z, r)}{\bar{N}} \bar{N}\, dg = 2\alpha L_r + o(r^2).$$

Dividing equation (30) by $2\overline{L_r}$ and observing that edge effects vanish as $r \to \infty$, we arrive at the result

$$\lim_{r\to\infty} \frac{1}{2\overline{L_r}} \overline{\sum_{j=1}^{M_r} F(s_j)} = \alpha. \tag{34}$$

Put

$$l = \lim_{r\to\infty} \frac{2\overline{L_r}}{\overline{M_r}}, \quad \beta = \lim_{r\to\infty} \frac{1}{\overline{M_r}} \overline{\sum_{j=1}^{M_r} F(s_j)}. \tag{35}$$

Then equation (34) may be written

$$\beta / l = \alpha. \tag{36}$$

In the ergodic case the limiting procedure in equation (31) serves to define the intuitive idea of the joint distribution of the 'arbitrary' interval I and the two adjacent angles ψ_1 and ψ_2 in the $\{P_i, \psi_i\}$ process (Figure 2). Formula (35) similarly defines the distribution of the length s of an 'arbitrary' side of a polygon of $\mathscr{T}$ (an 'arbitrary' edge of $\mathscr{T}$ if with probability 1 each side of each polygon in $\mathscr{T}$ is common to two polygons), and then l is its mean length.

We can now obtain the Laplace transform result mentioned in the introduction. Choose $f(u)$ as e^{-tu}. Then

$$f'(u) = -t\, e^{-tu}, \quad F(u) = \frac{1 - e^{-tu}}{t},$$

and, with $\mathbf{E}$ for 'expectation', equation (36) gives

$$\mathbf{E}\{e^{-tI}(1 - tI \cot\psi_1 \cot\psi_2)\} = \frac{1}{l} \frac{1 - \mathbf{E}(e^{-st})}{t}. \tag{37}$$

Assume that I, ψ_1 and ψ_2 are independent. Since $\mathscr{T}$ is homogeneous and isotropic, each ψ_i is distributed with density $\frac{1}{2}\sin\psi$, and $\mathbf{E}(\cot\psi_i) = 0$. Thus equation (37) simplifies to

$$\frac{1}{l} \frac{1 - \mathbf{E}(e^{-ts})}{t} = \mathbf{E}(e^{-tI}). \tag{38}$$

This may be inverted to give the relation

$$V(x) = 1 - l\frac{d}{dx}W(x) \tag{39}$$

between the distribution functions V of s and W of I. In particular, if I, ψ_1 and ψ_2 are independent and I is distributed exponentially, then s is also distributed exponentially and has the same mean. This is consistent with the results quoted in Section 3.2.1, for R. E. Miles has shown the same thing in the case of $\mathscr{T}$ generated by a Poisson process of lines in a plane [5].

Note, that equation (39) remains valid in the general case, if under $W(x)$ the inverse Laplace transform of the left-hand side of the equation (37) is understood. Since $V(0+)$ has to vanish, we have

$$l = \left[\frac{d}{dx}W(x)\right]_{x=0}^{-1}.$$

Let $\mathscr{P}$ be a random polygon of $\mathscr{T}$, in the sense in which each polygon of $\mathscr{T}$ is given equal weight. Consider the problem of finding the mean values of the random variables $H =$ the perimeter of $\mathscr{P}$, $A =$ the area of $\mathscr{P}$, A^2 and AH (see again [5]).

The integral geometry identities on which we rely are (domain $\mathscr{D}$ is fixed)

$$\pi A = \iint \chi\, dg, \tag{41}$$

$$3A^2 = \iint \chi^3\, dg, \tag{42}$$

$$AH = \frac{1}{2}\iint \chi^2[\sin^{-1}\psi' + \sin^{-1}\psi'']\, dg. \tag{43}$$

The equations (41) and (42) are well known (Blaschke, [6]), while (43) is derived by differentiation at $h = 0$ of (42) written for a 'parallel' domain $\mathscr{D}_h$.

The process of averaging, analogous to the one used above, leads to

$$\left.\begin{aligned} EA &= a \cdot EI, \\ EA^2 &= \frac{a}{3}EI^3, \\ EAH &= \frac{a}{2}EI^2[\sin^{-1}\psi_1 + \sin^{-1}\psi_2], \end{aligned}\right\}\quad a = EH.$$

At the same time

$$EH = l \cdot EN, \quad N = \text{number of sides in } \mathscr{P}.$$

EN is determined by the random pattern of arbitrary junction J (in the sense, that each junction is given equal weight) of $\mathscr{T}$. Indeed, for h.i. $\mathscr{T}$ with probability 1 $\mathscr{P}$ has exactly one left-hand point, as well as one right-hand point.

From this observation one easily finds, that

$$EN = \frac{2Em}{En}, \tag{44}$$

where the random variables m and n are defined as follows:

m = number of polygons, for which J serves as a vertex;

n = number of polygons, for which J serves as a vertex, and which are not intersected by a line of a fixed direction through J.

It had been shown in [2], how the distribution of the $\{\mathscr{P}_i, \psi_i\}$ process influences possible patterns of J. Here we note only, that in the case, when

$$\mathbf{P}\{J \text{ is of } T \text{ or } X \text{ types}\} = p_T + p_X = 1$$

application of equation (44) gives

$$EN = 2\frac{4p_X + 2p_T}{2p_X + p_T} = 4$$

so that EH, EA, EA^2 and EAH may be obtained in terms of the joint distribution of l, ψ_1 and ψ_2.

REFERENCES

1. Rollo Davidson, "Construction of line-processes. Second-order properties", 2.4 of this book.
2. R. V. Ambartzumian, "Random fields of segments and random mosaics on a plane", *Proceedings of the Sixth Berkley Symposium*, vol. 3.
3. ——, "Palm distribution and superpositions of independent point processes in R^2", *Stochastic Point Processes*, Wiley–Interscience (1972).
4. ——, "Метод инвариантного вложения в теории случайных прямых", Известия АН Арм ССР, Математика. T. V No. 3, (1970).
5. R. E. Miles, "Random polygons determined by random lines in a plane", *Proc. Nat. Acad. Sci. U.S.A.* **52** (1964), 901–907, 1157–1160.
6. W. Blaschke, *Vorlesungen über Integraleometrie*, Chelsea, New York (1949).

3.3

The Distance from a Given Point to the Nearest End of One Member of a Random Process of Linear Segments

R. COLEMAN

3.3.1 SUMMARY

Consider a process of straight lines embedded at random in the Euclidean space of k dimensions ($k = 1, 2, 3$), and having length distribution F. The distribution is found of the distance from a fixed point, chosen independently of the process of lines, to the nearest end of one of the lines.

3.3.2 INTRODUCTION

The length distribution of the recoil tracks left in nuclear track emulsion by hydrogen nuclei after collisions with a beam of fast neutrons can be used to reconstruct the energy distribution of the neutrons in the beam. The development of travelling microscopes which automatically record the coordinates of points under observation led Lehman and Brisbane [7] to construct search and measure sampling procedures for the particle tracks. A search is made for the end of a track closest to an arbitrary fixed point (chosen independently of the tracks); the track is measured; and its other end is used as the next fixed point. These procedures are considered in detail in Coleman [2]. Let G be the distribution of the distance from a fixed point, chosen independently of a process of straight lines embedded at random in R_k ($k = 1, 2, 3$), the Euclidean space of k dimensions, to the nearest end of one of the lines. In Coleman [2] the relationship is found between G and the joint distribution of the vector chain of search vectors and sampled lines, which is a realization of the path of the travelling microscope under the search and measure procedure. Here in this paper we develop the structure of the distribution G.

We shall need a model for a random process of straight lines. In Coleman [2] a model equivalent to the following is given for straight lines embedded at random in R_k $(k = 1, 2, 3)$.

Definition. If a set of straight lines obeys the following conditions it is a *Poisson process of lines of intensity* λ *in* R_k, *with lengths from the distribution* $dF(l)$.

Condition I. We consider a homogeneous Poisson process of intensity λ in R_k. This is an end from each line of the process. We shall call these ends the *first* ends.

Condition II. The lines have lengths from the distribution

$$dF(l)\ (0 < l < \infty).$$

Condition III. The orientations of the lines are from a uniform distribution over $S_k^*(1)$, where $S_k^*(u)$ is the surface of $S_k(u)$, a k-sphere of radius u in R_k. (Conditions II and III define the process of *second* ends of the lines.)

Condition IV. All the lengths and orientations of the lines are mutually independent.

We describe in the next section how the ends of the lines form a clustering process, and note here that other authors have examined the distances to the points of clustering processes. For example, Pielou [9] and Holgate [6] construct tests, which use these distances, for whether a clustering process is a Poisson process or not. Eberhardt [4] summarizes results of the use of these distances in estimating the intensity parameter of the process. In the particle track problem, however, we have a well-defined clustering process of ends of lines, which is not a Poisson process, and the intensity parameter of which is a nuisance parameter. Persson [8] and Holgate [5] come near to the spirit of this investigation in their studies of the distribution of the distance from a random point to the nearest point of the 'clustering' process which is a regular lattice of points.

3.3.3 THE PROBABILITY THAT A FIXED REGION CONTAINS NO ENDS OF LINES

We find the probability that any fixed Borel set, S, chosen independently of the process of lines, contains no ends of lines. We are thus treating the ends of the lines as a clustering process, and are finding the zero term of the clustering distribution of the number of ends of lines in S. This clustering distribution is a specialization of the distribution generated by the random scattering of seed about themselves by a homogeneous Poisson point-process of parent plants. These mechanisms and processes are considered for example by Thompson [11] and Bartlett [1, Section 2.3]. Here an arbitrary end of a line is a parent, and its other end is its offspring, and each parent has just one offspring isotropically about itself.

To find the probability that S contains no ends of lines, we calculate the probability for each line of the process that it has no ends in S. We therefore calculate, for each point r in R_k ($k = 1, 2, 3$), the joint probability that it is the first end of a line and that, if so, the line has at least one end in S. We then combine the complementary probabilities.

We partition R_k into small Borel sets $\{\delta r\}$. Each δr is so small that (by Condition I for the process of lines: that their first ends are a homogeneous Poisson point-process) the probability that it contains two or more first ends is $o(\delta r)$. We shall refine the partitioning so that in the limit δr is a point at r. We define the events:

$J(S)$: There are no ends of lines in S;

$A(\delta r)$: Borel set δr contains the first end of a line; and

$K(\delta r, S)$: There is the first end of a line in Borel set δr, and that line has at least one end in S.

Events $J(S)$ and $K(\delta r, S)$ are abbreviated to J and $K(\delta r)$ respectively, since S is kept fixed.

Then by Condition I for the process of lines

$$\operatorname{pr}\{A(\delta r)\} = \lambda\,\delta r + o(\delta r) = \lambda\,\delta r\{1 + o(1)\}. \tag{1}$$

Also, for every δr and $\delta r'$ ($\delta r \neq \delta r'$), the events $A(\delta r)$ and $A(\delta r')$ are independent. By Conditions I and IV, for every δr and $\delta r'$ ($\delta r \neq \delta r'$), the events $K(\delta r)$ and $K(\delta r')$ are independent. Therefore, for every δr,

$$\begin{aligned} \operatorname{pr}\{K(\delta r)\} &= \operatorname{pr}\{A(\delta r)\}\operatorname{pr}\{K(\delta r) \mid A(\delta r)\} \\ &= \lambda \operatorname{pr}_A\{K(\delta r)\}\,\delta r\{1 + o(1)\} \end{aligned} \tag{2}$$

by (1), where probabilities conditional on event A are denoted by pr_A. Then

$$\begin{aligned} \operatorname{pr}\{\overline{K(\delta r)}\} &= 1 - \lambda \operatorname{pr}_A\{K(\delta r)\}\,\delta r\{1 + o(1)\} \\ &= \exp[-\lambda \operatorname{pr}_A\{K(\delta r)\}\,\delta r\{1 + o(1)\}]. \end{aligned} \tag{3}$$

The event J will occur if event $K(\delta r)$ fails to occur for every δr; that is,

$$J = \bigcap_{\{\delta r\}} \overline{K(\delta r)}. \tag{4}$$

By the independence of the events on the right-hand side in (4),

$$\operatorname{pr}(J) = \prod_{\{\delta r\}} \exp\left[-\lambda \operatorname{pr}_A\{K(\delta r)\}\,\delta r\{1+o(1)\}\right]$$

$$= \exp\left[-\lambda \sum_{\{\delta r\}} \operatorname{pr}_A\{K(\delta r)\}\,\delta r\{1+o(1)\}\right]$$

$$\text{(5)} \qquad = \exp\left[-\lambda \int_{R_k} \operatorname{pr}_A\{K(r)\}\,dr\right]$$

(as we refine the partition $\{\delta r\}$ of R_k, letting $\max\|\delta r\| \to 0$). That is,

$$\text{(6)} \qquad \operatorname{pr}(J) = \exp\{-\lambda B(S)\},$$

where

$$B(S) = \int_{R_k} \operatorname{pr}_A\{K(r)\}\,dr.$$

The technique just applied is established more formally in Theorem 1 of Ryll-Nardzewski [10].

3.3.4 THE DISTRIBUTION OF THE DISTANCE TO THE NEAREST END OF A LINE

If the fixed Borel set, S, in R_k $(k = 1, 2, 3)$ is $S_k(u)$ (a k-sphere of radius u), then the event J, that S contains no ends of lines, is the same as the event 'the distance U from the centre of S to the nearest end of a line exceeds u'. Then the distribution function, $G(u)$, of the random variable U satisfies

$$\text{(7)} \qquad \operatorname{pr}(J) = \operatorname{pr}(U > u) = 1 - G(u).$$

Let us now write $B_k(2u)$ for $B(S_k(u))$. Then

$$\text{(8)} \qquad G(u) = 1 - \exp\{-\lambda B_k(2u)\} \quad (0 < u < \infty),$$

and

$$\text{(9)} \qquad dG(u) = \lambda \frac{dB_k(2u)}{du} \exp\{-\lambda B_k(2u)\}\,du \quad (0 < u < \infty),$$

where

$$\text{(10)} \qquad B_k(2u) = \int_{R_k} \operatorname{pr}_A\{K(r)\}\,dr \quad (0 < u < \infty).$$

We determine explicit formulae for $B_k(v)$ and $B'_k(v)$ $(k = 1, 2, 3)$ in the following subsections.

The R_1 case

We give the fixed interval, $S_1(u)$, of length $2u$ in the real axis the coordinates $(-u, u)$; then if a line has its first end at r and its second end at

$$r+l \quad (-\infty<l<\infty),$$

the set of lines, $\{(l, r)\}$, which would have at least one end in $(-u, u)$ is

$$\text{(11)} \qquad \mathscr{A}_1(u) \equiv \begin{Bmatrix} 0 \leqslant |r| \leqslant u \\ 0 \leqslant |l| < \infty \end{Bmatrix} \cup \begin{Bmatrix} u<|r|<\infty \\ -r-u \leqslant l \leqslant -r+u \end{Bmatrix};$$

that is, the union of those lines with their first ends in $(-u, u)$ and those whose first ends are not in $(-u, u)$, but whose second ends are.

We denote by $K^*(l, r; u)$ the indicator event

$$(l, r) \in \mathscr{A}_1(u),$$

so

$$\text{(12)} \qquad \mathrm{pr}_A\{K^*(l, r; u)\} = \begin{cases} 1 & ((l, r) \in \mathscr{A}_1(u)), \\ 0 & \text{(otherwise)}. \end{cases}$$

Then, if the lines come from a population with distribution of lengths $dF(l)$ $(0<l<\infty)$, for the event $K(r; u)$ ('a line has its first end at r and at least one end in $S_1(u)$') we have that

$$\text{(13)} \qquad \mathrm{pr}_A\{K(r; u)\} = \int_{l=-\infty}^{\infty} \mathrm{pr}_A\{K^*(l, r; u)\}\, dF^*(l),$$

where

$$dF^*(l) = \tfrac{1}{2} dF(|l|) \quad (-\infty<l<\infty).$$

Then if the first ends of the lines are a homogeneous Poisson process in the real axis, we have that

$$B_1(2u) = \int_{-\infty}^{\infty} \mathrm{pr}_A\{K(r; u)\}\, dr$$

$$= \int_{r=-\infty}^{\infty} \int_{l=-\infty}^{\infty} \mathrm{pr}_A\{K^*(l, r; u)\}\, dF^*(l)\, dr$$

$$= \iint_{(l,r) \in \mathscr{A}_1(u)} dF^*(l)\, dr$$

$$\text{(14)} \qquad = 2u \int_{l=0}^{\infty} C_1\left(\frac{l}{2u}\right) dF(l),$$

where

(15) $$C_1(w) = 1+\min(w, 1).$$

Thus

(16) $$B_1(v) = 2v - vF(v) + \int_{l=0}^{v} l\,dF(l),$$

and, trivially,

(17) $$B_1'(v) = \frac{dB_1(v)}{dv} = 2 - F(v).$$

The elaborate procedure of introducing set $\mathscr{A}_1$ and event K^* is not really necessary in the R_1 case, but it mimics the procedure which we shall adopt in the R_2 and R_3 cases, when otherwise we should find the changing of the orders of integration too unwieldy to be performed neatly.

The R_2 case

In R_2, for a line to have an end in disc $S_2(u)$, the axis on which it lies must cut circle $S_2^*(u)$ in two points, U and V say. We label the first end of the line with C, and suppose that distance CU exceeds distance CV. We take the centre O of $S_2(u)$ to be the origin of plane polar coordinates with fixed line $\varphi = 0$ $(0 \leqslant \varphi < 2\pi)$, so that C is at point (r, φ). We denote angle OCU by ψ $(0 \leqslant |\psi| < \frac{1}{2}\pi)$. (See Figure 1.) The line can then be described by coordinates $(\mathbf{l}, \mathbf{r}) \equiv ((l, \psi), (r, \varphi))$.

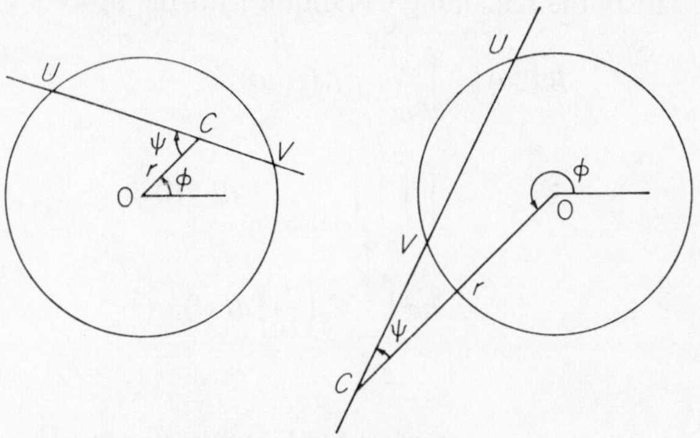

Figure 1. A coordinate system for a straight line having its first end at C and lying on the axis VU which cuts a circle centre O

Let us denote by $\mathscr{A}_2(u)$ the set of lines, $\{(\mathbf{l}, \mathbf{r})\}$, which can have at least one of their ends in $S_2(u)$. Then

(18) $$\mathscr{A}_2(u) \equiv \{\mathscr{A}(u;\, l, \psi, r) \cup \mathscr{A}(u;\, l, -\psi, r)\} \cap \{0 \leqslant \varphi < 2\pi\},$$

where

$$(19)\quad \mathscr{A}(u;\, l,\psi,r) \equiv \begin{Bmatrix} 0\leqslant l<\infty \\ 0\leqslant r\leqslant u \\ 0\leqslant \psi\leqslant \pi \end{Bmatrix} \cup \begin{Bmatrix} 0\leqslant l<2u \\ u\leqslant r\leqslant l+u \\ 0\leqslant \psi\leqslant \psi_{\max} \end{Bmatrix} \cup \begin{Bmatrix} 2u\leqslant l<\infty \\ l-u\leqslant r\leqslant l+u \\ 0\leqslant \psi\leqslant \psi_{\max} \end{Bmatrix},$$

and

$$(20)\qquad \psi_{\max} = \arccos\left(\frac{l^2+r^2-u^2}{2lr}\right) \quad \text{(p.v. in } [0,\pi]).$$

We denote by $K^*(\mathbf{l},\mathbf{r};\, u)$ the indicator event

$$(\mathbf{l},\mathbf{r})\in\mathscr{A}_2(u).$$

If the lines satisfy Conditions II, III and IV for the process, then the distribution of $\mathbf{l}$ is

$$(21)\qquad dF^*(\mathbf{l}) = dF(l)\frac{d\psi}{2\pi} \quad (\mathbf{l}\in R_2).$$

Then for event $K(\mathbf{r};\, u)$ ('a line has its first end at $\mathbf{r}$ and at least one end in $S_2(u)$') we have that

$$(22)\qquad \mathrm{pr}_A\{K(\mathbf{r};\, u)\} = \int_{\mathbf{l}\in R_2} \mathrm{pr}_A\{K^*(\mathbf{l},\mathbf{r};\, u)\}\, dF^*(\mathbf{l}).$$

Then if the first ends of the lines are a homogeneous Poisson process in R_2 (that is, satisfy the remaining Condition I for the process of lines),

$$B_2(2u) = \int_{R_2} \mathrm{pr}_A\{K(\mathbf{r};\, u)\}\, d\mathbf{r}$$

$$= \iiiint_{(\mathbf{l},\mathbf{r})\in\mathscr{A}_2(u)} dF^*(\mathbf{l})\, d\mathbf{r}$$

$$(23)\qquad = 2u^2\int_{l=0}^{\infty} C_2\left(\frac{l}{2u}\right) dF(l),$$

where

$$(24)\qquad C_2(w) = \begin{cases} \pi - \arccos w + w(1-w^2)^{\frac{1}{2}} & (0<w<1), \\ \pi & (1\leqslant w<\infty). \end{cases}$$

Thus

$$(25)\qquad B_2(v) = \tfrac{1}{2}v^2\left[\pi - \int_{l=0}^{v}\left\{\arccos\left(\frac{l}{v}\right) - \left(\frac{l}{v}\right)\left(1-\frac{l^2}{v^2}\right)^{\frac{1}{2}}\right\} dF(l)\right]$$

and

$$B_2'(v) = v\left\{\pi - \int_{l=0}^{v} \arc\cos\left(\frac{l}{v}\right) dF(l)\right\}. \tag{26}$$

The R_3 case

As for the R_2 case, for a line in R_3 to have an end in sphere $S_3(u)$, the axis on which it lies must cut $S_3^*(u)$ in two points, U and V say. We denote the first end of the line by C, and suppose that distance CU exceeds distance CV. We take the centre, O, of $S_3(u)$ to be the origin of spherical polar coordinates. Plane $OCUV$ cuts $S_3^*(u)$ in a grand circle $S_2^*(u)$. In plane $OCUV$, therefore, the system is that of the R_2 case, and this plane is taken to be at angle θ $(0 \leqslant \theta < 2\pi)$ with respect to axis OC. We let C be at $\mathbf{r} \equiv (r, \varphi, \beta)$. The angle OCU is denoted by ψ $(0 \leqslant \psi < \pi)$. Any line can then be described by the coordinates $(\mathbf{l}, \mathbf{r}) \equiv ((l, \psi, \theta), (r, \varphi, \beta))$.

Then $\mathscr{A}_3(u)$, the set of lines, $\{(\mathbf{l}, \mathbf{r})\}$, which would have at least one of their ends in $S_3(u)$, is

$$\mathscr{A}_3(u) \equiv \mathscr{A}(u; l, \psi, r) \cap \begin{Bmatrix} 0 \leqslant \varphi < 2\pi \\ 0 \leqslant \theta < 2\pi \\ 0 \leqslant \beta < \pi \end{Bmatrix}, \tag{27}$$

where $\mathscr{A}$ is defined in equation (19). Then, as for the R_1 and R_2 cases,

$$B_3(2u) = \int\!\!\int\!\!\int\!\!\int\!\!\int\!\!\int_{(\mathbf{l},\mathbf{r}) \in \mathscr{A}_3(u)} dF^*(\mathbf{l})\, d\mathbf{r} \tag{28}$$

$$= \frac{4\pi u^3}{3} \int_{l=0}^{\infty} C_3\left(\frac{l}{2u}\right) dF(l),$$

where

$$C_3(w) = \begin{cases} 1 + \frac{1}{2}w(3 - w^2) & (0 < w < 1), \\ 2 & (1 \leqslant w < \infty). \end{cases} \tag{29}$$

Thus

$$B_3(v) = \frac{1}{6}\pi v^3 \left\{2 - F(v) + \frac{1}{2}\int_{l=0}^{v} \left(\frac{l}{v}\right)\left(3 - \frac{l^2}{v^2}\right) dF(l)\right\} \tag{30}$$

and

$$B_3'(v) = \frac{1}{2}\pi v^2 \left\{2 - F(v) + \frac{1}{v}\int_{l=0}^{v} l\, dF(l)\right\} \tag{31}$$

$$(= \tfrac{1}{2}\pi v B_1(v)).$$

3.3.5 CONCLUSIONS

The results may be summarized in the following:

Theorem 1. *Consider a Poisson process of lines in* R_k $(k = 1, 2, 3)$, *having length distribution* $dF(l)$ $(0 < l < \infty)$ *and intensity* λ *of first ends of lines. The probability that the distance from a point chosen independently of the process of lines to the nearest end of a line exceeds u is*

$$\exp\{-\lambda B_k(2u)\} \quad (0 \leqslant u < \infty), \tag{32}$$

where

$$B_1(v) = 2v - vF(v) + \int_{l=0}^{v} l\, dF(l) \tag{33}$$

$$B_2(v) = \frac{1}{2}v^2\left[\pi - \int_{l=0}^{v}\left\{\arccos\left(\frac{l}{v}\right) - \left(\frac{l}{v}\right)\left(1 - \frac{l^2}{v^2}\right)^{\frac{1}{2}}\right\} dF(l)\right] \tag{34}$$

and

$$B_3(v) = \frac{1}{6}\pi v^3\left\{2 - F(v) + \frac{1}{2}\int_{l=0}^{v}\left(\frac{l}{v}\right)\left(3 - \frac{l^2}{v^2}\right) dF(l)\right\}. \tag{35}$$

It is of interest to note that the probability that there are no ends of lines of the process in $S_k(\frac{1}{2}v)$, a k-sphere of diameter v, depends only on the length distribution of those lines whose length does not exceed v; that is, on $F(l)$ $(0 < l \leqslant v)$. We can see this heuristically. By a generalization of a theorem of Doob [3, p. 404], since the first ends of the lines are a homogeneous Poisson process, so also are the second ends. The correlation between the two processes yields the clustering distribution of ends of lines. A line longer than v can have at most one end in $S_k(\frac{1}{2}v)$, and this end will be a point of a homogeneous Poisson process. The probability that an end of the line lies in $S_k(\frac{1}{2}v)$ will therefore not depend on the length of the line. For a line shorter than v, we must take into account the chance of both its ends lying in $S_k(\frac{1}{2}v)$.

We check the formulae of the theorem in the special case in which the lines degenerate into points. That is, distribution $dF(l)$ is an atom of probability of weight one at $l = 0$. Then by the above theorem

$$B_1(v) = v, \quad B_2(v) = \tfrac{1}{4}\pi v^2 \quad \text{and} \quad B_3(v) = \tfrac{1}{6}\pi v^3. \tag{36}$$

We recognize that these are the volumes of $S_k(\frac{1}{2}v)$ $(k = 1, 2, 3)$, k-spheres of diameter v. In this degenerate case the distance, U, from the fixed point to the nearest end of a line is the distance to the nearest first end of a line. The first ends are a homogeneous Poisson process of intensity λ, and the fixed point is chosen independently of the process of first ends, so

(37) $\text{pr}\{\text{there are no first ends in } S\} = \exp\{-\lambda \text{ volume } (S)\}$,

the zero term of a Poisson distribution; therefore

(38) $\text{pr}(U > u) = 1 - G(u) = \exp\{-\lambda \text{ volume } (S_k(u))\}$.

In this case therefore $B_k(2u) = \text{volume } (S_k(u))$, confirming equation (36).

Acknowledgements

I held a Science Research Council Studentship whilst carrying out the work for this contribution, which is based on a chapter of a Ph.D. thesis submitted to Cambridge University. My research supervisor was Mr. A. D. McLaren whose advice and encouragement are gratefully acknowledged.

REFERENCES

1. M. S. Bartlett, *Stochastic Population Models in Ecology and Epidemiology*, Methuen, London (1960).
2. R. Coleman, "Sampling procedures for the lengths of random straight lines", *Biometrika* **59** (1972), 415–426.
3. J. L. Doob, *Stochastic Processes*, Wiley, New York (1953).
4. L. L. Eberhardt, "Some developments in 'distance sampling' ", *Biometrics* **23** (1967), 207–216.
5. P. Holgate, "The distance from a random point to the nearest point of a closely packed lattice", *Biometrika* **52** (1965), 261–263.
6. ——, "Tests of randomness based on distance methods", *Biometrika* **52** (1965), 345–353.
7. R. L. Lehman and R. W. Brisbane, "Random-drift sampling—A study by computer simulation", *Nucl. Instr. and Meth.* **64** (1968), 269–277.
8. O. Persson, *Distance Methods*, Studia Forestalia Suecica, No. 15 (1964).
9. E. C. Pielou, "The use of point-to-plant distances in the study of pattern of plant populations", *J. Ecol.* **47** (1959), 607–613.
10. C. Ryll-Nardzewski, "Remarks on the Poisson stochastic process (III) (On a property of the homogeneous Poisson process)", *Studia Math.* **14** (1954), 314–318.
11. H. R. Thompson, "A note on contagious distributions", *Biometrika* **41** (1954), 268–271.

3.4

A Synopsis of 'Poisson Flats in Euclidean Spaces'

R. E. MILES

3.4.1 ABSTRACT

Homogeneous isotropic Poisson s-flats in Euclidean d-dimensional space $E^d\,(0 \leqslant s < d)$ are defined as a natural generalization of the standard linear Poisson process, for which $s=0$, $d=1$. Invariant densities of integral geometry enter naturally into the ergodic theory of n-subsets of s-flats in such systems. A wide class of ergodic gamma-type probability distributions, in a sense dual to the Poisson distributions for numbers of hits, is derived. Extensions of the theory to more general systems incorporating mixtures, anisotropy and associated cylinder sets are discussed, and mention is made of the fundamental role of the anisotropic case as a local limit when random s-dimensional varieties are superposed in E^d. Finally, the ergodic probability distributions of several associated random tessellations of convex polytopes are investigated.

3.4.2 INTRODUCTION

The present paper is based on two lectures given by the author at the Symposium on Integral Geometry and Geometrical Probability held at Oberwolfach in June 1969. The main object then was to present, within the limits of a paper of reasonable length, a wide-ranging account of Poisson flats in Euclidean spaces. Emphasis was placed on the underlying connection with integral geometry, and on the wide range of possible random geometric models encompassed by such a structure. As inevitable costs of such an ambitious project, the material was somewhat selective with few examples being included, full proofs were for the most part omitted, and questions of rigour were glossed over or even ignored. This paper has much the same defects, but on the other hand it is hoped it offers a relatively effortless and illuminating introduction to this area of random geometry.

Much of the material relating to the random tessellations determined by Poisson hyperplanes is taken from the author's unpublished Ph.D. thesis [13], while Theorem 2 was announced in [14]. Specialized accounts of the planar cases $s=0$, $d=2$ and $s=1$, $d=2$ are to be found in [17] and [15, 33] respectively. The author is preparing a series of papers in which the contents of the present paper are developed in full detail. The first two of these are [16] and [31].

3.4.3 PRELIMINARIES

First we introduce the Poisson and gamma probability distributions, which are of fundamental importance in this work. The Poisson distribution with parameter (= its expectation = its variance) λ has p.m.f.

(1) $$p_i = \mathrm{e}^{-\lambda}\,\lambda^i/i! \quad (i = 0, 1, \ldots).$$

The family of gamma distributions $\Gamma_\theta(\nu, \lambda)$ (θ, λ, ν all >0) has p.d.f.

(2) $$f(x) = \theta\lambda^{\nu/\theta}\,x^{\nu-1}\mathrm{e}^{-\lambda x^\theta}/\Gamma\left(\frac{\nu}{\theta}\right) \quad (x \geqslant 0),$$

and kth-order moment

(3) $$\mu_k = \left\{\Gamma\left(\frac{\nu+k}{\theta}\right)\Big/\Gamma\left(\frac{\nu}{\theta}\right)\right\}\lambda^{-k/\theta} \quad (k = 1, 2, \ldots).$$

If ν/θ is a positive integer, then the d.f. of $\Gamma_\theta(\nu, \lambda)$ is

(4) $$F(x) = q\left\{\frac{\nu}{\theta}-1, \lambda x^\theta\right\} \quad (x \geqslant 0),$$

where the tails of the Poisson (λ) distribution given by equation (1) are defined by

(5) $$p\{i, \lambda\} = 1 - q\{i, \lambda\} = \mathrm{e}^{-\lambda}\{1 + \ldots + (\lambda^i/i!)\}.$$

The cases $\theta = 1$ and $\theta = 1$, $\nu = 1$ yield the standard gamma distribution $\Gamma(\nu, \lambda)$ of index ν and parameter λ, and the exponential distribution of parameter λ, respectively.

The random systems we consider lie in E^d, in which $x, y, \ldots$ are points and o the origin ($d = 1, 2, \ldots$). The notation $|\ldots|$ is used for the modulus of a real number, the length of a vector, full dimensional Lebesgue measure and for determinants; the interpretation will be apparent from the context. The *ball* $|x| \leqslant q$ of centre o and radius q is denoted by Q_q; then its boundary ∂Q_q is the *sphere* $|x| = q$. The volume d-content $|Q_q|$ of Q_q is $v_d q^d$, where

$$v_d \equiv \pi^{d/2}/\Gamma(\tfrac{1}{2}d+1);$$

the surface $(d-1)$-content $|\partial Q_q|$ of Q_q is $\sigma_d q^{d-1}$, where

$$\sigma_d \equiv 2\pi^{d/2}/\Gamma(\tfrac{1}{2}d).$$

The weighted linear sum of the subsets $X_i \subset E^d$ with weights λ_i $(1 \leqslant i \leqslant n)$ is

$$\sum_1^n \lambda_i X_i = \left\{\sum_1^n \lambda_i x_i \colon x_i \in X_i,\ 1 \leqslant i \leqslant n\right\}. \tag{6}$$

We abbreviate '$X \cap Y \neq \varnothing$' to '$X \uparrow Y$', in words '$X$ *hits* Y'. The range space of a variable ... is often denoted by [...], while the subset comprising the single point ... is {...}. Finally, $\{x_*\}$ is shorthand for $\{x_i\}$ over the full range of i.

3.4.4 THEORY

An *s-flat* is the translate $J_s = \{x\} + J_{s(o)}$ of an s-dimensional linear subspace, or *s-subspace*, $J_{s(o)}$ in E^d $(0 \leqslant s < d)$. Examples: s: $0 \sim$ point, $1 \sim$ line, $2 \sim$ plane, $3 \sim$ (ordinary) space, ..., $d-1 \sim$ hyperplane. Thus, for a system of random s-flats in E^d, there are six cases of practical importance, viz. $0 \leqslant s < d \leqslant 3$; and ten if 'time' is included. Since an s-subspace in E^d has $s(d-s)$ degrees of freedom, an s-flat has $(s+1)(d-s)$ degrees of freedom and hence may be parametrized by a point $b \in [b] \subset E^{(s+1)(d-s)}$. This parametrization is trivial when $s = 0$. An s-flat hitting $X \subset E^d$ is termed an *s-secant* of X. Thus an arbitrary probability distribution concentrated on

$$B_X \equiv \{b \colon J_s(b) \uparrow X \subset E^d\} \tag{7}$$

(supposed measurable) 'is' the distribution of a random s-secant of X.

Suppose, for each measurable subset B of $[b]$,

$$\int_B f(b)\, db$$

is invariant with respect to the group of Euclidean motions in E^d; that is,

$$\int_B f(b)\, db = \int_{B'} f(b')\, db'$$

for all such motions. The existence and uniqueness, up to a constant factor, of f is ensured by the general theory. See, for example, Santaló [21, Part III] or Nachbin [18, Chap. III]. According to Nachbin (Example 6, pp. 143–144), the s-flats of E^d form a locally compact homogeneous space under the locally compact Euclidean transformation group, on which there is an invariant measure, unique up to a constant factor. We describe $f(b)$ as the *invariant density* in B with respect to the parametrization b, and

$$F(B) = \int_B f(b)\, db$$

as the corresponding *invariant measure*. An explicit form of the invariant density was first given by Blaschke, in a short monograph [2] which marked the birth of integral geometry as such; see also Petkantschin [19], Santaló [21; §24] and, more recently, Miles [32]. Suppose $u_1, \ldots, u_d$ are orthonormal vectors in E^d; $J_{s(o)}$ is spanned by $u_1, \ldots, u_s$; dO^s represents an $(s-1)$-dimensional volume element of the s-dimensional unit sphere with centre o through $u_1, \ldots, u_s$ (dO^d, which may be equated with du, is often written below simply as dO); and dx^{d-s} represents a $(d-s)$-dimensional volume element of the orthogonal complement of $J_{s(o)}$. In terms of these quantities, we have the intuitively apparent and convenient exterior differential relation

$$f(b)\, db = dJ_s \Big/ \int dJ_{s(o)} = dJ_{s(o)}\, dx^{d-s} \Big/ \int dJ_{s(o)} \tag{8}$$

where

$$dJ_{s(o)}\, dO^s \ldots dO^1 = dO^d \ldots dO^{d-s+1} \tag{9}$$

(dO^1 represents the 'volume element' of a measure concentrating unit mass on both ± 1), which implies

$$\int dJ_{s(o)} = \sigma_d \cdots \sigma_{d-s+1} / \sigma_s \cdots \sigma_1. \tag{10}$$

Apart from the constant factor $\int dJ_{s(o)}$, this is the form given by Santaló [22]. Examples: $s = 0$: $f(x) = 1$; $s = d-1$: if, in polar coordinates, (p, u) is the foot of the perpendicular from o to the hyperplane, then $f(p, u) = 2/\sigma_d$, the corresponding element being $(2/\sigma_d)\, dp\, dO$. Integrating equation (8) over B_X, and defining X_{d-s} to be the orthogonal projection of X onto the orthogonal complement of $J_{s(o)}$, we obtain

$$F(B_X) = \int |X_{d-s}|\, dJ_{s(o)} \Big/ \int dJ_{s(o)} = M_{d-s}\{X\}, \tag{11}$$

the *mean* $(d-s)$*-projection* of X. Examples (see [16, §2.3]): $M_d\{X\} = |X|$; for convex X, $M_{d-1}\{X\} = \{\Gamma(\frac{1}{2}d)/2\pi^{\frac{1}{2}}\Gamma(\frac{1}{2}[d+1])\}|\partial X|$; $M_o\{X\} = 1$. For $0 < M_{d-s}\{X\} < \infty$, the p.d.f.

$$f_X(b) = \begin{cases} f(b)/M_{d-s}\{X\} & (b \in B_X), \\ 0 & (b \notin B_X) \end{cases} \tag{12}$$

determines a *uniform isotropic* (random) *s-secant of X*. Such secants possess rather natural properties. Thus, if J_s is a uniform isotropic *s-secant* of X, then

(*a*) $\mathbf{P}(J_s \uparrow Y \subset X) = M_{d-s}\{Y\}/M_{d-s}\{X\}$, *independent of the 'position' of Y within X*;

(*b*) given $J_s \uparrow Y \subset X$, then J_s is uniform isotropic in Y; further,

(*c*) given that the flat of intersection of independent uniform isotropic secants in X hits X, this flat is a uniform isotropic secant of X of appropriate dimension.

Consider the random system comprising N independent uniform isotropic s-secants of X. The distribution of $\#_Y$, the number of secants hitting $Y \subset X$, is binomial $(N, M_{d-s}\{Y\}/M_{d-s}\{X\})$. As $N, M_{d-s}\{X\}$ both $\to \infty$ in such a way that $N/M_{d-s}\{X\} \to \rho$, the distribution of $\#_Y$ tends to Poisson $(\rho M_{d-s}\{Y\})$, in the usual way. *This and the relation* (11) *suggest the consideration of the stochastic flat process* $\mathfrak{P}(\rho; s, d)$ *in* E^d *corresponding to the (inhomogeneous) Poisson point process in* $[b]$ *of intensity* $\rho f(b)$. This definition ensures that $\mathfrak{P}(\rho; s, d)$ is stochastically invariant with respect to Euclidean motions. Thus it is both *homogeneous* (or 'strictly stationary'), i.e. stochastically invariant under translations; and *isotropic*, i.e. stochastically invariant under rotations. Accordingly, we describe $\mathfrak{P}(\rho; s, d)$ as *the isotropic homogeneous Poisson s-flat process of intensity* ρ *in* E^d. Immediate from equation (11) and the Poisson structure is

Theorem 1. *The number* $\#_X$ *of s-flats of* $\mathfrak{P}(\rho; s, d)$ *hitting* $X \subset E^d$ *has a Poisson* $(\rho M_{d-s}\{X\})$ *distribution. Further, given that* $\#_X = N$, *these N s-secants of X are independent uniform random s-secants of X.*

It is left to the reader to derive the p.g.f. of the multivariate Poisson distribution of $(\#_{X(1)}, \ldots, \#_{X(m)})$ for arbitrary $X(i) \subset E^d$. Examples: $\mathfrak{P}(\rho; 0, 1)$ is the standard linear Poisson point process, and $\mathfrak{P}(\rho; 0, d)$ the corresponding d-dimensional point process. For clarity, the 0-flats of $\mathfrak{P}(\rho; 0, d)$ are termed *particles*. Since a hyperplane partitions E^d into two separated half-spaces, $\mathfrak{P}(\rho; d-1, d)$ (which includes $\mathfrak{P}(\rho; 0, 1)$) has the effect of partitioning E^d into a tessellation $\mathscr{P}$ of random convex polytopes (see Section 3.4.6). (Thus a one-dimensional convex polytope is an interval.) The intensity ρ is also characterized as the mean s-content of s-flat of $\mathfrak{P}(\rho; s, d)$ per unit volume in E^d. We now state two of the fundamental properties of isotropic Poisson flat systems.

Independent superposition

If $\mathfrak{P}(\rho_1; s, d), \ldots, \mathfrak{P}(\rho_m; s, d)$ are independent, then

$$\bigcup_{i=1}^{m} \mathfrak{P}(\rho_i; s, d) \quad \text{is a} \quad \mathfrak{P}\left(\sum_{i=1}^{m} \rho_i; s, d\right).$$

Arbitrary section

If J_t is an arbitrary t-flat in E^d $(d-s \leqslant t < d)$, then $J_t \cap \mathfrak{P}(\rho; s, d)$ is a $\mathfrak{P}(\rho'; s+t-d, t)$, where

$$(13) \qquad \rho' = \left\{\Gamma\left(\frac{s+1}{2}\right)\Gamma\left(\frac{t+1}{2}\right)\bigg/\Gamma\left(\frac{d+1}{2}\right)\Gamma\left(\frac{s+t-d+1}{2}\right)\right\}\rho.$$

Henceforth, since ρ is mostly fixed, ρ is usually omitted from $\mathfrak{P}(\rho; s, d)$.

In this paragraph we forget $\mathfrak{P}(s, d)$ for the moment, and return to the integral geometry of s-flats in E^d. An *n-figure* of s-flats is defined to be an ordered set of n distinct s-flats (an unordered such set is termed an *n-set*). Parametrically,

$$(14) \qquad c = (b_1, \ldots, b_n) \in [c] \subset E^{n(s+1)(d-s)}.$$

The corresponding product invariant density in $[c]$ is

$$(15) \qquad f(c) = \prod_1^n f(b_i) \quad (c \in [c]).$$

However, we shall use instead a special 'structural' parametrization $\tilde{c}$ of an n-figure, best illustrated by an example. Taking $s = 0$, $d = 2$ and $b = (x, y)$, set

$$(16) \qquad \begin{cases} (x_1, y_1) = (x, y), \\ (x_2, y_2) = (x + l\cos\theta, y + l\sin\theta), \\ (x_i, y_i) = (x + \lambda_i l \cos\overline{\theta+\varphi_i}, y + \lambda_i l \sin\overline{\theta+\varphi_i}) \quad (3 \leqslant i \leqslant n). \end{cases}$$

Then

$$(17) \qquad \tilde{c} = (\underset{\text{location}}{x, y}; \quad \underset{\text{orientation}}{\theta}; \quad \underset{\text{scale}}{l}; \quad \underbrace{\lambda_3, \varphi_3, \ldots, \lambda_n, \varphi_n}_{\text{shape}})$$

is structural, the components contributing to the four elements of structure of the n-figure being indicated. Since in this case $f(c) = 1$,

$$(18)\qquad f(\tilde{c}) = \left|\frac{\partial c}{\partial \tilde{c}}\right| f(c) = l^{2n-3} \prod_{3}^{n} \lambda_i.$$

For general (s, d), the intersection of the n component s-flats of an n-figure is in general an $\{ns-(n-1)d\}$-flat. Attention is here restricted to the case in which this flat is either void or at most a point, i.e. $n \geqslant d/(d-s)$. Then a *centre* $z \in E^d$ specifying the location of the n-figure can be defined, and we may write $\tilde{c} = (z, \alpha)$. Thus $z = (x, y)$ in equation (17). Clearly the product invariant density factorizes to give $f(\tilde{c}) = 1 . f(\alpha)$. For $n > d/(d-s)$, there is in general a unique non-degenerate sphere hitting all n component s-flats and of minimal radius—the *minimal sphere* of the n-figure. If the minimal sphere has radius l, then we may write $\alpha = (l, \beta)$, in which the $\{n(s+1)(d-s)-(d+1)\}$-tuple β specifies *shape* and *orientation*. The 'characteristic length' l determines the *scale* of the n-figure. The reader should envisage the sequential construction of an n-figure by (*i*) z; (*ii*) β; (*iii*) l. Finally, integral geometric argument, which for reasons of space must be omitted (see result (4.19T) of [31]), serves to generalize equation (18) to

$$(19)\qquad f(\tilde{c}) = 1 . f(\alpha) = 1 . f(l, \beta) = l^{n(d-s)-(d+1)} f(1, \beta),$$

where $f(1, \beta) \equiv [f(l, \beta)]_{l=1}$.

Now reconsider $\mathfrak{P}(s, d)$. Define $\mathfrak{P}_n(s, d)$ to be the aggregate of n-figures generated by $\mathfrak{P}(s, d)$ $(n = 1, 2, \ldots)$. Thus each n-set consisting of members of $\mathfrak{P}(s, d)$ gives rise to $n!$ members of $\mathfrak{P}_n(s, d)$, and $\mathfrak{P}_n(s, d)$ may be regarded as a stochastic point process in $[c]$ or $[\tilde{c}]$. The a.s. (almost sure) countability of the members of $\mathfrak{P}(s, d)$ implies the same property for $\mathfrak{P}_n(s, d)$. Define $H_n\{X, \delta\alpha\}$ to be the number of members of $\mathfrak{P}_n(s, d)$ with centre $z \in X \subset E^d$ and with α-value lying in the notionally small $\{n(s+1)(d-s)-d\}$-dimensional interval $\delta\alpha$ in $[\alpha]$ with opposite vertices α and $\alpha + \delta\alpha$. In addition to its normal meaning as an increment in α, $\delta\alpha$ is also used for the above-mentioned interval and the $\{n(s+1)(d-s)-d\}$-content of this interval; the interpretation intended will be apparent from the context. The corresponding normalized 'empiric average' is

$$(20)\qquad h_q = H_n\{Q_q, \delta\alpha\}/|Q_q|.$$

Then

$$(21)\qquad \mathbf{E}(h_q) = |Q_q|^{-1}\, \mathbf{E} \int_{Q_q} H_n\{dz, \delta\alpha\} = |Q_q|^{-1} \int_{Q_q} \mathbf{E} H_n\{dz, \delta\alpha\}.$$

Appealing to the extreme 'Poisson' independence and the relation $f(c)\,dc = f(\tilde{c})\,d\tilde{c}$, we have

$$\mathbf{E}[H_n\{dz, \delta\alpha\}^k] = \rho^n f(\alpha)\,dz\,\delta\alpha\{1+O(\delta\alpha)\} \tag{22}$$

for an arbitrary positive integer k. Combining equations (21) and (22) with $k=1$ and applying the homogeneity of $\mathfrak{P}(s,d)$, we obtain

$$\mathbf{E}(h_q) = \rho^n f(\alpha)\,\delta\alpha\{1+O(\delta\alpha)\}. \tag{23}$$

Henceforth, for brevity, factors like $\{1+O(\delta\alpha)\}$ are usually omitted. Similarly,

$$\begin{aligned} \operatorname{Var} h_q &= |Q_q|^{-2}\int_{Q_q}\int_{Q_q}[\mathbf{E}\{H_n(dx,\delta\alpha)\,H_n(dy,\delta\alpha)\} \\ &\qquad - \mathbf{E}H_n(dx,\delta\alpha)\,\mathbf{E}H_n(dy,\delta\alpha)] \\ &\equiv |Q_q|^{-2}\int_{Q_q}\int_{Q_q} g(x-y,\delta\alpha)\,dx\,dy, \end{aligned} \tag{24}$$

say. Now g is zero except on a subset S of $Q_q \times Q_q$, where

$$|S|/|Q_q| \to a(\delta\alpha) < \infty$$

as $q\to\infty$. Further, application of Schwartz's inequality and equation (22) with $k=2$ shows that $g \leqslant b(\delta\alpha) < \infty$ on S. It follows that $\operatorname{Var} h_q \to 0$ as $q\to\infty$, and so

$$h_q \xrightarrow{\text{m.s.}} \rho^n f(\alpha)\,\delta\alpha, \quad \text{a constant,}$$

as $q\to\infty$. But, since $\mathfrak{P}(s,d)$ is homogeneous, Wiener's d-parameter ergodic theorem [30, Theorem II″] applies, giving

$$h_q \xrightarrow{\text{a.s.}} \text{a random variable,} \quad h \text{ say,}$$

as $q\to\infty$. Since m.s. and a.s. limits coincide (= in probability limit), we have

$$h_q \xrightarrow{\text{a.s.}} \rho^n f(\alpha)\,\delta\alpha$$

as $q\to\infty$. In fact, more generally

$$H_n\{X,\delta\alpha\}/|X| \xrightarrow{\text{a.s.}} \rho^n f(\alpha)\,\delta\alpha\{1+O(\delta\alpha)\} \tag{25}$$

as $|X|\to\infty$, where $X = X_q = qX_1$ is a bounded region of E^d star-shaped with respect to o and $|X|\to\infty$ is equivalent to $q\to\infty$. The significance of the limit (25) is emphasized by

$$H_n\{X,\delta\alpha_1\}/H_n\{X,\delta\alpha_2\} \xrightarrow{\text{a.s.}} f(\alpha_1)\,\delta\alpha_1\{1+O(\delta\alpha_1)\}/f(\alpha_2)\,\delta\alpha_2\{1+O(\delta\alpha_2)\} \tag{26}$$

as $|X|\to\infty$. *That is, $f(\alpha)$ is moreover the* (a.s.) *ergodic density of α for* $\mathfrak{P}_n(s,d)$. Ergodic densities, like invariant densities, are only defined up to a constant factor. It may be said that the ergodic density $f(\alpha)$ is the 'quotient' of the product invariant density by the uniform density of the centre: $f(\alpha)=f(\tilde{c})/f(z)$. Factorization of the ergodic density for disjoint sets of components of α means that these sets are ergodically independent (providing also, of course, that the joint range is the corresponding product range, as is usually the case). Thus l and β are ergodically independent. The ergodic density of l is not normalizable over its full range $[0,\infty)$, although it may of course be normalized over a truncated range. Generally speaking, as in equation (18), the orientation components of β are normalizable, whereas the shape components are not.

We now show how the ergodic density $l^{n(d-s)-(d+1)}$ may be normalized into an ergodic probability density in a rather natural manner. Consider a mapping

$$\alpha\to Y(\alpha)\subset E^d \tag{27}$$

where $Y(\alpha)$ may possibly be $\emptyset$, and it is supposed (writing, for brevity, $M(\alpha)$ for $M_{d-s}\{Y(\alpha)\}$):

(*a*) $M(\alpha)$ exists and is finite in $[\alpha]$;

(*b*) $M(\alpha)$ and, in a suitable sense, $Y(\alpha)$ itself are continuous in $[\alpha]$, except possibly on subvarieties of $[\alpha]$;

(*c*) $A\equiv\{\alpha\colon M(\alpha)$ is positive and continuous$\}$ has positive Lebesgue measure in $[\alpha]$.

The corresponding mapping

$$\tilde{c}=(z,\alpha)\to\{z\}+Y(\alpha) \tag{28}$$

is *translation invariant*. We shall be concerned with the aggregate

$$\mathscr{Y}_n=\{\{z\}+Y(\alpha)\colon(z,\alpha)\in\mathfrak{P}_n(s,d)\} \tag{29}$$

of random 'associated sets' generated by $\mathfrak{P}_n(s,d)$. Now define, for $\alpha\in A$, $H_n^{(m)}\{X,\delta\alpha\}$ to be the number of the $H_n\{X,\delta\alpha\}$ members of $\mathfrak{P}_n(s,d)$ with centre in X and α-value in $\delta\alpha$, whose associated set is hit by exactly m s-flats of $\mathfrak{P}(s,d)$, *excluding the n component s-flats.* Finally, write $\mathfrak{P}_n^{(m)}(s,d)$, $\mathscr{Y}_n^{(m)}$ for the corresponding subaggregates of $\mathfrak{P}_n(s,d)$, $\mathscr{Y}_n$.

The preceding ergodic theory may be repeated, utilizing the extreme Poisson independence properties of $\mathfrak{P}_n(s,d)$, the assumed continuity of $Y(\alpha)$, and Theorem 1, to show that

$$H_n^{(m)}\{X,\delta\alpha\}/H_n\{X,\delta\alpha\}\xrightarrow{\text{a.s.}}(\rho M(\alpha))^m\,\mathrm{e}^{-\rho M(\alpha)}/m!\quad(\alpha\in A) \tag{30}$$

as $|X| \to \infty$. Thus the ergodic density of $\mathfrak{P}_n^{(m)}(s, d)$ is

$$f^{(m)}(\alpha) = f(\alpha)\, M(\alpha)^m \, \mathrm{e}^{-\rho M(\alpha)} \quad (\alpha \in A). \tag{31}$$

The exponential factor clearly has a powerful effect rendering normalizable hitherto non-normalizable ergodic densities.

In applications, $Y(\alpha)$ is usually *homothetically invariant* in A, i.e.

$$Y(\alpha) = Y(l, \beta) = lY(1, \beta), \tag{32}$$

which implies

$$M(\alpha) = l^{d-s} M(1, \beta). \tag{33}$$

In this case, we may substitute $M(\alpha)$ for l in equations (19) and (31), i.e. $(l, \beta) = \alpha \to \alpha' = (M, \beta)$. There results

$$\frac{H_n^{(m)}\{X, \delta(M, \beta)\}}{|X|} \xrightarrow{\text{a.s.}} \frac{\rho^{m+n}}{m!\,(d-s)} \tag{34}$$
$$\underbrace{\underbrace{\times \frac{f(1, \beta)}{M(1, \beta)^{n-d/(d-s)}}}_{f^{(m)}(\beta)} \; \underbrace{M^{m+n-1-d/(d-s)} \, \mathrm{e}^{-\rho M} \, \delta M \, \delta\beta}_{f^{(m)}(M)}}_{f^{(m)}(M, \beta)}.$$

Thus M and β are ergodically independent, with densities as indicated. The ergodic distribution of β is in general rather complex, but on the other hand we have

Theorem 2. *In the homothetically invariant case, the ergodic distribution of the mean projections M_{d-s} of the members of $\mathscr{Y}_n^{(m)}$, given any 'meaningful' condition on β, and thus in particular no condition, is $\Gamma(m+n-d/(d-s), \rho)$.*

Theorem 2 yields a wide class of distributions corresponding to varying choices of s, d, $Y(\alpha)$ and conditions on β. Note that Theorems 1 and 2 are in a sense dual, since they give respectively $\mathbf{P}(\sharp \,|\, M_{d-s})$ and $\mathbf{P}(M_{d-s} \,|\, \sharp)$. In fact, Theorem 2 follows heuristically from the standard relation

$$\mathbf{P}(M_{d-s} \,|\, m, n) \propto \mathbf{P}(M_{d-s} \,|\, n)\, \mathbf{P}(m \,|\, M_{d-s}, n).$$

Examples of Theorem 2:

$(s, d) = (0, 1)$: the total lengths of sets of n consecutive inter-particle intervals are $\Gamma(n, \rho)$;

$= (0, 2)$: the areas of the 'empty' convex n-gons with particle vertices are $\Gamma(n-1, \rho)$;

$= (1, 2)$: the in-radii of the polygons of the random tessellation $\mathscr{P}$ are exponential (2ρ);

$= (1, 2)$: the perimeters of the n-gons of $\mathscr{P}$ are $\Gamma(n-2, \rho/\pi)$.

The reader is advised to verify these examples and perhaps construct others of his own. Note that, in applications of Theorem 2, $Y(\alpha)$ is usually *order invariant*, i.e. has the same value for all $n!$ n-figures corresponding to each n-set.

3.4.5 GENERALIZATIONS

We now consider three distinct generalizations of $\mathfrak{P}(s, d)$, which may all be simultaneously incorporated. As moreover s and d are arbitrary, a rather wide class of possible random models results.

(*a*) *Mixtures*. The *mixture*

$$\mathfrak{P}(\rho_0, \ldots, \rho_{d-1}; d) = \bigcup_{s=0}^{d-1} \mathfrak{P}(\rho_s; s, d), \tag{35}$$

where the $\mathfrak{P}(\rho_s; s, d)$ are supposed independent. Theorem 1 generalizes trivially, as do the results in Section 3.4.4 regarding independent superpositions and arbitrary sections. More interesting now is an n-figure, which comprises n_s s-flats

$$\left(0 \leqslant s < d; \quad \sum_{s=0}^{d-1} n_s = n\right).$$

Varga [27] gave various forms of the product invariant density (which he termed 'Crofton formulae') for some such mixed n-figures in E^2 and E^3. The n component flats intersect in a $\{\sum_s(sn_s) - (n-1)d\}$-flat, and there is a characteristic length l when $\sum_s(d-s)n_s > d$. The ergodic density is

$$f(l, \beta) = l^{\Sigma_s(d-s)n_s-(d+1)} f(1, \beta). \tag{36}$$

Since in general $M_1\{Y(\alpha)\}, \ldots, M_d\{Y(\alpha)\}$ are not simply interrelated, there is no satisfactory extension of Theorem 2. However, an important exception is the choice of $Y(\alpha)$ as a ball, in which case $M_{d-s} = v_{d-s} r^{d-s}$ ($0 \leqslant s \leqslant d$, r = radius). For example, the locally maximal empty balls in the interstices between the members of $\mathfrak{P}(\rho_0, \ldots, \rho_{d-1}; d)$ are 'tangential' to exactly $d+1$ flats; the radii of such balls touching n_s s-flats

$$(0 \leqslant s < d, \textstyle\sum_s n_s = d+1)$$

has ergodic p.d.f.

$$f(r) \propto r^{\Sigma_s(d-s)n_s-(d+1)} \exp\left[-\sum_s \rho_s v_{d-s} r^{d-s}\right], \tag{37}$$

a generalized gamma distribution. The radius distribution for the entire class of locally maximal empty balls is a weighted sum of distributions

(37); the weights being

(38)
$$W(n_*) = \int_{\substack{\text{locally}\\ \text{maximal}}} \lim_{|X|\to\infty} [H_{\{n_*\}}\{X, \delta r\, d\beta\}/|X|].$$

Clearly

$$W(n_*) \propto \prod_s \rho_s^{n_s}.$$

For further details, see relation (7.9) of [31].

(*b*) *Anisotropy*. Since a point has no orientation, this extension is only possible for $1 \leqslant s < d$. Parametrizing an s-subspace by the $s(d-s)$-tuple a, we have $b = (a, x^{d-s})$. Consider the anisotropic analogue

(39)
$$\hat{F}(db) = \Theta(da)\, dx^{d-s}$$

of equation (8), in which Θ is a general probability measure on $[a]$. The important property preserved from Section 3.4.4 is that $\hat{F}(B)$ is invariant under translations. The analogue of equation (11) is

(40)
$$\hat{F}(B_X) = \int |X_{d-s}(a)|\, \Theta(da) \equiv \hat{M}_{d-s}\{X\}.$$

A *uniform* Θ (random) *s-secant of* X has probability element

$$\hat{F}(db)/\hat{M}_{d-s}\{X\} \quad (b \in B_X).$$

$\mathfrak{P}(\rho; s, d; \Theta)$ is now defined in the natural way, $\sharp_X$ having a Poisson $(\rho \hat{M}_{d-s}\{X\})$ distribution (Theorem 1). The independent superposition

(41)
$$\bigcup_i \mathfrak{P}(\rho_i; s, d; \Theta_i) = \mathfrak{P}\left(\sum_i \rho_i; s, d; \sum_i \rho_i \Theta_i \middle/ \sum_i \rho_i\right),$$

and in general the section $J_t \cap \mathfrak{P}(\rho; s, d; \Theta)$ is a $\mathfrak{P}(\rho'; s+t-d, t; \Theta')$. One difference from the isotropic case should be noted. In $\mathfrak{P}(s, d)$, almost surely no two s-flats are parallel. But here, if a_0 is an atom of Θ, then a 'fraction' (in an ergodic sense) $\Theta(\{a_0\})$ of the members of $\mathfrak{P}(s, d; \Theta)$ have orientation a_0. The translational invariance means that $\mathfrak{P}(s, d; \Theta)$ is homogeneous, and so admits an ergodic theory. Turning to n-figures, the 'nth product' of equation (39) admits a similar decomposition to that in equation (19), in the sense that the independent $l^{n(d-s)-(d+1)}$ density carries over. Consequently Theorem 2 extends, $\Gamma(m+n-d/(d-s), \rho)$ now being the ergodic distribution of $\hat{M}_{d-s}$ for $\mathscr{Y}_n^{(m)}$ in the homothetically invariant case. In particular, since $\hat{M}_i\{Q_q\} = M_i\{Q_q\}$, ergodic 'ball' distributions are unchanged. Thus equation (37) extends unchanged when (*a*) and (*b*) are combined.

It emerges from Section 3.4.4 that the isotropic $\mathfrak{P}(s,d)$ is fundamental in the sense of integral geometry; on the other hand, $\mathfrak{P}(s,d;\Theta)$ is of fundamental importance as a *local limit*, as we now explain. Suppose V_0 is a mobile smooth s-dimensional variety in E^d, whose position is determined by a centre $z \in E^d$ and a d-frame $\{u_1, \ldots, u_d\}$ of orthonormal vectors emanating from z, fixed with respect to V_0. Randomizing by giving (z, u_*) the distribution $D(z, u_*)$ furnishes the *random image* V of V_0. We suppose $D(z \mid u_*)$ is continuous for all $\{u_*\}$ for which it is defined, and set

(42) $$Y = \{x \in E^d \colon 0 < \lim_{q \to 0} q^{s-d}\, \mathbf{P}[V \uparrow \{x\} + Q_q] < \infty\}.$$

Consider n independent random images $V^1, \ldots, V^N$ of V_0. It may be shown that, under the above conditions, the local limit of the system in the neighbourhood of $x \in Y$ as $N \to \infty$, under the local dilation

$$(y-x)' = N(y-x)$$

at x, is $\mathfrak{P}(\rho_x; s, d; \Theta_x)$. Both ρ_x and Θ_x depend on V_0 and D, as well as x.

Suppose that z is independent of $\{u_*\}$ and is uniform over some region R of E^d. Then, assuming edge effects near ∂R have been eliminated, the local limit is the same at all points of R. If, moreover, $\{u_*\}$ is uniform (i.e. normalized Haar measure on the d-dimensional rotation group) then the local limit is $\mathfrak{P}(s,d)$. The conditions for the local limit at x to be $\mathfrak{P}(\rho_x; s, d; \Theta_x)$ may be widened by, for instance, sampling the s-varieties from some distribution, and allowing them a certain degree of mutual dependence. Actually, $\mathfrak{P}(0,d)$ is equally fundamental as a local limit, having been discussed by Goldman [8].

(*c*) *Cylinders*. For each member J_s of $\mathfrak{P}(s,d)$, associate a random set W_{d-s} in the $(d-s)$-subspace orthogonal to J_s, the association being stochastically invariant with respect to translations of J_s. Consider the system of cylinder sets

(43) $$\mathscr{C} = \{J_s + W_{d-s} \colon J_s \in \mathfrak{P}(s,d)\}.$$

The case $s = 0$ has been considered by Takács [26] and Giger and Hadwiger [5], amongst others. If the W_{d-s} are independently and identically distributed, then the number of cylinders containing $x \in E^d$ has a Poisson $(\rho E|W_{d-s}|)$ distribution. It is then an ergodic result that the 'fraction of E^d' which is i-covered (in the sense of the limiting fraction of Q_q as $q \to \infty$) is

(44) $$p_i = (\rho \mathbf{E}|W_{d-s}|)^i \exp[-\rho \mathbf{E}|W_{d-s}|]/i! \quad (i = 0, 1, \ldots).$$

If the random sets W_{d-s} are almost surely convex, then mutual intersection probabilities of the members of $\mathscr{C}$ hitting arbitrary fixed convex sets of E^d

may be investigated by means of iteration of the complete system of kinematic formulae of integral geometry—see Streit [25]. This technique serves also to generalize Theorem 1 when, moreover, X itself is convex. Clearly independent superpositions and arbitrary sections yield corresponding Poisson cylinder systems, but equally clearly Theorem 2 does not extend. We conclude Section 3.4.5 by exploring a special case of (*c*).

Coverage and concentration

Consider N arbitrary subsets of a set X in a general space. Suppose $x \in X$ lies in $H(x)$ of these subsets, and define

(45) $$\underline{H}_X = \min_{x \in X} H(x), \quad \bar{H}_X = \max_{x \in X} H(x).$$

That is, the least- and most-covered regions of X are respectively $\underline{H}_X$- and $\bar{H}_X$-covered. Alternatively, $\underline{H}_X$ and $\bar{H}_X$ determine the overall *coverage* of X by the subsets, and the maximal *concentration* of the subsets in X, respectively. If the subsets are random, then we should like to know the joint p.m.f. of $(\underline{H}_X, \bar{H}_X)$.

Poisson disks

We now sketch the derivation of asymptotic probabilities of coverage and concentration for the special case of $\mathscr{C}$ in which disks of fixed radius r are centred at each particle of $\mathfrak{P}(0, 2)$.

(*i*) *Coverage*. A disk in E^2 is a *loc. max.*$^{(j-1)}$ disk with respect to $\mathfrak{P}(0, 2)$ if it contains $j+2$ particles, three of which lie in its perimeter circle in the form of an acute-angled triangle. Ignoring edge effects, which may be shown to be of negligible importance as $|X| \to \infty$, it is a geometrical identity that $\underline{H}_X \geqslant j$ if and only if every loc. max.$^{(j-1)}$ disk with centre in X has radius $< r$. Write $H^{(j-1)}\{X, \delta r\}$ for the number of loc. max$^{(j-1)}$ disks with centre in X and radius in $(r, r+\delta r)$. By a specialization of the theory of Section 3.4.4 it may be shown [17, Section 13] that, as $|X| \to \infty$,

$$H^{(j-1)}\{X, \delta r\}/|X| \xrightarrow{\text{a.s.}} 2\rho(\pi\rho)^{j+1} r^{2j+1} e^{-\pi\rho r^2} \delta r/(j-1)! \quad (j = 1, 2, \ldots).$$
(46)

Thus, asymptotically as $|X| \to \infty$, the aggregate of loc. max.$^{(j-1)}$ radii in X 'behaves like' an independent sample of size $j\rho|X|$ from a $\Gamma_2(2j+2, \pi\rho)$ distribution. Falsely assuming such behaviour we should have, for arbitrary positive θ,

(47) $$\mathsf{P}(\underline{H}_X \geqslant j) = (1 - p\{j, \pi\rho r^2\})^{j\rho|X|} \sim e^{-\theta}$$

as $|X| \to \infty$, provided $j\rho|X| p\{j, \pi\rho r^2\} = \theta$. This suggests

$$(48)\quad \mathbf{P}(\underline{H}_X \geqslant j) \sim \exp\left[-j\rho|X| e^{-\pi\rho r^2}\left\{1 + \ldots + \frac{(\pi\rho r^2)^j}{j!}\right\}\right] \quad (j = 0, 1, \ldots)$$

as $|X| \to \infty$. In fact, a rigorous *spatial* investigation of the homogeneous stochastic point-process of loc. max.$^{(j-1)}$ centres and their associated radii shows that our 'dependent sampling' is 'asymptotically sufficiently independent' for $\sup_{r>0}|$(left side − right side) in relation (48)$| \to 0$ as $|X| \to \infty$. The main argument is similar to that of Watson [28]. The formula (48) with $j = 1$ appears to be quite accurate even when $\rho|X|$ is as small as 100—see Gilbert [7, p. 330] for the results of a computer simulation.

(*ii*) *Concentration.* Since the method applying here may be regarded as the dual of that of (*i*), we give even sketchier details. A disk in E^2 is a *loc. min.*$_{(k-1)}$ disk with respect to $\mathfrak{P}(0, 2)$ if it contains $k+1$ particles, two of which lie in the perimeter circle at the ends of a diameter. Ignoring edge effects, $\bar{H}_X \leqslant k$ iff every loc. min.$_{(k-1)}$ and loc. max.$^{(k-2)}$ disk with centre in X has radius $> r$. Write $H_{(k-1)}\{X, \delta r\}$ for the total number of either of these types with radius in $(r, r+\delta r)$. Further specialization of Section 3.4.4 [17, Section 13] shows that, as $|X| \to \infty$,

$$(49)\quad H_{(k-1)}\{X, \delta r\}/|X| \xrightarrow{\text{a.s.}} 2(k+1)\,\rho(\pi\rho)^k r^{2k-1} e^{-\pi\rho r^2}\,\delta r/(k-1)! \qquad (k = 1, 2, \ldots).$$

Thus $H_{(k-1)}\{X, \delta r\}$ approximates to a $(k+1)\,\rho|X|$-sample from a $\Gamma_2(2k, \pi\rho)$ distribution. In analogy with equation (47),

$$(50)\qquad \mathbf{P}(\bar{H}_X \leqslant k) = (1 - q\{k-1, \pi\rho\})^{(k+1)\rho|X|},$$

suggesting

$$(51)\quad \mathbf{P}(\bar{H}_X \leqslant k) \sim \exp\left[-(k+1)\,\rho|X| e^{-\pi\rho r^2}\left\{\frac{(\pi\rho r^2)^k}{k!} + \ldots\right\}\right] \quad (k = 1, 2, \ldots).$$

This is in fact true in the same sense as the relation (48), and with a similar justification.

These results generalize to higher dimensions. For Poisson spheres in E^3,

$$(52)\quad \begin{cases} \mathbf{P}(\underline{H}_X \geqslant j) \sim \exp[-(3\pi^2/32)\,j(j+1)\,\rho|X| p\{j+1, 4\pi\rho r^3/3\}], \\ \mathbf{P}(\bar{H}_X \leqslant k) \sim \exp[-\{4 + (3/16)(\pi^2+16)(k-1) + (3\pi^2/32)(k-1) \\ \qquad (k-2)\}\,\rho|X| q\{k-1, 4\pi\rho r^3/3\}]. \end{cases}$$

It is hoped that the formulae (48), (51) and (52) may prove useful in statistical applications; for instance, in testing the hypothesis of independent uniformity for point-process data. A corresponding pair of formulae for general d may be derived. The interesting derivation utilizes relations (66) and (67) below together with iteration of the general kinematic formulae for spaces of constant curvature, given by Santaló [23]. Incidentally, this iteration implies Wendel's [29] formula giving the probability that N independent isotropic random hemispheres on a sphere in E^d completely cover that sphere ($N = 2, 3, \ldots$).

3.4.6 RANDOM TESSELLATIONS

A *tessellation* in E^d is defined to be an aggregate of convex polytopes which cover E^d without overlapping. (Henceforth we omit 'convex', since all polytopes considered are in fact convex.) In this final section we investigate certain natural random tessellations generated by homogeneous Poisson flat systems.

But first consider a general random tessellation $\mathscr{T}$ in E^d for which a probability space with all necessary regularity properties has been established. Our examples are all of this type, since each depends in a simple way upon its underlying $\mathfrak{P}(s, d)$. A natural desirable property of $\mathscr{T}$ is homogeneity, i.e. stochastic invariance under translations. This is implied by isotropy, i.e. stochastic invariance under rotations. Note that arbitrary sections by a flat are random tessellations with corresponding properties. For a polytope T, let $Z = (Z_1, \ldots, Z_m)$ be a partial (or even complete) *description* of its Euclidean invariant properties. For us, important possible components of Z are $\{M_*\}$; $\{N_*\}$, where N_s is the number of s-facets (s-dimensional polytope facets) in ∂T; $\{L_*\}$, where L_s is the sum of the s-contents of the N_s s-facets; and I, the in-radius. It is convenient to write $V \equiv L_d (= M_d)$, $S \equiv L_{d-1}$ and $N \equiv N_0 (= L_0)$. The polytope T is the convex hull of its set of N 0-facets or vertices. The standard reference on convex polytopes is Grünbaum [9].

We now sketch the general procedure required to establish ergodic distributions for $\mathscr{T}$. Suppose Z is an arbitrary description and that Z' is a 'particular value' in [Z]. Define H_q to be the number of polytopes of $\mathscr{T}$ 'within' Q_q, and $H_q(Z')$ to be the number of these for which $Z_i \leqslant Z_i'$ $(1 \leqslant i \leqslant m)$. We write 'within' here on the supposition that edge effects due to polytopes which hit ∂Q_q may be shown by *ad hoc* means to be of negligible importance as $q \to \infty$. The homogeneity of $\mathscr{T}$ implies the homogeneity of stochastic processes of the type $\{Y(x)\}$ $(x \in E^d)$, where $Y(x)$ is the value of the description Y for the polytope T_x of $\mathscr{T}$ containing

x. The homogeneity allows the application of Wiener's d-parameter ergodic theorem [30, Theorem II″] to the empiric averages

$$\int_{Q_q} Y(x)\,dx/|Q_q|$$

of such processes. Such application to $Y_1(x) = 1/V(x)$ and

$$Y_2(x) = K_{Z'}(x)/V(x),$$

where the indicator random variable $K_{Z'}(x)$ indicates the event

$$[Z_i(x) \leqslant Z_i' \quad (1 \leqslant i \leqslant m)]$$

ensures the existence of the almost sure limits $H, H(Z')$ of $H_q/|Q_q|$, $H_q(Z')/|Q_q|$, respectively, as $q \to \infty$. In the important metrically transitive case these limits, which are in general random, degenerate to constants. However, although in practice metrical transitivity is difficult to prove directly, the demonstration of the 'asymptotic independence of $\mathscr{T}$ in distant localities of E^d', a sort of d-dimensional mixing condition, suffices to ensure the constancy of H and $H(Z')$. Then

$$H_q(Z')/H_q \xrightarrow{\text{a.s.}} H(Z')/H \equiv F(Z'), \tag{53}$$

the value of the *ergodic d.f. of Z for $\mathscr{T}$* at the particular value Z'.

The general problem is the determination of $F(Z)$ for the important and natural descriptions Z. Write $G(Z)$ for the usually well-defined d.f. of Z for the polytope containing an arbitrary point of E^d, e.g. T_0. A by-product of the above ergodic theory is the basic relation

$$F(dV, dZ) = G(dV, dZ)/V \int V^{-1} G(dV), \tag{54}$$

which may be derived heuristically by regarding, on account of homogeneity, the origin o to be a 'random point in E^d'. This relation is important, if only because G is usually more accessible than F. Another way of expressing equation (54) is to say that T_0 is a 'Vf' polytope, where f represents the 'density' (i.e. combined p.m.f./p.d.f.) of F. Then a 'random polytope of $\mathscr{T}$' is an 'f' polytope. Another reassuring by-product of the ergodic theory is that the empiric mean of scalar Z (the quotient of the empiric averages of $Y_3(x) = Z(x)/V(x)$ and $Y_1(x)$) converges to the ergodic mean:

$$\mathsf{E}_q(Z) \xrightarrow{\text{a.s.}} \mathsf{E}(Z) = \int Z\,F(dZ) \quad \text{as } q \to \infty. \tag{55}$$

Example 1. *For $d = 2$, $\mathbf{E}(N) = 2\chi/(\chi-2)$ where χ is the mean number of sides meeting at each vertex of $\mathcal{T}$ (for further details, see Matachinski* [11] *and Miles* [17, §10]).

Example 2. *For $\mathscr{P}$ when $d = 2$,*

$$\mathbf{P}(\text{a random polygon is triangular}) = 2-(\pi^2/6) = 0.3551$$

while

$$\mathbf{P}(T_0 \text{ is triangular}) = (\pi^2/6)(25-36 \ln 2) = 0.0768$$

(see [33]).

The random tessellation $\mathscr{P}$.

Let us combine the elements (*b*), (*c*) of Section 3.4.5 in the case $s = d-1$. Suppose (b, w) represents the *hyperslab* containing all points of E^d whose perpendicular distance from the hyperplane b is at most w; b, w are its *mid-hyperplane* and *semi-thickness*, respectively. Let Φ be a general probability distribution in (a, w) for which $\mathbf{E}(w) < \infty$ and the marginal distribution of a is Θ. Generate in the usual way the system $\mathfrak{P}(\rho; d-1, d; \Phi)$ of hyperslabs. Thus its members are $\{(p_i, u_i, w_i)\}$ $(i = 1, 2, \ldots)$, where

(*i*) the perpendiculars $\{p_i\}$ from o to the mid-hyperplanes constitute a $\mathfrak{P}(0, 1)$ of intensity 2ρ on $[0, \infty)$;

(*ii*) independently of (*i*), $\{(u_i, w_i)\}$ are independently sampled from Φ.

Write $\mathscr{P}^+$ for the aggregate of polytope interstices between these hyperslabs. The system of mid-hyperplanes, which constitutes a $\mathfrak{P}(\rho; d-1, d; \Theta)$, has the effect of partitioning E^d into a random tessellation, $\mathscr{P}$ say. (Clearly, for $\mathscr{P}^+$ and $\mathscr{P}$ to exist, it is necessary that Θ be not 'too degenerate'.) Using Grünbaum's [9] terminology, almost surely every polytope of $\mathscr{P}^+$ and $\mathscr{P}$ is *simple*, i.e. each s-facet lies in the intersection of $d-s$ $(d-1)$-facets $(0 \leqslant s < d)$. Further, each s-facet of $\mathscr{P}$ lies in the boundaries of 2^{d-s} members of $\mathscr{P}$ $(0 \leqslant s < d)$. Now, given that o is not covered by any member of $\mathfrak{P}(d-1, d; \Phi)$, let T_0^+ be the member of $\mathscr{P}^+$ containing it. It may be shown that T_0^+ (under this condition) and T_0 (with no conditions imposed) have the same stochastic construction. Consequently, since equation (54) applies to both $\mathscr{P}^+$ and $\mathscr{P}$, *the ergodic distributions of $\mathscr{P}^+$ and $\mathscr{P}$ are identical*! Thus, for example, Theorem 2 and (*b*) of Section 3.4.5 imply that, for $\mathscr{P}^+$, the ergodic distribution of I is exponential (2ρ), and the conditional ergodic distribution of $\hat{M}_{d-s}$ given N_{d-s} is $\Gamma(N_{d-s}-d, \rho)$.

Although attention is restricted henceforth to $\mathscr{P}$ generated by $\mathfrak{P}(d-1, d)$, it should be borne in mind that the following results apply equally to associated $\mathscr{P}^+$. The invariant density element for a hyperplane in E^d is

$f(b)\,db = (2/\sigma_d)\,dp\,dO$. The corresponding product density element for a d-figure of hyperplanes is

$$(2/\sigma_d)^d \prod_1^d dp_i\,dO_i.$$

If z is their common point, then $p_i = z.u_i$ $(1 \leqslant i \leqslant d)$, and so

$$\left|\frac{\partial p_*}{\partial z}\right| = \text{modulus of } |u_1 \dots u_d| \equiv \Lambda_d(u_*) \tag{56}$$

say, the d-content of the parallelotope with edges $\{u_*\}$. Thus the product invariant density element is alternatively

$$(2/\sigma_d)^d \Lambda_d(u_*)\,dz \prod_1^d dO_i,$$

from which it follows that the ergodic p.d.f. of $\{u_*\}$ at the vertices of $\mathscr{P}$ is

$$\phi(u_*) = \Lambda_d(u_*) \Big/ \int \Lambda_d(u_*)\,dO_1 \dots dO_d. \tag{57}$$

The product invariant density element for a $(d+1)$-figure of hyperplanes is

$$(2/\sigma_d)^{d+1} \prod_0^d dp_i\,dO_i.$$

Write y, I for the in-centre and in-radius of the simplex so formed. Define $u_i' = \pm u_i$ so that the feet of the perpendiculars from y to the hyperplanes are $\{y + Iu_*'\}$. Then $|p_i| = y.u_i' + I$ $(0 \leqslant i \leqslant d)$, and so

$$\left|\frac{\partial p_*}{\partial(y, I)}\right| = \text{modulus of} \begin{vmatrix} 1 & \dots & 1 \\ u_0' & \dots & u_d' \end{vmatrix} \equiv \nabla_d(u_*'), \tag{58}$$

say, which is $d!$ times the d-content $\nabla_d(u_*')$ of the simplex with vertices at the points $u_i' \in \partial Q_1$. Thus the product invariant density element is alternatively

$$(2/\sigma_d)^{d+1} \nabla_d(u_*')\,dy\,dI \prod_0^d dO_i,$$

from which it follows that the in-simplices of $\mathscr{P}$ have ergodic p.d.f.

$$f(I, u_*) = 2\rho\,e^{-2\rho I} . \nabla_d(u_*) \Big/ \int_K \nabla_d(u_*)\,dO_0 \dots dO_d \tag{59}$$

restricted to

$$K = \{\{u_*\}\text{: there is no hemisphere of } \partial Q_1 \text{ containing all the } u_i\}.$$

An 'f' polytope may be constructed about its in-centre by means of equation (59), the remaining $\{N_{d-1}-(d+1)\}$ $(d-1)$-facets being determined by a $\mathfrak{P}(d-1, d)$ restricted so that none of its members hit the already determined in-ball. Actually, the joint orientation densities of equations (57) and (59) extend to the anisotropic case upon weighting by $\prod \Theta(du_i)$.

Suppose v is an arbitrary unit vector. Almost surely the relation 'extreme point of a polytope in the direction v' sets up a (1, 1) correspondence between the vertices and the members of $\mathscr{P}$. Hence, since each vertex is almost surely a vertex of 2^d polytopes, it is clear that $\mathbf{E}(N) = 2^d$. This essentially geometrical property extends to the anisotropic case. The stochastic construction of an 'f' polytope with respect to its 'extreme v point' is clear:

(*i*) construct $\mathfrak{P}(d-1, d)$;

(*ii*) independently construct d random hyperplanes through o with joint distribution (57) and, of the 2^d convex polytopal cones into which E^d is thereby partitioned, let C_0 be the a.s. unique one having o as extreme v point;

(*iii*) $T_0 \cap C_0$ is then an 'f' polytope.

Given the 2^d cones *up to a random rotation* (normalized Haar measure), they do not have equal chances of being C_0. Define the polar angle of the convex cone C (apex o) to be the angle of the convex polar cone

$$C^p = \{x^p : x^p \,.\, x \geqslant 0, \text{ all } x \in C\}.$$

Then the chance a given one of these cones is selected as C_0 in the random rotation is proportional to its polar angle. For example, when $d = 2$, the angles at a vertex have common p.d.f. $\frac{1}{2}\sin\theta$, whereas the polygon angle at its extreme v point has p.d.f. $\{1-(\theta/\pi)\}\sin\theta \, (0 \leqslant \theta \leqslant \pi)$.

Denote the t-facets of a polytope T by $T_{t,i}$ $(1 \leqslant i \leqslant N_t)$. Denote the mean s-projection of $T_{t,i}$, *with respect to the t-flat containing it*, by $M_{s,t}\{T_{t,i}\}$, and define

$$Y_{s,t}\{T\} = \sum_{i=1}^{N_t} M_{s,t}\{T_{t,i}\} \quad (0 \leqslant s \leqslant t \leqslant d). \tag{60}$$

Then the edge elements of the triangular array $\{Y_{s,t}\}$ are

$$Y_{s,s} = L_s, \quad Y_{s,d} = M_s, \quad Y_{0,s} = N_s \quad (0 \leqslant s \leqslant d). \tag{61}$$

Miles [13] has shown that, for $\mathscr{P}$ with respect to $\mathfrak{P}(\rho; d-1, d)$,

$$\mathbf{E}(Y_{s,t}) = \left\{2^{d+s-t}\binom{d}{t}\Gamma(\tfrac{1}{2}t+1)/\Gamma(\tfrac{1}{2}[t-s]+1)\right\} \times \{\Gamma(\tfrac{1}{2}[d+1])/\Gamma(\tfrac{1}{2}d)\rho\}^s \quad (0 \leqslant s \leqslant t \leqslant d), \tag{62}$$

and

$$\mathsf{E}(L_r L_s) = \frac{2^d \pi^{\frac{1}{2}}}{\Gamma(\frac{1}{2}[r+1])\,\Gamma(\frac{1}{2}[s+1])} \left\{\frac{\Gamma(\frac{1}{2}[d+1])}{\Gamma(\frac{1}{2}d)\,\rho}\right\}^{r+s}$$

$$\times \sum_{i=\max(r,s)}^{d} \binom{d}{i} \left(\frac{\pi}{2}\right)^i \frac{\Gamma(\frac{1}{2}[i+1])}{\Gamma(\frac{1}{2}i+1)} (i)_r (i)_s \quad (0 \leqslant r \leqslant s \leqslant d). \tag{63}$$

The first- and second-order ergodic moments (62) and (63) allow the variance–covariance matrix of $(L_0, \ldots, L_d)$, and thus in particular (N, S, V), to be evaluated. The moments $\mathsf{E}(N_*)$ in equation (62) may be obtained by a limiting process from a result of Cover and Efron [4, Theorem 1′].

Note that, for the mixture $\mathfrak{P}(\rho_0, \rho_{d-1}; d)$,

$$\mathsf{P}(\text{a 'random polygon of } \mathscr{P}\text{' contains no particles}) = \int_0^\infty e^{-\rho_0 V}\, dF_{\rho_{d-1}}(V), \tag{64}$$

the Laplace–Stieltjes transform of the ergodic d.f. of V for $\mathscr{P}$. This offers a possible combinatorial method of investigating this perhaps most important ergodic distribution of $\mathscr{P}$.

See [15, 33] for a more detailed discussion of the planar case, and [24] for the generalization of this to the hyperbolic plane.

The random tessellations $\mathscr{V}$, $\mathscr{D}$ *and* $\mathscr{V}_n$, $\dot{\mathscr{V}}_n$ $(n = 1, 2, \ldots)$ *generated by* $\mathfrak{P}(0, d)$

$\mathscr{V}$: Label the particles of $\mathfrak{P}(0, d)$ by y_*; for example, y_i might be the ith nearest particle to o $(i = 1, 2, \ldots)$.

$$T_i = \{x \in E^d : |x - y_i| \leqslant |x - y_j|, j \neq i\} \tag{65}$$

is almost surely a simple polytope, and $\mathscr{V} = \{T_*\}$ is a random tessellation—the *Voronoi* tessellation generated by $\mathfrak{P}(0, d)$ (see Rogers [20, Chap. 7]). Each s-facet lies in the s-flat of points which are equidistant from a set of $d-s+1$ particles. Thus, unlike $\mathscr{P}$, each s-facet lies in the boundaries of $d-s+1$ members of $\mathscr{V}$. In the 'practical' cases $d = 2, 3$, the first-order ergodic moments of V, S and $\{N_*\}$ were determined by Meijering [12], while Gilbert [6] evaluated the second-order moments of V by computer calculation of definite integrals.

Blaschke [3] and Petkantschin [19] independently obtained the form

$$dx_0 \ldots dx_s = \nabla_s(x_*)^{d-s}\, dJ_s\, dx_0^s \ldots dx_s^s \tag{66}$$

of the density of $(s+1)$-figures of points in E^d. Here $\nabla_s(x_*)$ is $s!$ times the s-content $\Delta_s(x_*)$ of the s-simplex with vertices $\{x_*\}$, J_s is the s-flat containing $\{x_*\}$, and $\{x_*^s\}$ are the coordinates of these points with respect to J_s. This served as the basis of a study by Kingman [10] of the random s-flat containing $s+1$ independent uniform random points of a convex body in E^d; in particular, he solved Sylvester's classical problem for a d-ball. In fact, by means of equation (66), all the moments $\mathbf{E}(\Delta_s^k)$ of the s-content Δ_s of the simplicial convex hull of an $(s+1)$-sample from certain spherically symmetric d-dimensional probability distributions (and distributions obtained by affine transformation from such distributions) may be determined; the $s+1$ sample points may even be dependent, but full spherical symmetry in E^d must be preserved. For example, if

$$x_1, \ldots, x_{r+s} \; (r \geqslant 0, s \geqslant 0, 2 \leqslant r+s \leqslant d+1)$$

are independent, $x_1, \ldots, x_r$ and $x_{r+1}, \ldots, x_{r+s}$ being uniformly distributed in Q_q and ∂Q_q, respectively, then

$$\mathbf{E}(\Delta_{r+s-1}^k) = \frac{q^{r+s-1}}{(r+s-1)!} \frac{\Gamma(\frac{1}{2}[r+s]\,[d+k]-s+1)}{\Gamma(\frac{1}{2}[(r+s)(d+k)-k]-s+1)}$$

$$\times \left\{\frac{\Gamma(\frac{1}{2}d+1)}{\Gamma(\frac{1}{2}[d+k]+1)}\right\}^r \left\{\frac{\Gamma(\frac{1}{2}d)}{\Gamma(\frac{1}{2}[d+k])}\right\}^s \prod_{i=d-r-s+2}^{d} \left\{\frac{\Gamma(\frac{1}{2}[k+i])}{\Gamma(\frac{1}{2}i)}\right\}. \tag{67}$$

Again, if $x_0, \ldots, x_s$ is an independent $(s+1)$-sample from the general d-dimensional normal distribution $\mathcal{N}(\mu, \Sigma)$, then

$$\mathbf{E}(\Delta_s^k) = \{(s+1)^{\frac{1}{2}}(2|\Sigma|^{1/d})^{s/2}/s!\}^k \prod_{i=d-s+1}^{d} \{\Gamma(\tfrac{1}{2}[k+i])/\Gamma(\tfrac{1}{2}i)\} \quad (1 \leqslant s \leqslant d). \tag{68}$$

A useful variant of equation (66) is

$$dx_1 \ldots dx_s = \Lambda_s(x_*)^{d-s}\, dJ_{s(o)}\, dx_1^s \ldots dx_s^s, \tag{69}$$

in which $\Lambda_s(x_*)$ is the s-content of the parallelotope with sides $\{x_*\}$, $J_{s(o)}$ is the s-subspace spanning $\{x_*\}$, and $\{x_*^s\}$ are the coordinates of these points with respect to $J_{s(o)}$. Using equation (69), it may similarly be proved that, if $\{x_*\}$ is an s-sample from $\mathcal{N}(\mu, \Sigma)$, $(s+1)^{\frac{1}{2}}$ times the s-content of the simplicial convex hull of o and $\{x_*\}$ also has the moments (68). Although this latter result is well known, equation (68) itself seems new. Of course (see, for example, Anderson [1, §7.5]) the common distribution is also the distribution of the 'generalized sample variance' of a $(d+1)$-sample from an s-dimensional normal distribution. The distributions of Δ_1 and Δ_2 in equation (68) are known. They are $\Gamma_2(d, \frac{1}{4}|\Sigma|^{1/d})$ and $\Gamma(d-1, 2/\sqrt{3}|\Sigma|^{1/d})$ respectively. The contents of this paragraph are fully discussed in [32].

We shall now show that

$$dx_0 \ldots dx_d = R^{d^2-1}\, \nabla_d(u_*)\, dz\, dR\, dO_0 \ldots dO_d, \tag{70}$$

where z, R are the circum-centre and circum-radius of $\{x_*\}$:

$$x_i = z + Ru_i \quad (0 \leqslant i \leqslant d).$$

If the foot of the perpendicular from o to the hyperplane (p_i, u_i) is x_i, then

$$dx_0 \dots dx_d = \prod_{i=0}^{d} p_i^{d-1}\, dp_i\, dO_i \tag{71}$$

which, by equation (58),

$$= \prod_{i=0}^{d} p_i(y, I, u_*)^{d-1}\, dO_i \,.\, \nabla_d(u_*)\, dy\, dI,$$

where y, I are the in-centre and in-radius, respectively, of the simplex formed by $\{(p_*, u_*)\}$,

$$\equiv \varphi(y, I, u_*)\, dy\, dI \prod_{i=0}^{d} dO_i,$$

say. But clearly

$$dx_0 \dots dx_d = \psi(R, u_*)\, dz\, dR \prod_{i=0}^{d} dO_i \tag{72}$$

for some function ψ, since z must be uniform and independent of (R, u_*). Now $y = o$ iff $z = o$, in which case $I = R$. Moreover, $dy = dz$ and $dI = dR$ at $y = z = o$. (This may be demonstrated geometrically in the case in which an arbitrary p_i is adjusted by dp_i, leaving the remaining p_i and all the u_i fixed. It follows that it is also true when all the p_i are adjusted by dp_i together.) Hence, by equations (71) and (72),

$$\psi(R, u_*) = \varphi(o, R, u_*) = R^{d^2-1}\, \nabla_d(u_*), \tag{73}$$

which completes the derivation of equation (70).

Combining equations (66) and (70), we obtain

$$dx_0 \dots dx_s = \nabla_s(u_*^s)^{d-s+1}\, R^{ds-1}\, dz\, dJ_{s(o)}\, dR\, dO_0^s \dots dO_s^s, \tag{74}$$

where z, R are the circum-centre and circum-radius of $\{x_*\}$ in the s-flat $\{z\} + J_{s(o)}$ containing these points. This relation, assisted by equation (67), is tailor-made for the determination of the ergodic moments

$$\mathbf{E}(L_s) = \frac{2^{d-s+1}\pi^{\frac{1}{2}(d-s)}\,\Gamma(\frac{1}{2}[d^2-ds+s+1])\,\Gamma(\frac{1}{2}d+1)^{d-s+s/d}\,\Gamma(d-s+s/d)}{(d-s)!\,d\Gamma(\frac{1}{2}[d^2-ds+s])\,\Gamma(\frac{1}{2}[d+1])^{d-s}\,\Gamma(\frac{1}{2}[s+1])\,\rho^{s/d}}$$

$$(0 \leqslant s \leqslant d) \tag{75}$$

of $\mathscr{V}$.

$\mathscr{D}$: Each vertex of $\mathscr{V}$ is the circum-centre of a set of $d+1$ particles of $\mathfrak{P}(0, d)$, the convex hull of which is a simplex. The aggregate of such random simplices is a tessellation—the *Delaunay* tessellation $\mathscr{D}$ (for a verification, see Rogers [20, Chap. 8]). On account of equation (70), the ergodic p.d.f. of $\mathscr{D}$

$$f(R; u_0, \ldots, u_d) \propto e^{-\rho v_d R^d} R^{d^2-1} \Delta_d(u_*). \tag{76}$$

Thus R is independent of $\{u_*\}$ and has a $\Gamma_d(d^2, \rho v_d)$ distribution. By means of the relation (76), the formula $V = R^d \Delta_d(u_*)$ and the moments (67), we obtain all the ergodic moments

$$\mathbf{E}(V^k) = \frac{(d+k-1)!\,\Gamma(\tfrac{1}{2}d^2)\,\Gamma(\tfrac{1}{2}[d^2+dk+k+1])\,\Gamma(\tfrac{1}{2}[d+1])^{d-k+1} \times \prod_{i=2}^{d+1}\{\Gamma(\tfrac{1}{2}[k+i])/\Gamma(\tfrac{1}{2}i)\}}{(d-1)!\,\Gamma(\tfrac{1}{2}[d^2+1])\,\Gamma(\tfrac{1}{2}[d^2+dk])\,\Gamma(\tfrac{1}{2}[d+k+1])^{d+1} \quad (2^d \pi^{\frac{1}{2}[d-1]} \rho)^k} \quad (k = 1, 2, \ldots) \tag{77}$$

of V for $\mathscr{D}$. For $d = 1$, $\mathscr{D} \equiv \mathscr{P}$ and V is exponential (ρ).

$\mathscr{V}_n$, $\dot{\mathscr{V}}_n$ $(n = 1, 2, \ldots)$: An arbitrary point of E^d almost surely possesses a set of n nearest particles; the points possessing the same set of n nearest particles form a simple polytope; the aggregate of such polytopes is defined to be $\mathscr{V}_n$. The random tessellation $\mathscr{V}_n$ is rather similar to $\mathscr{V} \equiv \mathscr{V}_1$, in that each s-facet is a facet of $d-s+1$ polytopes of $\mathscr{V}_n$. In fact, equation (74) implies that the joint orientation p.d.f. of the $d-s+1$ particles equidistant from an s-facet of $\mathscr{V}_n$ is given by

$$f(u_0^{d-s}, \ldots, u_{d-s}^{d-s}) \propto \nabla_{d-s}(u_*^{d-s})^{s+1} \quad (0 \leqslant s \leqslant d-2). \tag{78}$$

The joint orientation of the $\binom{d-s+1}{2}$ $(d-1)$-facets meeting at this s-facet may in theory be deduced from the relation (78).

Taking into account the *order* of the n nearest particles, we obtain $\dot{\mathscr{V}}_n$, a 'refinement' of, and more complex than, $\mathscr{V}_n$. In fact, denoting the union of polytope boundaries of a tessellation $\mathscr{T}$ by $\partial\mathscr{T}$, we have

$$\partial\dot{\mathscr{V}}_n = \partial\mathscr{V}_1 \cup \ldots \cup \partial\mathscr{V}_n. \tag{79}$$

It may be shown that the local limit (see Section 3.4.5) as $n \to \infty$ of $\dot{\mathscr{V}}_n$ at an arbitrary point of E^d is $\mathscr{P}$ with respect to $\mathfrak{P}(\rho_n; d-1, d)$, where

$$\rho_n = \{2^{3d-1} d!\,\Gamma(\tfrac{1}{2}d+1)^{2-1/d}/\pi^{\frac{1}{2}}(2d)!\}\,\rho^{1/d} n^{2-1/d}. \tag{80}$$

The values of each of $\mathbf{E}(V)$, $\mathbf{E}(S)$ and $\mathbf{E}(N)$ in the case $d = 2$ are given in [17, §10].

REFERENCES

1. T. W. Anderson, *An Introduction to Multivariate Statistical Analysis*, Wiley, New York (1958).
2. W. Blaschke *Integralgeometrie 1. Ermittlung der Dichten für lineare Unterräume im E_n*, Hermann, Paris (1935) (Act. Sci. Indust. No. 252).
3. ——, "Integralgeometrie 2. Zu Ergebnissen von M. W. Crofton", *Bull. Math. Soc. Roumaine des Sci.* **37** (1935), 3–11.
4. T. M. Cover and B. Efron, "Geometrical probability and random points on a hypersphere", *Ann. math. Statist.* **38** (1967), 213–220.
5. H. Giger and H. Hadwiger, Über Treffzahlwahrscheinlichkeiten im Eikörperfeld", *Ztschr. Wahrsch'theorie & verw. Geb.* **10** (1968), 329–334.
6. E. N. Gilbert, "Random subdivisions of space into crystals", *Ann. math. Statist.* **33** (1962), 958–972.
7. ——, "The probability of covering a sphere with *N* circular caps", *Biometrika* **52** (1965), 323–330.
8. J. R. Goldman, "Stochastic point-processes: limit theorems", *Ann. math. Statist.* **38** (1967), 771–779.
9. B. Grünbaum, *Convex Polytopes*, Wiley, New York (1967).
10. J. F. C. Kingman, "Random secants of a convex body", *J. Appl. Prob.* **6** (1969), 660–672.
11. M. Matschinski, "Considérations statistiques sur les polygones et les polyèdres", *Publ. Inst. Stat. Univ. Paris* **3** (1954), 179–201.
12. J. L. Meijering, "Interface area, edge length, and number of vertices in crystal aggregates with random nucleation", *Philips Res. Rep.* **8** (1953), 270–290.
13. R. E. Miles, *Random polytopes: the generalisation to n dimensions of the intervals of a Poisson process*, Ph.D. thesis, Cambridge University (1961).
14. ——, "A wide class of distributions in geometrical probability" (abstract), *Ann. math. Statist.* **35** (1964), 1407.
15. ——, "Random polygons determined by random lines in a plane", *Proc. Nat. Acad. Sci. U.S.A.* **52** (1964), 901–907; II, 1157–1160.
16. ——, "Poisson flats in Euclidean spaces. Part I: A finite number of random uniform flats", *Adv. Appl. Prob.* **1** (1969), 211–237.
17. ——, "On the homogeneous planar Poisson point-process", *Mathematical Biosciences* **6** (1970), 85–127.
18. L. Nachbin, *The Haar Integral*, van Nostrand, Princeton (1965).
19. B. Petkantschin, "Integralgeometrie 6. Zusammenhänge zwischen den Dichten der linearen Unterräume im *n*-dimensionalen Raum", *Abh. Math. Seminar Hamburg* **11** (1936), 249–310.
20. C. A. Rogers, *Packing and Covering*, Cambridge University Press (1964) (Math. Tract No. 54).
21. L. A. Santaló, *Introduction to Integral Geometry*, Hermann, Paris (1953) (Act. Sci. Indust. No. 1198).
22. ——, "Sur la mesure des espaces linéaires qui coupent un corps convexe et problèmes qui s'y rattachent", *Colloque sur les Questions de Réalité en Géométrie, Liège*, pp. 177–190, Georges Thone, Liège; Masson et Cie, Paris (1955).

23. ——, "Sobre la formula fundamental cinematica de la geometria integral en espacios de curvatura constante", *Mathematicae Notae* **18** (1962), 79–94.
24. ——, "Valores medios para poligonos formados por rectas al azar en el plano hiperbolico", *Universidad Nacional de Tucuman, Revista Ser. A* **16** (1966), 29–43.
25. F. Streit, "On multiple integral geometric integrals and their applications to probability theory", *Canadian J. Math.* **22** (1970), 151–163.
26. L. Takács, "On the probability distribution of the measure of the union of random sets placed in a Euclidean space", *Ann. Univ. Sci. Budapest., Eötvös Sect. Math.* **1** (1958), 89–95.
27. O. Varga, "Integralgeometrie 3. Croftons Formeln für den Raum", *Math. Z.* **40** (1935), 387–405.
28. G. S. Watson, "Extreme values in samples from *m*-dependent stationary stochastic processes", *Ann. math. Statist.* **25** (1954), 798–800.
29. J. G. Wendel, "A problem in geometric probability", *Math. Scand.* **11** (1962), 109–111.
30. N. Wiener, "The ergodic theorem", *Duke math. J.* **5** (1939), 1–18.

More recent references.

31. R. E. Miles, "Poisson flats in Euclidean spaces. Part II: Homogeneous Poisson flats and the complementary theorem", *Adv. Appl. Prob.* **3** (1971), 1–43.
32. ——, "Isotropic random simplices", *Adv. Appl. Prob.* **3** (1971), 353–382.
33. ——, "The various aggregates of random polygons determined by random lines in a plane", *Advances in Math.*, to appear.

3.5

On the Elimination of Edge Effects in Planar Sampling

R. E. MILES

3.5.1 SUMMARY

Practical weighting procedures are proposed to compensate for the edge effect bias inherent in sampling randomly shaped domains located at random in the plane. The family of parallel domains of a domain, and a dual such family for a smooth convex domain, arise naturally, yielding interesting connections with integral geometry.

3.5.2 INTRODUCTION

It appears that little or no attention has hitherto been paid to the problem of overcoming edge effects in planar sampling.

The underlying stochastic structure is a 'domain process', i.e. domains scattered as a random process throughout the plane with a constant intensity (number per unit area) ρ. The individual domains are themselves of random sizes and shapes, so that 'descriptions', i.e. Euclidean invariant quantities such as area, perimeter, etc., possess well-defined 'ergodic' distributions, with d.f.'s $F(\cdot)$. The general problem considered is the estimation of ρ and/or various $F(\cdot)$, or associated distributional quantities, from a limited record (or realization) of the process.

The 'natural' solution is to evaluate descriptions for those domains lying 'within' some arbitrarily chosen region, and to form the corresponding 'natural' estimates. However, edge effects are immediately evident since it must first be decided which domains to count: only those completely contained in the sampling region ('minus' sampling), those intersecting the sampling region ('plus' sampling), or some other choice. In both plus and minus sampling, the natural estimates have definite biases. In minus sampling a heuristic reason for this is as follows. Since the sampling region, being arbitrary, is essentially randomly located in the plane, the smaller domains have the best chance of being enclosed, and sufficiently large domains no chance. (Roughly speaking, in plus sampling this bias is reversed.) The emphasis in this article is on minus sampling, since it proved to be the best method in a recent application. For minus sampling in a circular region, a practical weighting procedure is proposed which compensates for the bias, and which characteristically yields consistent estimators under replication of sampling circles. Each domain is weighted by M^{-1}, where the description M is the 'R-measure' of the domain, R being the radius of the sampling circle. The general method employed may well find wider applicability to other problems involving edge effects.

Domains, domain descriptions and domain processes are introduced in Section 3.5.3. Working from two plausible assumptions, the basic 'unbiased' estimation formulae are established in Section 3.5.4. The R-measure M enters fundamentally into these formulae, and Section 3.5.5 is devoted to developing formulae for M which permit its value to be easily calculated for arbitrary domains. The two main areas of practical application, in Monte Carlo studies and to the analysis of sample data, are discussed in Section 3.5.6. The Poisson independent case is especially tractable, with unbiased estimators of the variances of certain estimators being readily available. The advantages of replication, and the consequent design problem of the size and location of the sampling circles, are considered; *ad hoc* considerations are often important. Section 3.5.7 treats plus sampling, and considers its advantages and disadvantages as compared with minus sampling. Formulae are given in Section 3.5.8 for the areas and perimeters of the family of exterior and interior parallel sets of a domain, and a dual such family for a smooth convex domain, which arise naturally in plus and minus sampling, respectively. Finally, these results lead to a conjecture regarding the kinematic measure of the positions of one mobile convex set contained in another fixed convex set in d dimensions.

Except for a portion of Section 3.5.8, we work entirely in the Euclidean plane E^2. For a subset X of E^2, X^c denotes its complement and ∂X its boundary; $\emptyset$ is the null set. The closed *disk* of centre x and radius R is denoted by $Q(x, R)$. Its bounding *circle* is $\partial Q(x, R)$, and $Q(x, 0)$ is simply $\{x\}$. Since frequent reference is made to $Q(o, R)$, where o is the origin, we usually write simply Q for this disk. Similarly, we use the notation $Q(R)$ for 'a disk of radius R'. The weighted linear *sum* of subsets X and Y, with weights λ and μ, is

$$\lambda X+\mu Y=\{\lambda x+\mu y\colon x\in X, y\in Y\}.$$

Note that, according to this notation, $X-Y\neq X\cap Y^c$.

$$X_R\equiv X+Q(o,R) \quad (R\geqslant 0)$$

is the *exterior parallel set* to X at distance R. If $K\subset E^2$ is convex, $A\{\cdot\}$ denotes area and $S\{\cdot\}$ denotes perimeter, then

$$A\{K_R\}=A\{K\}+S\{K\}R+\pi R^2 \tag{1}$$

and

$$S\{K_R\}=S\{K\}+2\pi R \quad (R\geqslant 0). \tag{2}$$

Equation (1) is the two-dimensional form of Steiner's formula (see Hadwiger [3], Section 6.1.8).

3.5.3 PLANAR DOMAIN PROCESSES

Domains

We now define a family $\mathscr{F}$ of 'admissible' classes of domains of E^2. Since an important application of our theory is to the entire class of convex polygons, we require in particular that this class belongs to $\mathscr{F}$. This consideration very much determines the character of the following definition.

First, we define a *domain* of E^2 to be a bounded region bounded by a simple closed curve comprising a finite number of smooth *arcs*, i.e. simple curves with continuous bounded curvature κ. The common point of two adjacent arcs is a *cusp*. For convenience, a domain is assumed to be closed, i.e. it contains its bounding curve. Finally, we assume each domain possesses a uniquely defined and Euclidean invariant *centre* point, i.e. if z is the centre of D, then $y+z$ is the centre of $\{y\}+D$ $(y\in E^2)$.

Let n be a positive integer, and $\mathscr{D}_n^0$ be an n-dimensional class of domains, each having centre o, parametrized by

$$\theta_n=(\theta_{n,1},\ldots,\theta_{n,n}).$$

Let

$$\mathscr{D}_n=\{\{z\}+D : z\in E^2, D\in\mathscr{D}_n^0\}.$$

Note that the class of ellipses is a $\mathscr{D}_3$. Given a sequence $\{\mathscr{D}_n^0\}$ $(n=1,2,\ldots)$, some members of which may be void, we define

$$\mathscr{D}^0=\bigcup_{n=1}^{\infty}\mathscr{D}_n^0,\quad \mathscr{D}=\bigcup_{n=1}^{\infty}\mathscr{D}_n.$$

The class of convex polygons may be represented by $\mathscr{D}$, with $\mathscr{D}_n$ non-void only for $n = 4, 6, 8, \ldots$. The domains of $\mathscr{D}_n^0$, $\mathscr{D}_n$, $\mathscr{D}^0$ and $\mathscr{D}$ are parametrized by θ_n, (z, θ_n), (n, θ_n) and $(z; n, \theta_n)$, respectively. By an *admissible* class of domains, we mean a class of the type $\mathscr{D}$. Clearly the family $\mathscr{F}$ of all admissible classes is sufficiently general for most practical requirements.

Domain descriptions

We are ultimately interested in Euclidean invariant *descriptions* of domains D, e.g. $A\{D\}$, $S\{D\}$, the in-radius $I\{D\} \equiv \sup$ radii of disks $\subset D$, the circum-radius $J\{D\} \equiv \inf$ radii of disks $\supset D$. The description (or rather one-dimensional family of descriptions) now defined is of immense importance in this work. Notationally, dependences on R are generally omitted, since (except for in Section 3.5.8) R is essentially a constant. Define, for each planar set X, the sets

$$X^{\alpha} = \{x \in E^2 : \{x\} + X \subset Q\}, \tag{3}$$

$$X^{\beta} = \{x \in E^2 : X \subset Q(x, R)\}. \tag{4}$$

If $J\{X\} > R$, then both X^{α} and X^{β} are $\emptyset$. Suppose that $J\{X\} \leqslant R$. Since $x \in X^{\alpha}$ iff $-x \in X^{\beta}$,

$$X^{\alpha} = -X^{\beta}.$$

Moreover, both X^{α} and X^{β} are convex. The *R-measure* of a domain D is defined as

$$\begin{cases} M\{D\} = A\{D^{\alpha}\} = A\{D^{\beta}\}, & R \geqslant J\{D\}, \\ \quad\quad\;\; = 0, & R < J\{D\}. \end{cases} \tag{5}$$

For example, for all $y \in E^2$, $M\{Q(y, r)\} = \pi(R-r)^2$ $(R > r)$ and 0 $(R \leqslant r)$ We write

$$\boldsymbol{Z} = (Z_1, \ldots, Z_m)$$

for a general vector description of a domain, each component being a scalar description. We emphasize that (n, θ_n) completely specifies a domain of $\mathscr{D}^0$, whereas in general $\boldsymbol{Z}$ does not.

Random domains and domain processes

Henceforth we restrict attention to one fixed admissible class $\mathscr{D}$, and to the associated $\mathscr{D}^0$. A tilde is occasionally used to distinguish random variables or vectors from the values they take. A random member $(\tilde{n}, \tilde{\theta}_n)$ of $\mathscr{D}^0$ is most simply specified by supposing $\tilde{n}$ has p.m.f. $\{p_n\}$ and $\tilde{\theta}_n$ given $\tilde{n} = n$ has conditional d.f. $G(\theta_n \mid n)$.

Assumption 1. All descriptions $\boldsymbol{Z}$ considered, regarded as functions $\boldsymbol{Z}(n, \theta_n)$ on $\mathscr{D}^0$, possess all necessary regularity properties. In particular,

$Z = Z(\tilde{n}, \tilde{\theta}_n)$, corresponding to the above random specification of $(\tilde{n}, \tilde{\theta}_n)$, has a well-defined joint d.f. $F(Z)$. Strictly speaking, we should write $F_{\tilde{Z}}(\cdot)$ rather than $F(Z)$. However, this abuse of notation simplifies many formulae without danger of confusion.

A stochastic point process in $\mathscr{D}$, i.e. $(z; n, \theta_n)$-space, induces a stochastic *domain process* in E^2. For a domain process, we define N_X to be the number of domains with centre in $X \subset E^2$; and $N_X(n, \theta_n)$ to be the number of these N_X-domains whose $\tilde{n}$-value equals n, and whose $\tilde{\theta}_n$-value is dominated by θ_n (i.e. $\tilde{\theta}_{n,i} \leqslant \theta_{n,i}$, $1 \leqslant i \leqslant n$). Again, we shall restrict attention to a single given domain process Δ.

Assumption 2. For Δ, for all measurable sets $X \subset E^2$ with Lebesgue measure $A\{X\} < \infty$,

$$\mathbf{E}N_X(n, \theta_n) = \rho A\{X\} p_n\, G(\theta_n \,|\, n),$$

where ρ is a positive constant, and p_n and $G(\cdot|\cdot)$ are as defined in Assumption 1.

Thus $\mathbf{E}N_X = \rho A\{X\}$. In fact our theory is only of real practical utility provided, for all $y \in E^2$, $N_{Q(y,R)}/\pi R^2$ converges, in some stochastic sense as $R \to \infty$, to ρ. We shall tacitly assume that this is so, even though the theory holds without it. Thus ρ should be thought of as the 'intensity' (or density) of Δ. A sufficient condition for Assumption 2 to hold is that the domain centres constitute an ergodic homogeneous point-process in E^2 with intensity ρ; and that (independently) the domains are independently and identically distributed about these centres like the random member of $\mathscr{D}^0$ specified above is about o. It is shown in my Ph.D. thesis (Cambridge University, 1961) that Assumptions 1 and 2 hold for the random tessellation $\mathscr{P}$ of convex polygons determined by ergodic homogeneous Poisson lines in E^2 (see Miles [4]), the centre of a convex polygon being defined, for example, as its extreme left-hand point. The theory of the present paper was actually developed in connection with a Monte Carlo study of that case by Mr. I. K. Crain and myself, currently being prepared for publication; this study is referred to below as *C & M*.

3.5.4 THE BASIC EXPECTATION FORMULAE IN MINUS SAMPLING

Write $d\theta_n$ for the infinitesimal n-dimensional interval with opposite vertices θ_n and $\theta_n + d\theta_n$. Then, in an obvious notation, Assumption 2 yields

$$\mathbf{E}N_X(n, d\theta_n) = \rho A\{X\} p_n\, G(d\theta_n \,|\, n). \tag{6}$$

Writing N_- for the number of members of Δ contained in Q, it follows from equations (3), (5) and (6) that

$$\mathbf{E}N_-(n, d\theta_n) = \rho M(n, \theta_n) p_n G(d\theta_n | n). \tag{7}$$

Summing and integrating equation (7) over the set $M < \tilde{M} \leqslant M + dM$, $Z < \tilde{Z} \leqslant Z + dZ$ and recalling Assumption 1, we obtain

$$\mathbf{E}N_-(dM, dZ) = \rho MF(dM, dZ). \tag{8}$$

Suppose $T(M, Z)$ is an arbitrary measurable function, and let $\sum_- Z_i$ denote the sum of the Z values of the members of Δ contained in Q. Multiplying equation (8) by $T(M, Z)/M$ and summing and integrating totally, we obtain

$$\mathbf{E} \textstyle\sum_- M_i^{-1} T(M_i, Z_i) = \rho \mathbf{P}(\Theta)\, \mathbf{E}\{T(M, Z) | \Theta\}, \tag{9}$$

where

$$\Theta \equiv [M > 0] = [J < R]$$

and $\mathbf{E}$ and $\mathbf{P}$ on the right-hand side of equation (9) are with respect to the d.f.s $F(\cdot)$. Our minus sampling estimation formulae stem directly from the 'master formula' (9). Its statistical utility derives from the fact that

$$\hat{T}(M, Z) \equiv \textstyle\sum_- M_i^{-1} T(M_i, Z_i) \tag{10}$$

is an unbiased estimator of the right-hand side of equation (9). The following are some important cases of equation (9).

$$\mathbf{E} \sum_{-\,Z_i \leqslant Z} M_i^{-1} = \rho \mathbf{P}(\Theta) F(Z | \Theta), \tag{11}$$

$$\mathbf{E} \textstyle\sum_- M_i^{-1} = \rho\, \mathbf{P}(\Theta), \tag{12}$$

$$\mathbf{E} \textstyle\sum_- Z_i = \rho\, \mathbf{P}(\Theta)\, \mathbf{E}(MZ | \Theta), \tag{13}$$

$$\mathbf{E}N_- = \rho\, \mathbf{P}(\Theta)\, \mathbf{E}(M | \Theta). \tag{14}$$

Conversely, it is seen from equation (13) that the 'natural' estimates are biased by a factor M. Suppose the members of Δ are almost surely bounded, i.e. $F(J) = 1$ for J sufficiently large, and let

$$J_0 = \inf_{F(J)=1} J.$$

Then, for $R > J_0$, $\mathbf{P}(\Theta) = 1$ and the right-hand side of equation (9) simplifies to $\rho\, \mathbf{E}T(M, Z)$; in particular, $\sum_- M_i^{-1}$ becomes an unbiased estimator of ρ.

By Assumption 2, the expected total area of domains with centre in X is $\rho A\{X\}\, \mathbf{E}(A)$. Hence the expected number of domains covering each point of E^2 is $\rho\,\mathbf{E}(A)$ and so, if every point of E^2 lies in m and only m domains, then

$$\rho\,\mathbf{E}(A) = m.$$

For a random (polygonal) tessellation, $m = 1$.

3.5.5 FORMULAE FOR M

Having demonstrated the importance of the description M in eliminating edge effects by minus sampling, we now develop formulae for it. Consider a domain D with $J\{D\} < R$, and define

$$D\dagger = \bigcap_{y:Q(y,R)\supset D} Q(y, R).$$

Clearly $D \subset D\dagger$, and $D\dagger$ is convex and closed. In practice, $D\dagger$ is easily calculated by 'rolling' a $Q(R)$ 'around' D. Inclusion implies that

$$M\{D\} \geqslant M\{D\dagger\}.$$

If this inequality were strict, then there would exist a $Q(R)$ containing D but not $D\dagger$, a contradiction. Hence

$$M\{D\dagger\} = M\{D\}.$$

Now consider $\partial D\dagger$, i.e. $\partial(D\dagger)$. Clearly $\kappa \geqslant R^{-1}$ at all points of $\partial D\dagger$. More precisely, $\partial D\dagger$ is the disjoint union

$$\bigcup_1^3 \Omega_i,$$

where

Ω_1 comprises arcs of $Q(R)$s, the points of which may or may not belong to ∂D $(\kappa = R^{-1})$;

Ω_2 comprises arcs of ∂D, with $R^{-1} < \kappa < \infty$ at all points; and

Ω_3 comprises a finite number of cusps of ∂D $(\kappa = +\infty)$.

An arbitrary convention may be adopted regarding the end-points of arcs of Ω_1 and Ω_2: the theory is unaffected. Disregarding such end-points, the condition for a point y of ∂D to belong to a member of Ω_2 is that $\kappa(y) > R^{-1}$ and D is contained in one of the two $\partial Q(R)$s tangential to D at y; a similar condition applies to Ω_3. Since $\kappa \geqslant R^{-1}$ on $\partial D\dagger$, we have $S\{D\dagger\} \leqslant 2\pi R$. Hence both Ω_1 and Ω_2 possess at most countable numbers of members of positive length. Let the associated *chord* lengths of the arcs of Ω_1 be $\{L_i\}$ $(i = 1, 2, \ldots)$, and the sum of the actual *arc* lengths of Ω_2 be L_0.

Now define D^*, the *R-hull* of D, to be the closed convex set bounded by these chords associated with Ω_1 and the arcs of Ω_2; thus $D^* \subset D\dagger$,

$$S\{D^*\} = \sum_0^\infty L_i,$$

and clearly

$$M\{D^*\} = M\{D\dagger\} = M\{D\}.$$

The R-hull of an n-gon is the m-gon convex hull of m of its cusps, where $2 \leqslant m \leqslant n$. Note that, for $R = \infty$, both $D\dagger$ and D^* are the convex hull of D. If $D^* = D$, we say D is *R-convex*. This generalizes the concept of convexity, which is ∞-convexity.

Lemma 1. *If D is R-convex, if U, V are any two distinct points of $\Omega_2 \cup \Omega_3$ and if $H_\pm$ are the two closed half-spaces whose common boundary is the line UV, then $H_\pm \cap D$ are also R-convex.*

Proof. The reader might be well advised to draw a figure. Write $Q_\pm$ for the two $Q(R)$s whose bounding circles contain U and V, with centres in $H_\pm$, respectively. Since $\kappa \geqslant R^{-1}$ on $\partial D\dagger$, we have

$$(H_\pm \cap D)\dagger = (H_\pm \cap D\dagger) \cup (H_\pm \cap Q_\pm),$$

from which it follows that

$$(H_\pm \cap D)^* = (H_\pm \cap D\dagger)^* = H_\pm \cap D.$$

Theorem 1. *If D is R-convex, and $V_1, \ldots, V_m$ are arbitrary distinct points of $\Omega_2 \cup \Omega_3$, then the polygonal convex hull of $V_1, \ldots, V_m$ is also R-convex.*

Proof. By repeated application of Lemma 1.

Next follow two geometrical lemmas.

Lemma 2. *The smaller of the two areas into which a $Q(R)$ is partitioned by a chord of length $2l$ is $R^2 \sin^{-1}(l/R) - l\sqrt{(R^2 - l^2)}$ $(l \leqslant R)$.*

Lemma 3. *Consider an R-convex triangle UVW specified by two of its side lengths, $2u$ and $2v$, and by the enclosed angle W. Suppose, beginning with UW as a chord of Q, the triangle is moved* without rotation *and with W in contact with ∂Q, until it reaches the position $U'V'W'$ in which $V'W'$ is a chord of Q (see Figure 1). Then the area swept out by VW in this movement is $u\{\sqrt{(R^2-u^2)} + \sqrt{(R^2-v^2)}\cos W - v \sin W\}$.*

Proof. The required area equals the area of the parallelogram $VWW'V'$, i.e.

(15) $$4ul \sin\varphi.$$

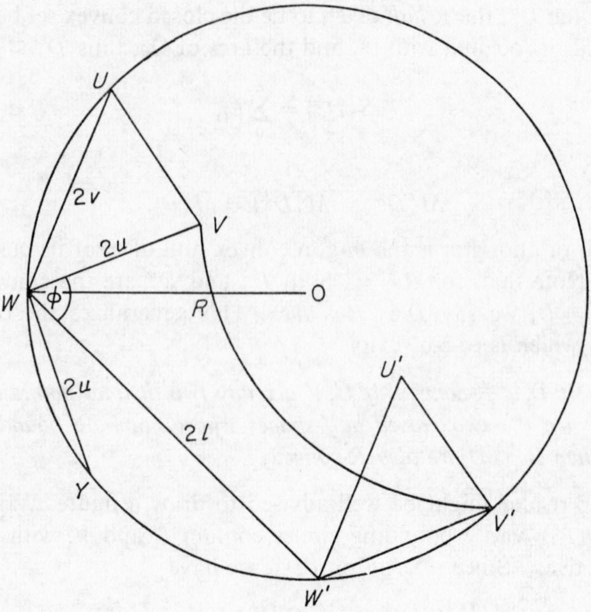

Figure 1

The angle of rotation about 0,

$$\angle WOW' = \angle VWY \equiv \theta,$$

say. Hence, considering angles at W, we have

$$\theta = \cos^{-1}\frac{u}{R} + \cos^{-1}\frac{v}{R} - W \tag{16}$$

and

$$\varphi = \frac{\pi}{2} - \frac{\theta}{2} + \cos^{-1}\frac{v}{R} - W. \tag{17}$$

The result follows upon eliminating l, θ and φ from

$$l = R\sin\frac{\theta}{2},$$

and equations (15), (16) and (17).

Lemma 4. *For a line segment of length L ($L \leqslant 2R$),*

$$M = 2R^2 \sin^{-1}\frac{\sqrt{(4R^2-L^2)}}{2R} - \left(\frac{L}{2}\right)\sqrt{(4R^2-L^2)}.$$

Proof. Apply equation (3) and Lemma 2.

Lemma 5. *For an R-convex triangle for which $J \leqslant R$, with sides L_i $(i = 1, 2, 3)$ and area A,*

$$M = \left(\pi - \sum_{i=1}^{3} \sin^{-1}\frac{L_i}{2R}\right) R^2 - \frac{1}{4}\sum_{i=1}^{3} L_i \sqrt{(4R^2 - L_i^2)} + A.$$

Proof. Let V be the vertex of the triangle opposite to the side of length L_2. The subset of positions of V such that the translated triangle is contained in Q is outlined by the broken line in Figure 2. Its area as given is

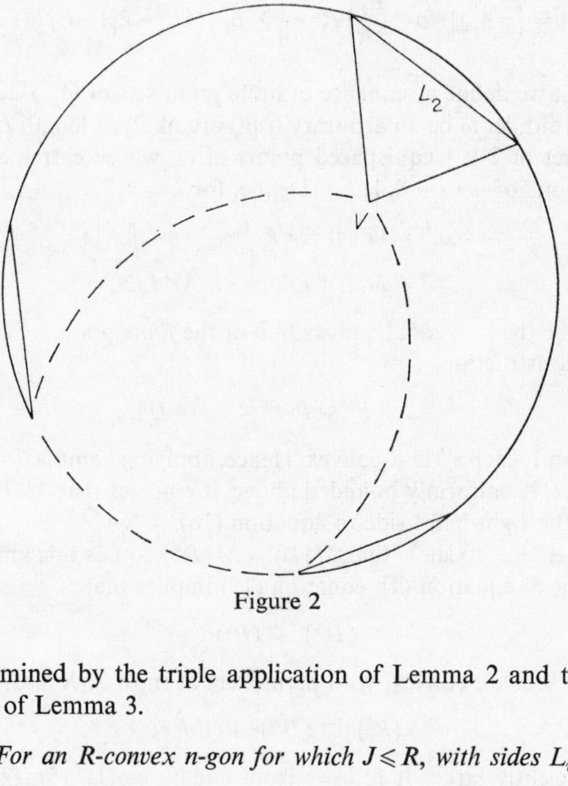

Figure 2

easily determined by the triple application of Lemma 2 and the double application of Lemma 3.

Lemma 6. *For an R-convex n-gon for which $J \leqslant R$, with sides L_i $(1 \leqslant i \leqslant n)$ and area A,*

$$M = \left(\pi - \sum_{i=1}^{n} \sin^{-1}\frac{L_i}{2R}\right) R^2 - \frac{1}{4}\sum_{i=1}^{n} L_i \sqrt{(4R^2 - L_i^2)} + A \quad (n = 3, 4, \ldots).$$

Proof. We proceed by induction, writing H_n for the proposition of the lemma for a given n. Lemma 5 states that H_3 is true. Suppose V_{j-1}, V_j and V_{j+1} are three adjacent vertices of an R-convex $(m+1)$-gon P_{m+1}, write P_3

for the triangle $V_{j-1}V_jV_{j+1}$ and P_m for the m-gon $P_{m+1}\cap P_3^c$. By Lemma 1, both P_3 and P_m are R-convex. Clearly

$$M\{P_m\}-M\{P_{m+1}\} = M\{\text{segment } V_{j-1}V_{j+1}\}-M\{P_3\}.$$

The value of the right-hand side may be deduced from Lemmas 4 and 5, and it is then only a matter of algebra to show that H_m implies H_{m+1}.

Theorem 2. *If D is a domain with* $J\{D\}\leqslant R$ *and* $\{L_i\}$ $(i = 0, 1, \ldots)$ *defined as above, then*

$$(18)\quad M\{D\} = \left(\pi-\sum \sin^{-1}\frac{L_i}{2R}\right)R^2-\frac{1}{4}\sum L_i\sqrt{(4R^2-L_i^2)}-L_0R+A\{D^*\}.$$

Proof. First we define a sequence of finite point sets of $\Omega_2\cup\Omega_3$. Assuming Ω_2 is non-void, let ω be an arbitrary (full) arc of Ω_2 of length L. Define ω_j to be the set of 2^j+1 equispaced points of ω, whose extreme points are the end-points of ω $(j = 0, 1, \ldots)$. Define, for $k = 1, 2, \ldots,$

$$\begin{aligned}\omega^k &= \varnothing \quad \text{if } L<k^{-1}\\ &= \omega_j \quad \text{if } L/2^{j+1}<k^{-1}\leqslant L/2^j,\end{aligned}$$

and P^k to be the polygonal convex hull of the *finite* point set $(\bigcup \omega^k)\cup\Omega_3$. Then, by construction,

$$(19)\qquad \ldots\subset P^k\subset P^{k+1}\subset \ldots \subset D^*.$$

By Theorem 1, each P^k is R-convex. Hence, applying Lemma 6 and recalling that κ on Ω_2 is uniformly bounded above, it emerges that $M\{P^k\}$ tends, as $k\to\infty$, to the right-hand side of equation (18).

It only remains to show that $M\{D\} = M\{D^*\}$ equals this limiting value. On account of equation (4), equation (19) implies that

$$(20)\qquad (D^*)^\beta \subset (P^k)^\beta.$$

Both these sets are convex, with perimeters at most $2\pi R$ and, given $\varepsilon>0$,

$$(21)\qquad (P^k)^\beta \subset (D^*)^\beta + Q(o, \varepsilon)$$

for k sufficiently large. It follows from equations (1), (5), (20) and (21) that

$$M\{D\}\leqslant M\{P^k\}\leqslant M\{D\}+2\pi\varepsilon R+\pi\varepsilon^2$$

for k sufficiently large.

Example 1. *Consider a domain which is the union of a rectangle with sides 2l and 2r and two semi-disks of radii r (see Figure 3). For M to be positive,* *it is necessary and sufficient that* $l+r<R$. *It is left to the reader to show that*

$$\text{(22)}\qquad \begin{cases} L_0 = 2r\left(\pi - 2\sin^{-1}\dfrac{l}{R-r}\right), \\ L_i = \dfrac{1}{2}\left[l + r\sqrt{\left\{1-\left(\dfrac{l}{R-r}\right)^2\right\}}\right] \quad (i=1,2). \end{cases}$$

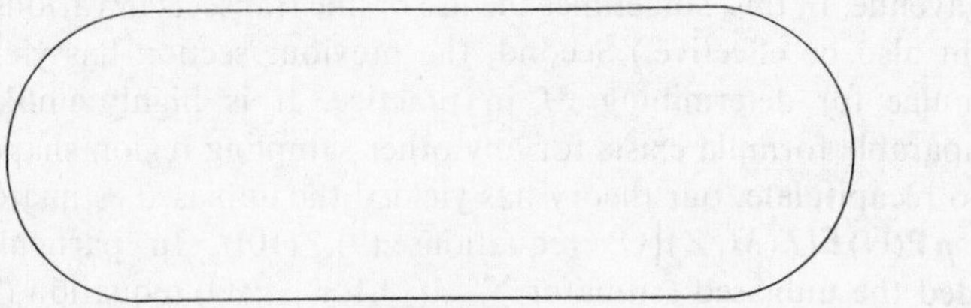

Figure 3

The case of an ellipse with semi-axes a, b $(R>a>b)$ *is much more difficult. However, simplification occurs in the R-convex case when* $\kappa \geqslant R^{-1}$ *everywhere, i.e. when* $a^2 \leqslant bR$. Then $L_i = 0$ $(i = 1, 2, \ldots)$ *and*

$$L_0 = 4aE\left(\frac{\pi}{2}, \sqrt{\left(1-\frac{b^2}{a^2}\right)}\right),$$

where $E(\cdot,\cdot)$ *is an Elliptic Integral of the Second Kind* (Dwight [2], p. 179; Table, p. 322).

Next we give a useful approximation in Theorem 2 for small sets. For fixed R,

$$\text{(23)}\qquad \pi R^2 + A\{(\lambda D)^*\} - M\{\lambda D\} \sim \lambda RS\{(\lambda D)^*\}$$

as $\lambda \to 0$. Omitting the asterisks and λs, and replacing $\sim$ by $=$, in the relation (23) yields the approximate formula actually utilized in the computer mensuration *C & M* of convex polygons with mean area approximately $\pi R^2/10^4$.

We end this section by posing the open problem of developing similar formulae for the R-measure (with respect to spheres) of domains in E^3.

3.5.6 ESTIMATION

General

As indicated above, our aim is to compensate for edge effects in the sampling of those domains of a domain process enclosed within a sampling circle (or circles).

There are two important reasons for using a circle rather than any other shape. First, for any other shape in general the (ergodic) orientation distribution of the domains will enter into the formulae. Since we are here only concerned with the distributions of Euclidean invariant domain descriptions, this would be a totally unnecessary additional complication. (Actually, our M^{-1} weighting method, cf. equation (11), may well extend to the determination of orientation distributions, but we shall not explore this avenue. In this, sometimes the use of line transects in various directions might also be effective.) Second, the previous section has yielded useful formulae for determining M in practice. It is highly unlikely that a comparable formula exists for any other sampling region shape.

To recapitulate, our theory has yielded the unbiased estimator $\hat{T}(M, Z)$ for $\rho\,\mathbf{P}(\Theta)\,\mathbf{E}\{T(M,Z)\,|\,\Theta\}$ (equations (9), (10)). In particular, it has yielded the unbiased estimator $\sum_{-} M_i^{-1}$ for $\rho\,\mathbf{P}(\Theta)$ (equation (12)). These two estimators suggest

$$\hat{t}(M,Z) \equiv \textstyle\sum_{-} M_i^{-1}\, T(M_i, Z_i) / \sum_{-} M_i^{-1}$$

as a natural, although in general *biased*, estimator of $\mathbf{E}\{T(M,Z)\,|\,\Theta\}$, where $T(\cdot,\cdot)$ is an arbitrary measurable function (for examples of $T(\cdot,\cdot)$, see equations (11)–(14)).

The variance of the estimators in the Poisson independent case

Naturally we should like to know something about the variances of the estimators $\hat{T}$ and $\hat{t}$. This only seems possible for $\hat{T}$ when the centres constitute a homogeneous Poisson process (or constant intensity ρ, say) and the domains are independently and identically distributed about these centres. In this case, N_{-} is Poisson distributed with mean value given by equation (14), and

$$\mathbf{E}(N_{-}^2) = (\mathbf{E}N_{-})^2 + \mathbf{E}(N_{-}).$$

For any $T \equiv T(M, Z)$,

$$\text{(24)}\qquad \mathbf{E}(\hat{T}^2) = \mathbf{E}\left(\textstyle\sum_{-} \dfrac{T_i^2}{M_i^2}\right) + 2\mathbf{E}\left(\textstyle\sum_{-\atop i\neq j} \dfrac{T_i\,T_j}{M_i\,M_j}\right)$$

$$= \rho\,\mathbf{P}(\Theta)\,\mathbf{E}\left(\frac{T^2}{M}\,\Big|\,\Theta\right) + \{\mathbf{E}(N_{-}^2) - \mathbf{E}(N_{-})\}\,\mathbf{E}\left(\frac{T_i}{M_i}\right)\mathbf{E}\left(\frac{T_j}{M_j}\right).$$

The Poisson independence implies that

$$\mathbf{E}\left(\textstyle\sum_{-} \dfrac{T_i}{M_i}\right) = (\mathbf{E}N_{-})\,\mathbf{E}\left(\dfrac{T_i}{M_i}\right)$$

which, applying equations (9) and (14), yields the interesting relation

$$\mathbf{E}\left(\frac{T_i}{M_i}\right) = \frac{\mathbf{E}(T|\Theta)}{\mathbf{E}(M|\Theta)}.$$

Substituting into equation (24), we obtain

$$\operatorname{Var}\hat{T} = \rho\,\mathbf{P}(\Theta)\,\mathbf{E}\left(\frac{T^2}{M}\Big|\,\Theta\right) = \mathbf{E}\left(\Sigma_{-}\frac{T_i^2}{M_i^2}\right).$$

Thus an unbiased estimator of the variance of $\hat{T}$ is given by

$$\Sigma_{-}\frac{T_i^2}{M_i^2}.$$

Replication

Now consider the not necessarily independent estimates $\hat{T}_1, \ldots, \hat{T}_n$ derived from observing Δ in n distinct $Q(R)$s.

$$\hat{\hat{T}} \equiv \sum_1^n \lambda_j \hat{T}_j \quad \left(\sum_1^n \lambda_j = 1\right)$$

is again an unbiased estimator of $\rho\,\mathbf{P}(\Theta)\,\mathbf{E}(T|\Theta)$. If the identically distributed $\hat{T}_j$ are independent with finite variance, then the $\hat{\hat{T}}$ with minimum variance occurs for $\lambda_j = n^{-1}$ $(1 \leqslant j \leqslant n)$. In this combined estimator each sampling disk, and hence each domain within the disks, contributes with equal weight, a fact which may well simplify computations. Of course, with independent replication, the sample variance of $\hat{T}_1, \ldots, \hat{T}_n$ is an unbiased estimator of the variance of $\hat{T}$, and further standard statistical procedures are likewise applicable. If the disk realizations are dependent, then minimum variance of $\hat{\hat{T}}$ is in general achieved with unequal λ_j.

If the realizations in a sequence of disks are 'sufficiently independent', then it is to be expected that, as $n \to \infty$,

$$\frac{1}{n}\sum_1^n \hat{T}_j \to \rho\,\mathbf{P}(\Theta)\,\mathbf{E}(T|\Theta) \tag{25}$$

in some stochastic sense (e.g. this will commonly occur if the disks lie in a lattice). If convergence in probability prevails in formula (25), then we should have, as $n \to \infty$,

$$\hat{\hat{t}} \equiv \sum_1^n \hat{T}_j \Big/ \sum_1^n \hat{h}_j \to \mathbf{E}(T|\Theta)$$

in probability, where $\hat{h} \equiv \Sigma_{-} M_i^{-1}$. These formulae offer perhaps the strongest argument for adopting M^{-1} weighted estimators: the fact that under reasonable conditions they are usually consistent.

Estimation in practice

We now briefly consider the two main types of situation in which the above theory may be applied.

(*a*) It may be possible to simulate Δ within a circle, on a computer say.

(*b*) A sample of Δ, e.g. an aerial photograph, may require analysis.

But first some general preliminary remarks. Generally speaking, in experimental situations, the value of ρ may often be known in advance, whereas information regarding the d.f.s $F(\cdot)$ is usually lacking. As regards measurement, we implicitly assume that there is no ambiguity in the separate identities of individual domains (observational problems may arise, for example, with high values of $\rho\,\mathbf{E}(A)$), and that all required Z-values for each domain may be determined without undue difficulty. This was so in *C & M*. Although stiuations come to mind where one might wish to estimate the Θ-dependent quantities on the right-hand sides of equations (9) and (11)–(14) for some given R-value, we shall ignore this case and concentrate on the estimation of the corresponding unconditioned quantities which occur when $\mathbf{P}(\Theta) = 1$. Intuition suggests that R should be chosen as large as possible, or at least sufficiently large that $\mathbf{E}(N_{-})$ is reasonably large. If this is not feasible, then replication will be desirable, to amass a sufficient sample. Choosing R large also tends to avoid the undesirable occurrence of small M-values in the estimator denominators. Expressed another way, the minimum observed value of $M/\pi R^2$ should be as close to unity as possible. If the domains are almost surely bounded and J_0 (see Section 3.5.4) is known, then efforts should clearly be made to choose R at least as large as J_0; if not, then it is only known that $\mathbf{P}(\Theta) \uparrow 1$ as $R \to \infty$, itself however a good reason for choosing R large. Of course, nothing may be inferred from minus sampling in $Q(R)$s regarding domains with $J > R$. In practice, other considerations may lead to hypotheses regarding $J > R$. For example, it might be assumed that certain (ergodic) domain densities are 'smoothly' monotone decreasing, even exponentially decreasing, beyond some point. Within each $Q(R)$ there will also be portions of domains severed by $\partial Q(R)$. We do not consider here how the measurement of these portions might be incorporated to improve estimates, although the theory of Section 3.5.7 below may shed some light on this.

(*a*) *Monte Carlo simulation.* For definiteness, we assume this is performed on a computer. This is no real limitation, as the considerations below will generally extend to non-computer simulation. Often a restriction on the size of R will be imposed by the computer's capacity. Naturally, with such a maximal R, one would replicate independently as many times as one's budget allows. However, there is an important exception. If there is a high dependence between neighbouring domains of Δ, then it may be advisable to replicate more times with a smaller R-value, in order to get a sufficiently varied sample. In *C & M*, the R-value was computer decided, ρ was known, and the budget permitted approximately seventy independent replications.

(*b*) *Sample data.* Here the usual circumstance is an *excess* of data, and the main problem is to decide what subsample to measure. Often it may be feasible to count the total number of domains in, but not the Z-values for, the entire data. In this case it is probably best to estimate ρ by simply

$$\hat{\rho} = \frac{\text{total number of domains}}{\text{total area}},$$

using some *ad hoc* procedure to account for boundary-severed domains, e.g. weight them by the apparent fraction inside the sample region. By using the entire data rather than the relative paucity contained in the sampling disks, the inevitable bias will usually be more than compensated for by a greatly increased precision. In $F(\cdot)$ estimation, the best plan may well be to select the centres of the sampling disks on an equilateral triangular lattice, of side length l, say $(l > 2R)$. Of course, this lattice should extend throughout the sample, and its position and orientation should be arbitrary, i.e. independent of Δ. Since the number of sampling disks is roughly proportional to l^{-2}, the work of mensuration is roughly proportional to $(R/l)^2$. Thus the ratio R/l is usually predetermined. Within this restriction, above considerations will suggest a minimum value for R. The final choice of R and l may be partially decided by the apparent degree of 'dependence' between neighbouring localities of the sample. The more [less] dependence, the smaller [larger] will be both R and l. Often, a mere glance at the data will indicate for which values of R negligible error is committed in assuming $\mathbf{P}(\Theta) = 1$.

3.5.7 PLUS SAMPLING

Should it be possible to measure all domains of Δ intersecting Q, then plus sampling, as defined in Section 3.5.4, may be preferable to minus sampling as described in Sections 3.5.4–3.5.6. The theory closely parallels that above, so we only indicate the salient formulae and differences.

Writing N_+ for the number of members of Δ intersecting Q, the counterpart of equation (8) is

$$\mathbf{E}N_+(dM_+, d\mathbf{Z}) = \rho M_+ F(dM_+, d\mathbf{Z}),$$

where

$$M_+\{D\} \equiv A\{D_R\}$$

(in a consistent notation, M would be M_-). The 'master formula' is now

$$\mathbf{E}(\hat{T}) = \rho\, \mathbf{E}T(M_+, \mathbf{Z}), \tag{26}$$

where $\hat{T} \equiv \sum_+ M_{+,i}^{-1} T(M_{+,i}, \mathbf{Z}_i)$. Note that no condition, like θ in equation (9), complicates the right-hand side of equation (26). The derivation from equation (26) of the formulae corresponding to formulae (11)–(14) is left to the reader. As before, in the 'Poisson independent' case, $\sum_+(T_i^2/M_{+,i}^2)$ is an unbiased estimator of the variance of $\hat{T}$. However, the advantage of the right-hand side being unconditioned is counterbalanced by two often serious shortcomings.

First, only for a restricted class of domains is there a formula for $M_+\{\cdot\}$ analogous to that for $M\{\cdot\}$ given in Theorem 2, as we shall now see. Suppose D has a smooth boundary, for which clearly

$$-\infty < \inf_{\partial K} \kappa \equiv \kappa_- \leqslant J^{-1} \leqslant I^{-1} \leqslant \kappa_+ \equiv \sup_{\partial K} \kappa < \infty.$$

Note that $\kappa_+ > 0$ and that D is convex iff $\kappa_- \geqslant 0$. For $y \in \partial D$, write y_R for the point distant R along the outward normal at y. The locus of y_R as y traverses ∂D is a closed curve $C_R(D)$ such that, except at points of inflection of ∂D, the radii of curvature of D at y and of $C_R(D)$ at y_R differ by R. If $C_R(D)$ is simple, then it is ∂D_R, and elementary computations yield

$$M_+\{D\} = A\{D\} + S\{D\}R + \pi R^2. \tag{27}$$

The validity of equation (27) clearly extends to domains with outward pointing cusps (at which $\kappa = +\infty$). In particular, equation (27) is valid for all convex sets, when it is Steiner's formula (1). Thus the validity of Steiner's formula is wider than is commonly supposed (see also Theorem 3 below).

Second, there may be difficulties in observing domains not enclosed in Q. For example, it is relatively easy to simulate the restriction of Poisson lines to Q and then to measure the N_- enclosed polygons of $\mathscr{P}$. But it is much more difficult to extend this simulation sufficiently far outside Q in order to be able to measure all N_+ intersecting polygons. This is the main reason why minus sampling was utilized in *C & M*. Often there will be problems in the measurement of large domains intersecting Q. Such domains may actually contain Q, and may extend outside the sampling region itself. As an illustration, suppose that the domains are all disks, with random radii $\tilde{\theta}$. Taking $T(M_+, Z) = M_+$ in equation (26), we have

$$\mathbf{E}(N_+) = \rho\, \mathbf{E}(M_+) = \rho\{\pi R^2 + R\, \mathbf{E}(\tilde{\theta}) + \pi\, \mathbf{E}(\tilde{\theta}^2)\}.$$

Hence, for any meaningful theory, we must have $\mathbf{E}(\tilde{\theta}^2) < \infty$. It would certainly help matters if the boundary of every intersecting domain also intersects Q, i.e. $I < R$ almost surely.

We shall not dwell upon sampling procedures in this case as, by and large, previous remarks carry over. Whenever feasible, it would seem good sense to carry out both a minus and a plus sampling, since they complement each other to some extent.

3.5.8 CONNECTIONS WITH INTEGRAL GEOMETRY

In this final section, we regard R no longer as a positive constant, but as a continuous parameter taking both positive and negative values.

For a general planar subset X, the *interior parallel set* at distance R is defined as

$$X_{-R} = \{x \in X\colon Q(x, R) \subset X\} \quad (0 < R \leqslant I).$$

Again, consider a domain D with smooth boundary, with $y_R[y_{-R}]$ the point distant R along the outward [inward] normal at $y \in \partial D$ $(R \geqslant 0)$. Let $B_+ > 0$ and $B_- < 0$ be the supremum and infimum, respectively, of the values of R for which the locus $C_R(D)$ of y_R is a *simple* closed curve. Note that, if D is convex, then $B_- = -\kappa_+^{-1}$ and $B_+ = \infty$. Since $C_R(D) = \partial D_R$ for $B_- \leqslant R \leqslant B_+$, the family $\{D_R\}$ $(B_- \leqslant R \leqslant B_+)$ of parallel domains satisfies

$$(28) \qquad (D_{R_1})_{R_2} = D_{R_1+R_2} \quad (B_- \leqslant R_1 \leqslant B_+, B_- \leqslant R_1 + R_2 \leqslant B_+),$$

with any member of the family determining the entire family. Relations (1), (2) and (27) extend easily to

Theorem 3. *For a domain D with smooth boundary,*

$$A\{D_R\} = A\{D\} + S\{D\}\, R + \pi R^2$$

and

$$S\{D_R\} = S\{D\} + 2\pi R \quad (B_- \leqslant R \leqslant B_+).$$

Strangely, there appears to be no reference to this simple result, an essential companion of Theorem 4 below, in the literature.

Next, suppose $\kappa_- > 0$, in which case D is convex, $B_- = -\kappa_+^{-1}$ and $B_+ = \infty$. Accordingly, we now write K instead of D. Recalling the definition of $X_R^\beta \equiv X^\beta$ in equation (4), it is readily seen that, for $R \geqslant \kappa_-^{-1}$, $C_{-R}(K) = \partial K_R^\beta$.

On account of the simple relationship between the radii of curvature at corresponding boundary points of parallel domains, we have, for $R_1 \geqslant \kappa_-^{-1}$, the basic relations

$$(K^\beta_{R_1})_{R_2} = K^\beta_{R_1+R_2} \quad (R_2 \geqslant \kappa_-^{-1} - R_1) \tag{29}$$

and

$$(K^\beta_{R_1})^\beta_{R_2} = K^\beta_{R_2-R_1} \quad (R_2 \geqslant R_1 - \kappa_+^{-1}) \tag{30}$$

corresponding to equation (28). In view of equations (29) and (30), $\{K^\beta_R\}$ $(R \geqslant \kappa_-^{-1})$ may be regarded as the family of parallel domains *dual* to $\{K_R\}$ $(R \geqslant -\kappa_+^{-1})$.

Theorem 4. *For a convex domain K with* $\kappa_- > 0$,

$$A\{K^\beta_R\} = A\{K\} - S\{K\}R + \pi R^2 \tag{31}$$

and

$$S\{K^\beta_R\} = 2\pi R - S\{K\} \quad (R \geqslant \kappa_-^{-1}). \tag{32}$$

Proof. Theorem 2 implies equation (31). Taking $R_1 = R_2$ in equation (30) and applying equation (31), we have $A\{K\} = A\{K^\beta_R\} - S\{K^\beta_R\}R + \pi R^2$. Equation (32) results upon combining this relation with equation (31).

We next extend the result (31) into a rather more general conjecture in E^d. For convex planar sets the fundamental formula of Blaschke takes the form (see Santaló [5], p. 36))

$$\int_{K^0 \cap K \neq \emptyset} dK = 2\pi(A^0 + A) + S^0 S, \tag{33}$$

where K^0 is a fixed, and K a mobile, convex domain and dK is the kinematic density element (Santaló [5], p. 22). Poincaré's formula (Santaló [5], p. 31) implies that

$$\int_{\partial K^0 \cap \partial K \neq \emptyset} \nu \, dK = 4S^0 S,$$

where ν is the total number of points comprising $\partial K^0 \cap \partial K$. If ν may assume only the values 0 and 2 (except possibly on a set of zero kinematic measure), then

$$\int_{\partial K^0 \cap \partial K \neq \emptyset} dK = 2S^0 S. \tag{34}$$

This condition is satisfied when

$$\inf_{\partial K} \kappa \geqslant \sup_{\partial K^0} \kappa,$$

in which case it follows from equations (33) and (34) that

$$\int_{K \subset K^0} dK = 2\pi(A^0 + A) - S^0 S. \tag{35}$$

Theorem 2, with $L_1 = L_2 = \ldots = 0$, is a special case of equation (35) in which $K^0 = Q$.

Now the analogue of equation (33) in d dimensions is

$$\int_{K^0 \cap K \neq \emptyset} dK = \lambda_d \sum_{i=0}^{d} \binom{d}{i} W_i\{K^0\} W_{d-i}\{K\}, \tag{36}$$

where

$$\lambda_d \equiv d!\, \pi^{d(d-1)/4} \Big/ 2 \prod_{j=1}^{d-1} \Gamma(\tfrac{1}{2}j + 1),$$

and $W_i\{K\}$ is the ith 'Quermassintegrale' of K (see Hadwiger [3], Section 6.1.6); $W_i\{K\}$ has the important stochastic interpretation as

$$\pi^{\frac{1}{2}} \Gamma(\tfrac{1}{2}[d-i]+1)/\Gamma(\tfrac{1}{2}d+1)$$

times the mean $(d-i)$-dimensional content of the orthogonal projection of K onto an isotropic random $(d-i)$-subspace in E^d (see Busemann [1], p. 46). Equations (33), (35) and (36) suggest the generalization

$$\int_{K \subset K^0} dK = \lambda_d \sum_{i=0}^{d} (-1)^i \binom{d}{i} W_i\{K^0\} W_{d-i}\{K\}, \tag{37}$$

which is easily verified to be true when K^0 and K are hyperspheres. Thus, finally, we make the

Conjecture. *The formula* (37) *holds for convex sets K^0, K of E^d if K may be positioned so as to be contained in K^0, and $(\partial K^0) \cap K$ is, for all positions of K, either $\emptyset$ or a connected set.*

REFERENCES

1. H. Busemann, *Convex Surfaces*, Wiley, New York (1958).
2. H. B. Dwight, *Tables of Integrals and other Mathematical Data*, 4th ed., Macmillan, New York (1961).
3. H. Hadwiger, *Vorlesungen über Inhalt, Oberfläche und Isoperimetrie*, Springer-Verlag (1957).
4. R. E. Miles, "Random polygons determined by random lines in a plane", *Proc. Nat. Acad. Sci.* (*U.S.A.*) **52** (1964), 901–907; II, 1157–1160.
5. L. A. Santaló, *Introduction to Integral Geometry*, Hermann, Paris (1953). (Act. Sci. Indust., No. 1198).

3.6

Line-processes, Road and Fibres

ROLLO DAVIDSON

The line-processes considered are stochastic processes of lines in the plane. If we choose an origin O and direction Ox in the plane, any line L may be represented by coordinates (p, θ), where p is the perpendicular distance from O to L and θ is the angle between the perpendicular and Ox. We may thus think of the line-process as a point-process on the (p, θ) cylinder. We consider only those line-processes which

(*a*) are locally square-summable (i.e. the mean-square number of lines cutting any circle is finite), and possess second-order product-moment densities (*g-functions*) (see Bartlett [1]) $g(p, \theta; p', \theta')$ which, for simplicity, we assume continuous;

(*b*) have a.s. no pairs of parallel lines;

(*c*) are second-order stationary under the rotations and translations of the plane.

We make assumption (*b*) because without it the mathematical theory is quite different and much weaker; but it is natural to think of the lines of the process being thrown down at random, when parallelism will not occur. It should be noted that the stationarity in (*c*) carries over to stationarity of the associated point-process on the (p, θ) cylinder with respect to the rotations and *shears—not* translations—of the cylinder.

Under the assumptions (*a*), (*b*) and (*c*) we have

Theorem 1. (*i*) *$g(p, \theta; p', \theta')$ is an even function of $\omega = \theta - \theta'$ alone; (ii) $g(\omega)$ is positive-definite.*

(Note that the analogue of (*ii*) for the g-functions of point-processes on the line does *not* hold (Bartlett [2]).)

The archetypal stationary line-process is the Poisson process. This may be constructed as a Poisson process on the (p, θ) cylinder; or equivalently on the plane as follows. Let O, $t_1, t_2, \ldots$ be successive points of a Poisson process on $\{t \geqslant 0\}$. With origin O place lines in the plane at perpendicular distances t_i from O and with the perpendiculars independently and uniformly distributed about O. More generally, we may construct a non-negative mean-square continuous stochastic process $V(\theta)$ on the circle, invariant under its rotations, and then put a Poisson process on the cylinder with local rate $V(\theta)$ at (p, θ); in this way we obtain the *most general* doubly stochastic Poisson line-process. There are other line-processes satisfying (*a*), (*b*) and (*c*), but it is not known if any of these are strictly stationary.

The practical idea behind looking at line-processes is that they provide a model for the structure of paper. Paper may be regarded as a (closely packed) random process of long fibres. It seems reasonable to regard the fibres as line in the plane, and to suppose that the line-process we obtain satisfies (*a*), (*b*) and (*c*). Now the strength of paper surely lies in the density of matting of its fibres; it is thus of interest to find what line-processes have the greatest specific density of intersections per unit area. Clearly we have to make our measure of specific density invariant under changes in the density of the process itself, that is, under changes in the thickness of the paper.

Let $k(P)$ be the mean number of intersections of lines of the line-process P in a unit area. Let $g(\omega)$ have Fourier series

$$g(\omega) = \sum_{n=0}^{\infty} a_n \cos(2n\omega).$$

Then all the *a*s are non-negative since g is positive-definite (Theorem 1); and we have

Theorem 2.

$$k(P) = a_0 - \pi \sum_{n=1}^{\infty} a_n / 2(4n^2 - 1).$$

The normalizing factor, to obtain invariance under change of density of the process, we take to be the mean number $N(A)$ of pairs of distinct lines of P in some given area A. The area A has to be specified in advance because $N(A)$ is not an additive function of A; but whatever A is, we find that

$$\mathrm{N}(A) = \sum_{n=0}^{\infty} a_n c_n,$$

where all the *c*s are non-negative. If now we introduce the mixed Poisson process, which is just the ordinary Poisson process with a random change of global density, we may state

Theorem 3. *The mixed Poisson process has the greatest specific intersection-density,* $k(P)/N(A)$.

Proof. From the formulae for $k(P)$ and $N(A)$, their quotient is largest when all the *a*s, except a_0, vanish; for all the *a*s are non-negative (and for a non-vacuous process a_0 must be positive). But the mixed Poisson process has this property.

It should be observed that the *a*s with positive suffices measure the tendency of the lines to have similar orientations, that is, the degree of departure from independent uniform distributions of orientation. Thus the content of Theorem 3 is really that to make your paper as strong as possible you should avoid bunching the fibres in particular directions; which is, indeed, intuitively obvious.

The conclusions from the application of line-processes to roads, however, are not perhaps quite what one would expect. The problem is to put a network of (conceptually) infinite straight roads in the plane so as to make the average road journey as short as may be. The lines will form a process which we assume to be strictly stationary and ergodic (in practical terms, this last means that the amount of asphalt we will lay is given); parallel roads are now permitted. Such a network might occur in the design of a new large city.

We have now to define the efficiency of the network. We take a random point A on the network, and consider all the points B on the network and at distance, as the crow flies, d from A. Let $r(d, A)$ be the mean distance of the points B from A via the roads of the network. We define the *coefficient of inefficiency* $m(d)$ to be the mean value, over A, of $r(d, A)$, divided by d. Thus $m(d) \geqslant 1$ always, and we want $m(d)$ to be as near 1 as possible.

The classical street plan is, of course, the rectangular grid. We calculate its efficiency in

Theorem 4. *For the rectangular grid,* $m(d) \geqslant (2+\pi)/4$, *and approaches this limit as* $d \to \infty$.

Now we should be able to do better than the value, about 1.3, given in this theorem; so we advance the idea of putting down the roads in a Poisson line-process. Taking the point A at random on a road then means constructing the process with A as fixed origin and a road going through it. For the Poisson process we have

Theorem 5. (*i*) *For all d, $m(d)$ is finite;* (*ii*) *$m(d) \to 1$ as $d \to \infty$.*

In the proof of (*i*), one shows that the distribution of the road distance from *A* to a typical *B* has exponential tail, and so possesses moments of all orders. In the proof of (*ii*), we look at *A* and a typical *B*, and invert the problem: instead of letting $d \to \infty$, we keep *d* fixed and let the process become infinitely dense. The proof uses the independence properties of the Poisson process, but, from the structure theory of doubly stochastic Poisson processes and considering what happens in the case of the rectangular grid, we think that the theorem would hold for doubly stochastic Poisson processes.

What is nice, of course, is the statement (*ii*). This says that in the case of long journeys, we should much prefer the Poisson process to the rectangular grid.

REFERENCES

1. M. S. Bartlett, *An Introduction to Stochastic Processes*, Cambridge University Press, 1966.
2. ——, "The spectral analysis of point processes", *J. statist. Soc. B* **25** (1963), 264–281.

For a thorough treatment of the Poisson line-process, see

3. R. E. Miles, "Random polygons determined by random lines in the plane, I and II", *Proc. Nat. Acad. Sci. U.S.A.* **52** (1964), (4) 901–907 and (5) 1157–1160.

3.7

On an Unfinished Manuscript of R. Davidson's

F. PAPANGELOU

(1) Shortly before his death R. Davidson started writing a paper on 'Processes of lines and marked points' and there are more than strong indications in it that by that time he had settled the following question: given a stationary process of cars-with-velocities on a road, can it be proved that the number of cars in a given stretch does not converge in probability to zero as time goes to infinity? The manuscript is unfortunately unfinished; it gives a proof for the case of independent velocities and breaks off at the point where treatment of the general case is begun. Though the independent case is only a special one, we think Davidson's ingenious solution is worth presenting here, as is of course the general problem itself, which is of interest not only in its own right but also in connection with some processes of lines in the plane, as will be seen. We describe below in condensed and rewritten form the contents of his manuscript.

(2) Let Z be a stationary process of cars-with-velocities. More specifically, cars are distributed on a road in a stationary marked point-process, the marks being their velocities. The numbers of cars in a fixed stretch of road are observed at general times t; it is required to show that these numbers do not converge to zero in probability (they clearly cannot converge to zero a.s.). This problem is easily seen to be equivalent to the following: with the same situation, an observer is placed at a fixed point and counts the numbers of cars passing in successive seconds; one has to prove that these numbers do not converge in probability to zero.

More formally, Z is a point-process $\{x_i, v_i\}$ in the plane, stationary under 'horizontal' shifts $x \to x+t$ and a.s. locally finite (only finitely many x_is in any finite interval). It is assumed that there are no multiple points.

Let $[a, b]$ be a fixed interval on the x-axis and for each $t \in (-\infty, \infty)$ let Z_t be the number of points $\{x_i, v_i\}$ such that $a \leqslant x_i + v_i t \leqslant b$. It is required to show that $\mathbf{P}(Z_t \geqslant 1)$ does not converge to zero as $t \to \infty$.

This same problem arises also in connection with some line-processes. Suppose we construct a process of lines as a process N of points-with-directions in the plane $\{x_i, y_i, \theta_i\}$, strictly stationary under the rigid motions of the plane. (Not all line processes are of this type.) We assume that N has a.s. some points; then it is trivial that the expected number of lines of N cutting any circle (or line segment) is infinite. Is it possible to strengthen this statement to say that the actual number cutting any circle is infinite? The answer to this is easily seen to be 'No'. However, one may conjecture the following:

Conjecture A. (For ergodic processes only.) *Given any non-zero radius, there exists with probability* 1 *a circle with that radius and with infinitely many lines of N through it.*

Conjecture B. *Given any line segment, with positive probability infinitely many lines of N pass through it.*

We discuss Conjecture B. Split N up into processes N_m $(m = 0, \pm 1, \ldots)$ corresponding to points $\{x_i, y_i, \theta_i\}$ of N such that the perpendicular distance of (x_i, y_i) from our fixed line segment I is between m and $m+1$. If $N(I)$ is the number of lines of N intersecting I, then

$$N(I) = \sum_{-\infty}^{\infty} N_m(I)$$

and so

$$\mathbf{P}(N(I) = \infty) \geqslant \mathbf{P}\left(\bigcap_{n=0}^{\infty} \bigcup_{m=n}^{\infty} \{N_m(I) \geqslant 1\}\right) \geqslant \limsup_{m \to \infty} \mathbf{P}(N_m(I) \geqslant 1).$$

Instead of considering N_m as a process of marked points in a band, we may section N_m by the centre line of the band (i.e. the line parallel to I and at distance $m + \frac{1}{2}$ from it), obtaining a process $\{z_j^m, \varphi_j^m\}$ of marked points (= lines) on a line. Furthermore, rather than consider the various processes N_m, we can, by stationarity, consider only N_0 and move instead the segment I, without changing its orientation, through a distance m in the direction perpendicular to it. Let Z_m be the number of lines of N_0 passing through this position of I. It is sufficient to prove

$$\limsup_{m \to \infty} \mathbf{P}(Z_m \geqslant 1) > 0.$$

If we write the process N_0 in the form $\{x_j, v_j\}$ with $x_j = z_j^0$ and $v_j = \cot \varphi_j^0$, we see the relationship with the cars-with-velocities problem.

Proposition 1. *If in the process $Z=\{x_i, v_i\}$ the velocities v_i are (identically distributed and) independent of each other and of the starting positions $\{x_i\}$ of the cars, then there is $\alpha>0$ such that $\mathbf{P}(Z_t\geqslant 1)\geqslant\alpha$ for all t.*

Proof. First thin out Z by deleting (x_i, v_i) whenever $x_{i+1}-x_i<d$. Clearly, there exists a $d>0$ such that the resulting process Z' is non-empty with positive probability. This process is also stationary under 'horizontal' translations and this implies that the measure m in R^2 defined by $m(B)=$ expected number of points of Z' in B (B Borel subset of R^2) is a product measure of the form $\mu\otimes F$, where μ is a multiple of the Lebesgue measure and F the distribution of the velocities v_i.

Let now $I=[a, b]$ be an interval with $b-a<d$ and Z_t' the number of points (x_i, v_i) of Z' such that $a\leqslant x_i+v_i t\leqslant b$. The expectations

$$\mathbf{E}Z_t' \quad (-\infty<t<\infty)$$

are all equal since $\mathbf{E}Z_t'=m(B_t)=\mu([a, b])$ by Fubini's theorem

$$(B_t=\{(x, v)\colon a\leqslant x+vt\leqslant b\}).$$

By the Schwarz inequality

$$\mathbf{P}(Z_t'\geqslant 1)\geqslant\frac{(\mathbf{E}Z_t')^2}{\mathbf{E}(Z_t'^2)},$$

and the proposition will be established if we show $\mathbf{E}(Z_t'^2)\leqslant 2$ for all t.

Let $p(x)$ be the probability that a car starting at x will have a velocity making it pass through I at time t. Obviously,

$$p(x)=F\left(\left[\frac{a-x}{t}, \frac{b-x}{t}\right)\right].$$

Given the pattern $\{x_i\}$ of initial positions, let Y_i be 1 or 0 according as v_i is in $[(a-x_i)/t, (b-x_i)/t]$ or not. Then

$$\begin{aligned}
\mathbf{E}(Z_t'^2) &= \mathbf{E}(\mathbf{E}(Z_t'^2 \,|\, \text{given } \{x_i\}))\\
&= \mathbf{E}\left(\mathbf{E}\left(\left(\sum_i Y_i\right)^2 \Big| \text{ given } \{x_i\}\right)\right)\\
&= \mathbf{E}\left(\mathbf{E}\left(\sum_i Y_i^2+2\sum_{i<j} Y_i Y_j \Big| \text{ given } \{x_i\}\right)\right)\\
&= \mathbf{E}\left(\sum_i p(x_i)+2\sum_{i<j} p(x_i)p(x_j)\right)\\
&\leqslant \mathbf{E}\left(\sum_i p(x_i)+\left(\sum_i p(x_i)\right)^2\right).
\end{aligned}$$

Now note that for each realization of $\{x_i\}$,

$$\sum_i p(x_i) = \sum_i F\left(\left[\frac{a-x_i}{t}, \frac{b-x_i}{t}\right]\right) = F\left(\bigcup_i \left[\frac{a-x_i}{t}, \frac{b-x_i}{t}\right]\right) \leqslant 1,$$

the intervals being disjoint because $b-a<d$ and no two xs are within distance d of each other. This proves that $\mathbf{E}(Z_t'^2) \leqslant 2$.

For the case of independent velocities it is known that if the common distribution of the velocities is absolutely continuous relative to the Lebesgue measure, then the distribution of Z_t converges to a mixed Poisson distribution (see C. Stone, 'On a theorem by Dobrushin', *Ann. Math. Stat.* **39** (1968), 1391–1401, and the references cited there).

4

STOCHASTIC TREE-STRUCTURES

4.1

The Probabilities of the Shapes of Randomly Bifurcating Trees

E. F. HARDING

4.1.1 INTRODUCTION

If an entity grows by bifurcation at the ends of its 'free' or terminal branches, starting from a single branch, then after a given number of bifurcations there will be a variety of possible shapes. When the free end which bifurcates on a particular occasion is a random one of those available, the different possible shapes will acquire a probability distribution which can be calculated once the set of possible shapes is specified. This distribution is of interest in population genetics, as described in [3], in polymer science [1], in the study of tributary-systems of rivers [4] and in a diversity of other applications. It might also have contributed to the investigation of bronchial structure which Rollo Davidson had embarked on. The growth of such an entity may be regarded as a multiplicative regenerative phenomenon, since the events dependent on any bifurcation are stochastically identical to those dependent on the very first bifurcation. Thus the present topic lies on the borders of both Stochastic Analysis and Stochastic Geometry—rather, Stochastic Topology—with ramifications into many other fields.

The material given here is selected from that in [3], and I am grateful to the Editor of *Advances in Applied Probability* for permission to use it. Much more detail may be found in the original article, as well as proofs omitted here; this article is intended to draw attention to the topic and to set the scene for that by Hammersley and Grimmett which follows it in this book. These authors give a proof of a result conjectured in [3] and make some further progress on an obscure and difficult problem raised there, by means of a most interesting and elegant technique.

I dedicate this contribution to the memory of Rollo Davidson, colleague and companion, in regret for all he has ceased to partake in.

Formal specification of unlabelled tree-shapes

We now define the fundamental entity—the *unlabelled rooted bifurcating tree-shape*.

Definition 1. An *unlabelled rooted bifurcating tree-shape*, or *shape*, is a type of linear graph satisfying

(*i*) . and $\overset{\cdot}{\wedge}$ are shapes. The shape . consists of a single *node* alone which (in this case) is both *root* and *terminus*. The shape $\overset{\cdot}{\wedge}$ consists of three nodes of which the highest is the root, the other two being termini or *ends*, and two *branches* connecting the root to the termini.

(*ii*) If $\dot{P}$ and $\dot{Q}$ are shapes, then so is the entity

$$\overset{\cdot}{\widehat{\dot{P}\quad\dot{Q}}}$$

constructed by superimposing the roots of $\dot{P}$ and $\dot{Q}$ upon the termini of $\overset{\cdot}{\wedge}$ whose root becomes the root of the composite shape. The termini of the composite shape are all the termini of $\dot{P}$ and $\dot{Q}$ taken together.

(*iii*) If $\dot{P}$ and $\dot{Q}$ are shapes, then

$$\overset{\cdot}{\widehat{\dot{P}\quad\dot{Q}}} \quad \text{and} \quad \overset{\cdot}{\widehat{\dot{Q}\quad\dot{P}}}$$

are the same shape (i.e. in their planar representations shapes are *commutative*).

(*iv*) Every shape except . can be expressed as

$$\overset{\cdot}{\widehat{\dot{P}\quad\dot{Q}}}$$

where $\dot{P}$ and $\dot{Q}$ are shapes.

Definition 2. (*i*) The shape $\overset{\cdot}{\wedge}$ is called a *fork*, having root and ends as in Definition 1.

(*ii*) A node *A* is *ancestor* to a node *B* if *A* is the root of a fork, one of whose ends is the root of a shape containing *B*. Conversely, *B* is called a *descendant of A*.

(*iii*) The unique node in a shape that is ancestor to all the other nodes of the shape is called the *root* of the shape.

(*iv*) The nodes of a shape that have no descendants are called *ends* or *termini* of the shape.

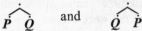

(*v*) The shape . is called **1**.

(*vi*) The shape  is called **2**.

(*vii*) If two shapes, $\dot{\boldsymbol{P}}$ and $\dot{\boldsymbol{Q}}$, are combined as in Definition 1(*ii*), then the resulting shape is called the *sum* of $\dot{\boldsymbol{P}}$ and $\dot{\boldsymbol{Q}}$ and is written $(\boldsymbol{P}+\boldsymbol{Q})$.

(The operation of addition thus defined is commutative but not associative—$((\boldsymbol{P}+\boldsymbol{Q})+\boldsymbol{R}) \neq (\boldsymbol{P}+(\boldsymbol{Q}+\boldsymbol{R}))$.)

(*viii*) The root of a shape is said to be of *generation* 0. Any other node is of generation n, where n is 1 more than the highest generation number of any of its ancestors.

(*ix*) If, for shapes $\boldsymbol{P}, \boldsymbol{Q}, \boldsymbol{R}$, we have $\boldsymbol{P} = \boldsymbol{Q}+\boldsymbol{R}$, than $\boldsymbol{Q}$ and $\boldsymbol{R}$ are called *subshapes of the first order of* $\boldsymbol{P}$. A shape is a subshape of the nth order of $\boldsymbol{P}$, if it is a subshape of the first order of a subshape of the $(n-1)$th order of $\boldsymbol{P}$. The order of a subshape is equal to the generation number of its root.

Definition 3. (*i*) The *degree*, $\delta(\boldsymbol{P})$, of a shape $\boldsymbol{P}$ is the number of its termini. It obeys

$$\delta(\mathbf{1}) = 1, \quad \delta(\boldsymbol{P}+\boldsymbol{Q}) = \delta(\boldsymbol{P})+\delta(\boldsymbol{Q}). \tag{1}$$

It is easily seen that a shape of degree δ has altogether $2\delta-1$ nodes.

(*ii*) The *height*, $h(\boldsymbol{P})$, of a shape $\boldsymbol{P}$ is the maximum of the generation numbers of its nodes. It obeys

$$h(\mathbf{1}) = 0, \quad h(\boldsymbol{P}+\boldsymbol{Q}) = 1+\max\,[h(\boldsymbol{P}), h(\boldsymbol{Q})], \tag{2}$$

and is also the maximum of the distances from the root to the termini.

(*iii*) A node which is not a terminus is called *balanced* if the shape of which it is the root has identical subshapes of the first order, and it is otherwise called *unbalanced*.

(*iv*) The *symmetry*, $\sigma(\boldsymbol{P})$, of a shape $\boldsymbol{P}$ is the number of its balanced nodes. It obeys

$$\begin{cases} \sigma(\mathbf{1}) = 0, \quad \sigma(\boldsymbol{P}+\boldsymbol{Q}) = \sigma(\boldsymbol{P})+\sigma(\boldsymbol{Q}) \quad \text{if } \boldsymbol{P} \neq \boldsymbol{Q}, \\ \qquad\qquad\quad \sigma(\boldsymbol{P}+\boldsymbol{P}) = 1+2\sigma(\boldsymbol{P}). \end{cases} \tag{3}$$

The *mutability*, or *asymmetry*, $\mu(\boldsymbol{P})$, is the number of unbalanced nodes. It is complementary to the symmetry, and obeys

$$\begin{cases} \mu(\boldsymbol{P})+\sigma(\boldsymbol{P}) = \delta(\boldsymbol{P})-1, \\ \qquad \mu(\mathbf{1}) = 0, \quad \mu(\boldsymbol{P}+\boldsymbol{P}) = 2\mu(\boldsymbol{P}), \\ \mu(\boldsymbol{P}+\boldsymbol{Q}) = 1+\mu(\boldsymbol{P})+\mu(\boldsymbol{Q}) \quad \text{if } \boldsymbol{P} \neq \boldsymbol{Q}. \end{cases} \tag{4}$$

Formal specification of labelled tree-shapes

Addition of unlabelled shapes is commutative because the nodes can be told apart only by how they are connected to other nodes in the shape. In some applications the terminal nodes are distinguished by labels, and this labelling makes distinct shapes that, unlabelled, would be identical.

Definition 4. (*i*) Given a shape $\boldsymbol{P}$ of degree n, and a set of n distinct symbols $(L_1, \ldots, L_n)$ called *labels*, any assignment of the labels one by one to the termini of $\boldsymbol{P}$ is called a labelling, and the combination of the shape $\boldsymbol{P}$ with any particular labelling is called a *labelled shape.*

(*ii*) If $\boldsymbol{Q}$ is a subshape of $\boldsymbol{P}$, the particular assignment of $\delta(\boldsymbol{Q})$ labels to $\boldsymbol{Q}$ is called the *sublabelling assigned to* $\boldsymbol{Q}$. The sublabellings assigned to the two identical subshapes descending from a balanced node may be exchanged without changing the labelled shape.

4.1.2 THE ENUMERATION OF LABELLED AND UNLABELLED SHAPES

Let k_n denote the number of labelled shapes of degree n. Then

$$k_n = \frac{(2n-2)!}{2^{n-1}(n-1)!} = 1.3.5.\ldots.(2n-3). \tag{5}$$

Let S_n denote the number of unlabelled shapes of degree n. Then

$$S_n = \begin{cases} S_1 S_{n-1} + S_2 S_{n-2} + \ldots + S_{m-1} S_m & \text{if } n = 2m-1, \\ S_1 S_{n-1} + S_2 S_{n-2} + \ldots + S_{m-1} S_{m+1} + \tfrac{1}{2} S_m(S_m+1) & \text{if } n = 2m. \end{cases} \tag{6}$$

From this recurrence equation S_n may be calculated for any given value of n, since clearly $S_1 = S_2 = 1$. But, in contrast with the case for labelled shapes, no explicit formula for S_n as a function of n has ever been obtained despite many studies at different times. Such a formula would almost certainly illuminate some of the difficulties described later in studying the probability distribution on the set of unlabelled shapes generated by random bifurcation.

4.1.3 PRELIMINARIES TO THE PROBABILITY THEORY OF SHAPES GENERATED BY RANDOM BIFURCATION

We now turn to consider the probability distribution induced on the set by the process of random bifurcation. Section 4.1.3 sets out a special notation for shapes, and an introductory account of the determination of the probability distribution, which will be useful later. Sections 4.1.4 and 4.1.5 present, for unlabelled and labelled shapes respectively, the results obtained on the probability distributions.

A notation for unlabelled shapes

There are S_n unlabelled shapes of degree n. If these shapes were indexed in a definite way by the integers $1, 2, \ldots, S_n$ then the notation $\boldsymbol{n}_j$ would be a specific, unique notation by which the jth in this sequence could be recovered. By describing a natural ordering of all the shapes of given degree we can specify a notation which allows any given shape to be reconstructed from its representative symbols without the necessity of referring to the complete sequence of shapes of given degree.

Convention. If any unlabelled shape is expressed as the sum of two first-order subshapes, e.g. $\boldsymbol{P} = \boldsymbol{Q}+\boldsymbol{R}$, it will be understood that $\delta(\boldsymbol{Q}) \leqslant \delta(\boldsymbol{R})$ and $\boldsymbol{Q}$ will be called the *left-hand subshape*, with explicit reference to the possibility of representing a shape on paper in a canonical way. If the subshapes are of the same degree, then the subshape earliest in the ordering to be defined will be the left-hand subshape.

The ordering of the shapes of given degree is (informally) a 'dictionary' ordering,

firstly with respect to increasing degree of left-hand subshape,
secondly with respect to the order of the right-hand subshape,
thirdly with respect to the order of the left-hand subshape.

Formally, we define an ordering of the shapes of degree n, in correspondence with the integers $1, 2, \ldots, S_n$, as follows.

(*a*) There is only one shape of degree 1 and it is denoted by $\boldsymbol{1}_1$.

(*b*) The ordering of shapes of any higher degree is defined recursively in terms of the orderings of shapes of lower degree, and the resulting sequence of shapes of degree n is denoted by $\boldsymbol{n}_1, \boldsymbol{n}_2, \ldots, \boldsymbol{n}_{S_n}$, where

$$\boldsymbol{n}_i = \boldsymbol{r}_k + \boldsymbol{s}_m \quad (r+s = n, r \leqslant s, 1 \leqslant k \leqslant S_r, 1 \leqslant m \leqslant S_s)$$

precedes

$$\boldsymbol{n}_j = \boldsymbol{t}_p + \boldsymbol{u}_q \quad (t+u = n, t \leqslant u, 1 \leqslant p \leqslant S_t, 1 \leqslant q \leqslant S_u)$$

if and only if

$$r \leqslant t \text{ and either } r < t$$
$$\text{or } r = t \text{ and } m < q$$
$$\text{or } r = t,\ m = q, \text{ and } k < p,$$

in which case we have $i < j$. This definition clearly specifies the order of any shape.

With reference to the notation introduced in parts (*v*), (*vi*), and (*vii*) of Definition 2, we can now construct the set of unlabelled shapes, in a series ordered overall by degree, as follows.

$$\begin{aligned}
S_1 = 1: &\quad \mathbf{1}_1 = \mathbf{1},\\
S_2 = 1: &\quad \mathbf{2}_1 = \mathbf{1}_1 + \mathbf{1}_1 = \mathbf{2},\\
S_3 = 1: &\quad \mathbf{3}_1 = \mathbf{1}_1 + \mathbf{2}_1 = \mathbf{1} + \mathbf{2},\\
S_4 = 2: &\quad \mathbf{4}_1 = \mathbf{1}_1 + \mathbf{3}_1 = (\mathbf{1} + (\mathbf{1} + \mathbf{2}))\\
&\quad \mathbf{4}_2 = \mathbf{2}_1 + \mathbf{2}_1 = \mathbf{2} + \mathbf{2},\\
S_5 = 3: &\quad \mathbf{5}_1 = \mathbf{1}_1 + \mathbf{4}_1 = \mathbf{1} + (\mathbf{1} + (\mathbf{1} + \mathbf{2}))\\
&\quad \mathbf{5}_2 = \mathbf{1}_1 + \mathbf{4}_2 = \mathbf{1} + (\mathbf{2} + \mathbf{2})\\
&\quad \mathbf{5}_3 = \mathbf{2}_1 + \mathbf{3}_1 = \mathbf{2} + (\mathbf{1} + \mathbf{2}),
\end{aligned}$$

and so on, in accordance with the ordering defined above.

It is also straightforward to determine explicitly the composition of the shape corresponding to any named symbol, say $\boldsymbol{n}_i$, by a direct method which does not require the entire set of shapes of degree n to be considered. An example in [3] shows how this may be done.

Generation of unlabelled shapes by random bifurcation

Let us explicitly trace out the set of possible sequences of bifurcation for $\delta = 1, 2, 3, 4$, which is shown graphically in Figure 1. For $\delta = 1$ we have

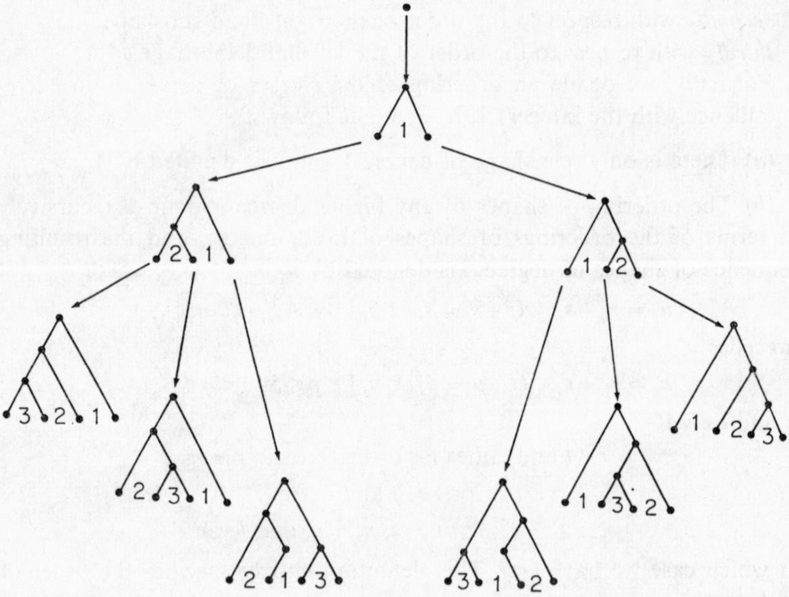

Figure 1

the initial ancestral stock before any split has occurred and there is only one possibility. For $\delta = 2$ we have the (again unique) case which occurs after the first bifurcation. Thus both these cases have probability, conditional on degree, of unity. (From now on, 'probability conditional on degree' will be called simply 'probability'.) At this second stage there are now two termini at which bifurcation may occur, with (by hypothesis) probability $\frac{1}{2}$ for each. Either possibility gives rise to the same shape, namely $\mathbf{3}_1$, which therefore also has probability 1. Similarly, choosing with uniform probability from the available termini that which shall next split, we obtain the 6 possibilities with $\delta = 4$ that are shown in Figure 1, each of which has probability $\frac{1}{6}$. But it can be seen that these 6 cases include the shape $\mathbf{4}_1$ 4 times and the shape $\mathbf{4}_2$ twice, so that their respective probabilities are $\frac{2}{3}$ and $\frac{1}{3}$. The arrows in the figure join termini to the shapes that arise by splitting at these termini.

4.1.4 THE PROBABILITY OF A GIVEN UNLABELLED SHAPE

By $\mathsf{P}[\boldsymbol{R}]$ we denote the probability that a given shape $\boldsymbol{R}$ will be reached by the process of random bifurcation, as the result of $\delta_{\boldsymbol{R}}-1$ splits. Thus, although the degree of a shape is not explicitly mentioned, it is understood to be given.

Theorem 1. If $\boldsymbol{n}$, any shape of degree n, is the sum $\boldsymbol{r}+\boldsymbol{s}$ of two subshapes $\boldsymbol{r}$ and $\boldsymbol{s}$, say, then

$$\mathsf{P}[\boldsymbol{n}] = \begin{cases} 2(n-1)^{-1}\,\mathsf{P}[\boldsymbol{r}]\,\mathsf{P}[\boldsymbol{s}] & \text{if } \boldsymbol{r} \neq \boldsymbol{s}, \\ (n-1)^{-1}(\mathsf{P}[\boldsymbol{r}])^2 & \text{if } \boldsymbol{r} = \boldsymbol{s}. \end{cases} \tag{7}$$

Proof. Let the two termini resulting from the first split be arbitrarily labelled L and R. It is then easily seen that the probability that, after $n-1$ splits, L is the root of a subshape of degree r is $1/(n-1)$ $(r = 1, \ldots, n-1)$; the process is in fact a Pólya Urn Scheme. It is also clear that, conditional on the value r, the first-order subshapes descending from L and R are independent of each other. Now delete the labels L and R and use the commutative property of unlabelled-shape addition, and equation (7) follows. (This is a simpler proof than either of the two in [3], which still have, however, some independent interest.)

The probability distribution on the set of unlabelled shapes

With the help of Theorem 1 it would be possible, in principle, to construct the complete probability distribution on the set of shapes induced by random bifurcation, by the simple direct method of evaluating, for each n, $\mathsf{P}[\boldsymbol{n}_i]$ for $i = 1, 2, \ldots, S_n$. However, when n is large, S_n is very large and such an approach is not practicable. In any case, in practice one is usually interested in the following, more limited information.

(*a*) What are the k largest probabilities of shapes of degree n.

(*b*) What are their respective multiplicities.

(*c*) What (if needed) are the corresponding shapes.

Theorem 2. To obtain the probabilities, multiplicities and identities of the most probable, second most probable, ..., kth most probable shapes of degree n it is sufficient to know, for shapes of each degree m $(1 \leqslant m < n)$ the probabilities, multiplicities and identities respectively of the most probable, second most probable, ..., and $[k + \log_2(n/m)]$th most probable shapes of degree m (where the notation $[x]$ denotes the largest integer not exceeding x).

Proof. See [3].

4.1.5 THE PROBABILITY DISTRIBUTION ON THE SET OF LABELLED SHAPES

The allocation of n labels to the termini of a shape of degree n, at random and independently of the shape, gives rise to a set of labelled shapes (already enumerated in Section 4.1.2) and to a probability distribution on this set.

Denote by $L(\boldsymbol{P})$ the number of distinct labellings of the termini of a given shape $\boldsymbol{P}$, that is, the number of assignments of n distinct labels to the termini which can be distinguished in accordance with Definition 4(*ii*). For example, the shape $\boldsymbol{4}_2 = (\boldsymbol{1}+\boldsymbol{1})+(\boldsymbol{1}+\boldsymbol{1})$ can be labelled in only 3 different ways with labels A, B, C, D, namely $((A, B), (C, D))$, $((A, C), (B, D))$, $((A, D), (B, C))$ all the remaining permutations of A, B, C, D being obtainable from these by exchanging identical subshapes with balanced roots.

Suppose $\boldsymbol{P} = \boldsymbol{Q} + \boldsymbol{R}$, where the degrees of $\boldsymbol{P}$, $\boldsymbol{Q}$ and $\boldsymbol{R}$, are n, r and s respectively $(r + s = n)$. In any labelling of $\boldsymbol{P}$, there are $\binom{n}{r}$ choices of the r labels to be assigned to $\boldsymbol{Q}$ leaving s labels to be assigned to $\boldsymbol{R}$, and these two subshapes may be labelled in $L(\boldsymbol{Q})$ and $L(\boldsymbol{R})$ ways respectively with the labels assigned to them. If $\boldsymbol{Q} = \boldsymbol{R}$ then the labels assigned to $\boldsymbol{Q}$ may be exchanged with those assigned to $\boldsymbol{R}$ without changing the labellings. It is therefore clear that

$$L(\boldsymbol{P}) = \begin{cases} \binom{n}{r} L(\boldsymbol{Q})\, L(\boldsymbol{R}) & \text{if } \boldsymbol{Q} \neq \boldsymbol{R}, \\ \frac{1}{2}\binom{n}{r} (L(\boldsymbol{Q}))^2 & \text{if } \boldsymbol{Q} = \boldsymbol{R}. \end{cases} \tag{8}$$

Taking labels to be assigned at random to the termini of a shape, we can find the probability of a particular labelled shape by dividing the probability of the unlabelled shape by the number of distinct labellings of that shape, since all labellings of a given shape are equally likely. Thus, if we denote by $\mathbf{P}_L[\boldsymbol{P}]$ the probability of each of the $L(\boldsymbol{P})$ equally likely labelled shapes corresponding to a given unlabelled shape, we have

$$\mathbf{P}_L[\boldsymbol{P}] = \frac{\mathbf{P}[\boldsymbol{P}]}{L(\boldsymbol{P})}, \tag{9}$$

so that, combining equations (7) and (8), we obtain the analogue for labelled shapes of the formula (7) for unlabelled shapes, namely

$$\mathbf{P}_L[\boldsymbol{P}] = \frac{2}{n-1} \frac{r!\,(n-r)!}{n!} \mathbf{P}_L[\boldsymbol{Q}]\, \mathbf{P}_L[\boldsymbol{R}] \tag{10}$$

which is true whether $\boldsymbol{Q} = \boldsymbol{R}$ or not, and is therefore free of the factor $\frac{1}{2}$ whose occurrence in equation (7) in the case $\boldsymbol{Q} = \boldsymbol{R}$ is responsible for the fact that the probability distribution of unlabelled shapes is so obscure.

The computation of the probabilities $\mathbf{P}_L[.]$ is much facilitated by the following device. For a shape $\boldsymbol{P}$ of degree n define $M(\boldsymbol{P})$ by

$$M(\boldsymbol{P}) = n!\, \mathbf{P}_L[\boldsymbol{P}]. \tag{11}$$

Then, from equation (10),

$$M(\boldsymbol{P}) = 2(n-1)^{-1} M(\boldsymbol{Q})\, M(\boldsymbol{R}) \tag{12}$$

for all shapes $\boldsymbol{P}$ whose first-order subshapes are denoted by $\boldsymbol{Q}$ and $\boldsymbol{R}$ and equation (12) is again true whether $\boldsymbol{Q} = \boldsymbol{R}$ or not.

The search for the most probable labelled shape of degree n met with more success than the search for the most probable unlabelled shape. I offered the following conjecture, with the challenge to make an honest theorem of it.

Conjecture. *The most probable labelled shape of degree n is*

$$\boldsymbol{n}^* = \boldsymbol{r}^* + (\boldsymbol{n} - \boldsymbol{r})^*, \tag{13}$$

The most probable unlabelled shapes of degrees up to 160. The table gives the degrees (r, s) of the two first-order subshapes into which the most probable shape of degree n splits. In all cases the two first-order subshapes are themselves the most probable shapes of their respective degrees, and from this information the complete shape can be constructed. Further data on the probability distribution of unlabelled shapes (second, third, etc., most probable shapes, and actual probabilities) are available, and may be obtained from me if required.

n	(r, s)	n	(r, s)	n	(r, s)	n	(r, s)
1	1, 0	41	20, 21	81	34, 47	121	59, 62
2	1, 1	42	15, 27	82	35, 47	122	55, 67
3	1, 2	43	15, 28	83	35, 48	123	60, 63
4	1, 3	44	17, 27	84	32, 52	124	58, 66
5	2, 3	45	20, 25	85	35, 50	125	58, 67
6	2, 4	46	19, 27	86	31, 55	126	59, 67
7	3, 4	47	20, 27	87	32, 55	127	60, 67
8	3, 5	48	20, 28	88	35, 53	128	62, 66
9	4, 5	49	21, 28	89	42, 47	129	62, 67
10	3, 7	50	23, 27	90	35, 55	130	63, 67
11	4, 7	51	23, 28	91	32, 59	131	62, 69
12	5, 7	52	25, 27	92	42, 50	132	55, 77
13	5, 8	53	25, 28	93	38, 55	133	66, 67
14	5, 9	54	23, 31	94	42, 52	134	60, 74
15	7, 8	55	27, 28	95	40, 55	135	58, 77
16	7, 9	56	27, 29	96	44, 52	136	59, 77
17	8, 9	57	25, 32	97	42, 55	137	60, 77
18	8, 10	58	27, 31	98	43, 55	138	59, 79
19	7, 12	59	27, 32	99	47, 52	139	62, 77
20	8, 12	60	28, 32	100	45, 55	140	63, 77
21	8, 13	61	27, 34	101	42, 59	141	67, 74
22	7, 15	62	27, 35	102	47, 55	142	55, 87
23	8, 15	63	31, 32	103	48, 55	143	66, 77
24	9, 15	64	27, 37	104	49, 55	144	67, 77
25	12, 13	65	32, 33	105	50, 55	145	55, 90
26	11, 15	66	31, 35	106	47, 59	146	59, 87
27	12, 15	67	32, 35	107	52, 55	147	60, 87
28	13, 15	68	32, 36	108	53, 55	148	58, 90
29	12, 17	69	34, 35	109	50, 59	149	59, 90
30	13, 17	70	32, 38	110	52, 58	150	60, 90
31	15, 16	71	35, 36	111	52, 59	151	67, 84
32	15, 17	72	35, 37	112	52, 60	152	62, 90
33	15, 18	73	35, 38	113	55, 58	153	63, 90
34	15, 19	74	32, 42	114	55, 59	154	67, 87
35	15, 20	75	35, 40	115	55, 60	155	58, 97
36	15, 21	76	34, 42	116	55, 61	156	59, 97
37	17, 20	77	35, 42	117	55, 62	157	67, 90
38	15, 23	78	35, 43	118	55, 63	158	59, 99
39	19, 20	79	32, 47	119	59, 60	159	62, 97
40	15, 25	80	38, 42	120	58, 62	160	63, 97

where $\mathbf{r}^*$ *and* $(\mathbf{n}-\mathbf{r})^*$ *are the most probable labelled shapes of degrees r and* $(n-r)$ *respectively and, for* $n > 1$,

$$(14) \qquad r = 2.2^{[\log_2 \frac{1}{3}(n-1)]}$$

(*where, as usual,* $[x]$ *denotes the integral part of* x).

It is clear that if the formula for r is right then the first-order subshapes are the most probable of their degrees. To verify the conjecture it is sufficient to verify that equation (14) gives the solution of

$$(15) \qquad (n-1)\, M(\mathbf{n}^*) = \max_{1 \leqslant r \leqslant [n/2]} [2M(\mathbf{r}^*)\, M((\mathbf{n}-\mathbf{r})^*)],$$

which is what Hammersley and Grimmett have achieved in their study of a much more general version of this problem, [2].

The problem of finding the most probable unlabelled shape of degree n is much more difficult. In [3] I had nothing to offer. Hammersley and Grimmett find that this problem, too, belongs to the class of maximal solutions of the subadditive inequality, and they go a little way towards its solution. For values of n up to 160 the most probable shape of degree n was found by the method of Theorem 2, and the results are given in the table opposite (to which Hammersley and Grimmett also refer).

REFERENCES

1. M. Gordon, T. G. Parker and W. B. Temple, "On the number of distinct orderings of a vertex-labelled graph when rooted on different vertices", *J. Combin. Th.* **11** (1971), 142–156.
2. J. M. Hammersley and G. R. Grimmett, "Maximal solutions of the generalized subadditive inequality", 4.2 of this book.
3. E. F. Harding, "The probabilities of rooted tree-shapes generated by random bifurcation", *Adv. Appl. Prob.* 3 (1971), 44–77.
4. R. L. Shreve, "Statistical laws of stream numbers", *J. Geol.* **74** (1966), 17–37.

4.2

Maximal Solutions of the Generalized Subadditive Inequality

J. M. HAMMERSLEY and G. R. GRIMMETT

4.2.1 INTRODUCTION

In an enquiry on maximum likelihoods of genetic histories, Harding [1] met a function defined recursively for $n = 1, 2, \ldots$ by means of

$$(1) \qquad M(1) = 1, \quad M(n) = \max_{1 \leqslant r \leqslant \frac{1}{2}n} \left[\frac{2M(r)\,M(n-r)}{(n-1)} \right] \quad (n \geqslant 2).$$

Writing $\rho(n)$ for the value of r which maximized the right-hand side of equation (1) he conjectured that either $\rho(n)$ or $n-\rho(n)$ was always a power of 2 lying between $\frac{1}{3}n$ and $\frac{2}{3}n$; and he verified this conjecture on a computer for $n \leqslant 100$. More specifically, his conjecture amounts to

$$(2) \qquad \rho(n) = \begin{cases} 2^{k-1} & \text{if } 1 \leqslant n/2^k \leqslant \frac{3}{2}, \\ n - 2^k & \text{if } \frac{3}{2} \leqslant n/2^k \leqslant 2. \end{cases}$$

We shall prove this conjecture in the wider setting (with other applications, perhaps) of *generalized subadditive functions* [2], namely sequences

$$\varphi = \{\varphi(n)\}_{n=1,2,\ldots}$$

satisfying the inequality

$$(3) \qquad \varphi(n) \leqslant \varphi(r) + \varphi(n-r) + g(n) \quad (1 \leqslant r < n \geqslant 2),$$

where $g = \{g(n)\}_{n=1,2,\ldots}$ is a given sequence. If $\varphi(n)$ is a solution of (3), so is $\varphi(n) - \lambda n$ for any constant λ; and hence, taking $\lambda = \varphi(1)$, we may confine ourselves to the *canonical solution* of (3), namely the solution with $\varphi(1) = 0$. We may also assume $g(1) = 0$, because (3) only involves $g(n)$ for $n \geqslant 2$. We shall write S for the space of all real sequences $\{s(n)\}_{n=1,2,\ldots}$ with $s(1) = 0$. Thus we are looking for solutions $\varphi \in S$ of (3) for given $g \in S$.

We can regard S as a complete topological space over the reals, with vector addition $\alpha s+\beta t=\{\alpha s(n)+\beta t(n)\}$ as usual, and some convenient metric, say

$$(4) \qquad d(s,t)=\lim_{n\to\infty}\frac{\Sigma_n}{1+\Sigma_n}, \quad \Sigma_n=\sum_{i=1}^{n}|s(i)-t(i)|.$$

Indeed S only fails to be a Banach space by lacking $d(\alpha s,0)=|\alpha|d(s,0)$.

The inequality (3) is equivalent to

$$(5) \qquad \varphi(n)\leqslant \min_{1\leqslant r\leqslant \frac{1}{2}n}[\varphi(r)+\varphi(n-r)]+g(n) \quad (n\geqslant 2).$$

Hence, if we define $f\in S$ recursively by means of

$$(6) \qquad f(1)=0, \quad f(n)=\min_{1\leqslant r\leqslant \frac{1}{2}n}[f(r)+f(n-r)]+g(n) \quad (n\geqslant 2),$$

then f is a solution of inequality (3). Moreover it is the maximal solution of (3) in the sense that $f(n)\geqslant\varphi(n)$ for all n and any solution φ of inequality (3). This follows at once by induction on n. Harding's relation (1) is equivalent to equation (6) in the particular case

$$(7) \qquad f(n)=-\log M(n), \quad g(n)=\log \tfrac{1}{2}(n-1).$$

We propose to study equation (6) in general, and we define $\sigma(n)=\sigma(n,f)$ to be the set of integers r in $1\leqslant r\leqslant \frac{1}{2}n$ which minimize $f(r)+f(n-r)$. Let $\rho=\{\rho(n)\}_{n=2,3,\ldots}$ be any given sequence of integers such that $1\leqslant\rho(n)\leqslant\frac{1}{2}n$, and define S_ρ to be the set of all $f\in S$ such that $\sigma(n,f)=\rho(n)$ for $n=2,3,\ldots$. (*Note:* we regard a sequence as a one-valued function of n; so this definition is framed to exclude f from every S_ρ if the set $\sigma(n,f)$ contains more than one member for some n.) Any S_ρ is either empty or else a convex positive cone of S, and S is the union (over all ρ) of the closures of the S_ρ.

[To sketch a proof of the closure assertion, suppose $f\in S$ and prescribe $\varepsilon>0$. Let $h=\{h(i)\}$ be a sequence of independent random variables $h(i)$, with $h(i)$ uniformly distributed on the interval $[f(n)-2^{-n}\varepsilon, f(n)+2^{-n}\varepsilon]$. Then, with probability 1, $\sigma(n,h)$ is single-valued for all $n\leqslant n_0$ for any prescribed n_0, and hence for all n. Since any such sequence h satisfies $d(f,h)<\varepsilon$, there exists certainly (not merely with probability 1) some sequence $h^*\in\bigcup_\rho S_\rho$ such that $d(f,h^*)<\varepsilon$.]

From equation (6) we see that f is a continuous transform of g, which we may write

$$(8) \qquad f=Tg.$$

Indeed, recursive solution of equation (6) will yield $f(n)$ as a linear combination

$$f(n) = \sum_{m \leqslant n} t_{nm} g(m), \tag{9}$$

where the t_{nm} are integers.

The inverse transform

$$g = T^{-1} f \tag{10}$$

is also continuous and, when $f \in S_\rho$, can be written as

$$g(n) = f(n) - f[\rho(n)] - f[n - \rho(n)]. \tag{11}$$

However, the transformation T is not linear on S. It is merely positive-homogeneous

$$T(\alpha g) = \alpha Tg \quad (\alpha \geqslant 0) \tag{12}$$

and superadditive

$$T(g+h) \geqslant Tg + Th, \tag{13}$$

because $Tg + Th$ is a solution of inequality (3) with $g(n) + h(n)$ in place of $g(n)$. In matrix form T and T^{-1} are both lower triangular with unit diagonal.

While the foregoing exhibits the general structure (in fashionable but equally superficial jargon), the heart of our problem lies in specifying the appropriate sequence ρ. Specifically, given g, to which region S_ρ, if any, does Tg belong? Once this question is answered, we can compute f from (9). We have only answered the question in the three simplest cases:

Case (*i*): If g is strictly decreasing, then $\rho(n) = 1$.

Case (*ii*): If g is increasing and strictly convex, then $\rho(n) = [\frac{1}{2}n]$, the integer part of $\frac{1}{2}n$.

Case (*iii*): If g is increasing and strictly concave, then $\rho(n)$ is given by equation (2).

Proofs appear in the next sections, together with the corresponding explicit solutions (20), (27) and (41) for f.

There are sequences g such that Tg belongs to no S_ρ. (For example, if $g = (0, 0, \ldots)$ then $f = Tg = (0, 0, \ldots)$ and $\sigma(n)$ clearly consists of all integers in $[1, \frac{1}{2}n]$. Thus $\sigma(n)$ is not single-valued and Tg is a limit point of every S_ρ but a member of none.) Every Tg must be a limit point of some S_ρ (because $S = \bigcup_\rho \overline{S_\rho}$), and (in principle) we can choose

$$g^i = \{g^i(n)\}_{n=1,2,\ldots} \to g$$

as $i \to \infty$ such that $Tg^i \in S_\rho$. This will yield Tg via

$$Tg = \lim_{i\to\infty} Tg^i.$$

For example, if $g(n) = n-1$ we may find an increasing convex sequence converging to g and use case (*ii*). Alternatively we may find an increasing concave sequence converging to g and use case (*iii*).

For which ρ is S_ρ empty? We have no general answer to this; but such ρ certainly exist, for example

(14) $$\rho = (\rho(2), \rho(3), \dots) = (1, 1, 1, 2, 2, \dots).$$

If S_ρ were not empty for (14) we should have the obviously incompatible inequalities

(15) $$\begin{cases} f_1+f_3<f_2+f_2, \\ f_2+f_3<f_1+f_4, \\ f_2+f_4<f_3+f_3. \end{cases}$$

Similarly any sequence which commences like

(16) $$\rho = (1, 1, 2, 1, 3, \dots)$$

is not allowed, and it is easy to produce plenty of other forbidden sequences.

Hammersley [2] proved an asymptotic result for any solution φ of inequality (3) whenever g belonged to a certain subset of S. Namely, if g is non-decreasing for all $n \geqslant \xi > 0$, then the finite convergence of $\sum g(n)/n^2$ is equivalent to the existence of the limit

$$l = \lim_{n\to\infty} \varphi(n)/n,$$

where $-\infty \leqslant l < \infty$. In particular this holds for the maximal solution $f = Tg$ of inequality (3). For example, Harding's case of

$$g(n) = \log \tfrac{1}{2}(n-1) \quad (n>2)$$

is clearly non-decreasing, and $\sum g(n)/n^2 < \infty$. So

$$l = \lim_{n\to\infty} f(n)/n$$

exists and may be easily calculated. By case (*iii*), $f \in S_\rho$, where ρ is given by equation (2), and iterated solution of equation (6) yields

(17) $$2^{-n} f(2^n) = 2^{-1} g(2) + 2^{-2} g(2^2) + \dots + 2^{-n} g(2^n).$$

So

$$l = \lim_{n\to\infty} 2^{-n} f(2^n) = -\log 2 + \sum_{i=2}^{\infty} 2^{-i} \log(2^i - 1) \tag{18}$$

$$= 0.253\ldots.$$

4.2.2 PROOF OF RESULTS

Proof of Case (*i*)

The proof is by induction on n, the result being trivially true for $n = 2$. Suppose it holds for $m = 1, 2, \ldots, n-1$. Iteration of

$$f(m) = f(1) + f(m-1) + g(m) \quad (m < n) \tag{19}$$

yields

$$f(m) = \sum_{i=1}^{m} g(i). \tag{20}$$

So

$$f(n) = f[\rho(n)] + f[n - \rho(n)] + g(n) \tag{21}$$

$$= \sum_{i=1}^{\rho(n)} g(i) + \sum_{i=1}^{n-\rho(n)} g(i) + g(n)$$

$$\geqslant \sum_{i=1}^{n} g(i)$$

with equality if and only if $\rho(n) = 1$, since g is strictly decreasing. This completes the proof.

Notation for Cases (*ii*) *and* (*iii*)

The iterative decompositions of

$$f(n) = f[\rho(n)] + f[n - \rho(n)] + g(n) \tag{22}$$

may be represented diagrammatically by a tree with a simple *root* vertex with *weight n*, giving birth in the first generation to two vertices weighted $\rho(n)$, $n - \rho(n)$, each of which can create similarly two next generation vertices with suitably representative weights. A vertex fails to divide if it is weighted one. To the two vertices born of the same vertex, with weight w say, we assign the smaller weight $\rho(w)$ to the left-hand vertex of the pair and the larger weight $w - \rho(w)$ to the right-hand vertex. Clearly the weight of vertex A is precisely the number of terminal vertices generated by A and its descendants.

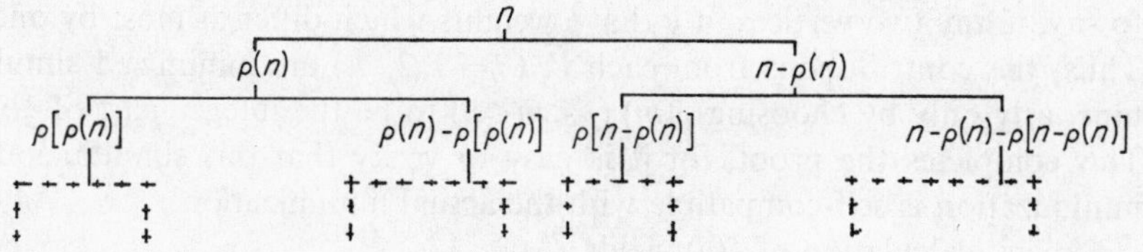

(23)

The solution $f = Tg$ is given by

(24) $$f(n) = \sum_{i \in I} g(a_i),$$

where the summation includes each vertex in the diagram (23), the weight of the ith vertex being a_i.

Vertex A is an *ancestor* of vertex B if B belongs to the subtree of which A is the root. The *depth* of A is the number of ancestors which it possesses, and the *depth* of a subtree is the depth of its root. A tree is *plenary* if the weights of any two vertices of equal depth are the same.

Proof of Case (*ii*)

We may extend the diagram (23) by allowing any vertex of zero or unit weight to give birth to vertices weighted (zero, zero) or (zero, one) respectively. We set $g(0) = 0$. This does not affect the strict convexity of g, since g is increasing and $g(1) = 0$. Let V_j be the set of depth-j vertices in the extended tree. For each j, V_j has 2^j members with total weight n. Their contribution of

$$\sum_{i=1}^{2^i} g(a_i)$$

to equation (24), where a_i is the weight of the ith vertex, is not changed by the extension. Furthermore we may extend the definition of g from a strictly convex increasing function on the non-negative integers to a convex non-decreasing function on the reals by

(25) $g(x) = (1-x+i)\,g(i+1)+(x-i)\,g(i) \quad (x \in (i, i+1)$ for some integer $i)$.

Then

(26) $$\sum_{i=1}^{2^i} g(a_i) \geqslant 2^j g\Big(2^{-j}\sum_{i=1}^{2^j} a_i\Big) = 2^j g(2^{-j}n)$$

with equality if and only if $a_i = 2^{-j}n$ $(i = 1, 2, \ldots, 2^j)$ whenever this is integer valued, or otherwise all the a_i lie in the interval of unit length with integer end-points embracing $2^{-j}n$. This is only possible with integer solutions if the spread of the weights of V_j is as small as possible; that is to say, if any two vertices in V_j have weights which differ at most by one. Thus, the contributions from each V_j ($j = 1, 2, \ldots$) are minimized simultaneously only by choosing $\rho(m)$ ($1 < m \leqslant n$) to be the integer part of $\frac{1}{2}m$. This completes the proof, for it is easy to verify that this simultaneous minimization is self-compatible with the actual minimization.

Explicit calculation of $f(n)$ yields

$$f(n) = \sum_{i=0}^{[\log_2 n]} 2^i(g([2^{-i}n]) + \{2^{-i}n\}\Delta g([2^{-i}n])), \tag{27}$$

where [ ] and { }, are the integer and fractional parts of their content and $\Delta g(x) = g(x+1) - g(x)$.

Proof of Case (iii)

We order the vertices of a tree as follows: a vertex α is *earlier* than a vertex β (written $\alpha < \beta$) when the depth of α is less than the depth of β, or else when α and β are of equal depth and α lies to the right of β in the same row of the tree. We shall first prove the lemma:

Lemma. *When the vertices of an optimally weighted tree are taken in increasing order of lateness, the weights of the vertices form a non-increasing sequence.*

Proof. Suppose the lemma is not true; and let β_0 be the earliest vertex which is heavier than some still earlier vertex, and let α_0 be the earliest vertex which is lighter than β_0. Writing $w(\alpha)$ for the weight of the vertex α, we can then assert

$$w(\alpha) \geqslant w(\beta) \quad \text{whenever } \alpha \leqslant \beta < \beta_0, \tag{28}$$

while

$$w(\alpha_0) < w(\beta_0), \quad \alpha_0 < \beta_0. \tag{29}$$

Any vertex is both lighter and later than any of its ancestors; so the inequality (29) ensures that α_0 cannot be an ancestor of β_0 nor vice versa. Therefore, α_0 and β_0 have a latest common ancestor γ distinct from both. Let $\alpha_1, \alpha_2, \ldots, \alpha_p$ be the successive ancestors of α_0 leading up to $\alpha_p = \gamma$; and let $\beta_1, \beta_2, \ldots, \beta_q$ be corresponding ancestors of β_0 leading up to $\beta_q = \gamma$. Since $\alpha_0 < \beta_0$, we see that α_0 is not deeper than β_0, which implies

$$p \leqslant q. \tag{30}$$

If $q = 1$, then $p = 1$, and γ is the immediate common ancestor of both α_0 and β_0, which implies

$$w(\beta_0) = \rho[w(\gamma)] \leqslant w(\gamma) - \rho[w(\gamma)] = w(\alpha_0) \tag{31}$$

in contradiction of the inequality (29). Hence

$$q \geqslant 2. \tag{32}$$

If α and β are two vertices satisfying $\alpha \leqslant \beta$, the immediate ancestor of α is not later than the immediate ancestor of β. Hence $\alpha_0 < \beta_0$ implies $\alpha_1 \leqslant \beta_1 < \beta_0$, which implies successively

$$\alpha_i \leqslant \beta_i < \beta_{i-1} < \ldots < \beta_0 \quad (i = 1, 2, \ldots, p). \tag{33}$$

Writing

$$a_i = w(\alpha_i), \quad b_i = w(\beta_i), \tag{34}$$

we obtain from the inequalities (28), (29) and (33)

$$a_0 < b_0, \quad a_i \geqslant b_i \quad (i = 1, 2, \ldots, p). \tag{35}$$

Now consider the effect of interchanging the two subtrees with roots at α_0 and β_0. Most of the weights in the tree will merely be shifted in position without changing their magnitudes; and the only weights to suffer a change of magnitude will be those at $\alpha_1, \alpha_2, \ldots, \alpha_{p-1}$ and $\beta_1, \beta_2, \ldots, \beta_{q-1}$. The former will be changed from $a_1, a_2, \ldots, a_{p-1}$ to $a_1 + \delta, a_2 + \delta, \ldots, a_{p-1} + \delta$, and the latter from $b_1, b_2, \ldots, b_{q-1}$ to $b_1 - \delta, b_2 - \delta, \ldots, b_{q-1} - \delta$, where

$$\delta = b_0 - a_0 > 0. \tag{36}$$

Thus the total score (24) for the new tree will exceed the total score for the original tree by an amount

$$\Delta = \sum_{i=1}^{p-1} [g(a_i + \delta) - g(a_i) + g(b_i - \delta) - g(b_i)] + \sum_{i=p}^{q-1} [g(b_i - \delta) - g(b_i)]. \tag{37}$$

In equation (37), we interpret a void sum

$$\left(\text{such as } \sum_{i=1}^{0} \quad \text{or} \quad \sum_{i=p}^{p-1}\right)$$

as zero if it should occur. However, the first sum on the right of equation (37) can only be void if $p = 1$, in which case the inequality (32) prevents the second sum from being void. So at least one of the sums in equation (37) is non-void, and we shall be able to conclude that

$$\Delta < 0 \tag{38}$$

provided that each individual term actually occurring on the right of equation (37) is strictly negative. And this is true; because the inequality (36) implies

$$g(b_i - \delta) - g(b_i) < 0 \tag{39}$$

inasmuch as g is a strictly increasing function. Also, since

$$b_i - \delta < b_i \leqslant a_i < a_i + \delta,$$

the chord of the strictly concave function g from $b_i - \delta$ to $a_i + \delta$ must lie strictly below the chord from b_i to a_i, and therefore the mid-point of the former chord lies strictly below the mid-point of the latter chord. Hence

$$g(a_i + \delta) + g(b_i - \delta) - g(a_i) - g(b_i) < 0. \tag{40}$$

This establishes the inequality (38). However (38) contradicts the assumption that the original tree was optimally weighted in the sense of having a minimum total score; and this contradiction proves the lemma.

Now consider an optimally weighted tree. Descend from its root, always taking the left-hand branch, until a vertex V of unit weight is reached (at depth D, say). Let v be any other vertex of depth D. Then $v < V$, and the lemma gives $w(v) \geqslant 1$. Suppose $w(v) > 2$. Then v has a right-hand descendant v' at depth $D+1$ such that $v' > V$ and

$$w(v') \geqslant \tfrac{1}{2}w(v) > 1 = w(V),$$

contradicting the lemma. So $1 \leqslant w(v) \leqslant 2$. The vertices at depth D, read as a row from right to left, have non-increasing weights. Hence, if the weights in the right-hand half of the row are not all 2, the weights in the left-hand half must all be 1. In either case, one of the two depth-1 subtrees of the original tree will have equal weights of unity in its last row, and so will be plenary. The weight of its root will be a power of 2, say 2^i. The weight of the other depth-1 subtree must be at least 2^{i-1} and at most 2^{i+1}. If the left-hand subtree is plenary, we get $2.2^i \leqslant n \leqslant 3.2^i$; and if the right-hand subtree is plenary, we get $\frac{3}{2}.\ 2^i \leqslant n \leqslant 2.2^i$. In either case, $\frac{1}{3}n \leqslant 2^i \leqslant \frac{2}{3}n$ holds. This completes the proof of equation (2).

Explicit calculation of $f(n)$ now yields

$$f(n) = \sum_{k=0}^{i} (2^k - 1) g(2^{i-k}) + \left[\frac{n-2^i}{2^i - k}\right] (g(2^{i-k+1}) - g(2^{i-k})) \tag{41}$$

$$+ g\left(2^{i-k}\left(1 + \left\{\frac{n-2^i}{2^{i-k}}\right\}\right)\right),$$

where $1 < 2^{-i}n \leqslant 2$, and [ ] and { } indicate the integer and fractional parts of their content.

4.2.3 UNLABELLED SHAPES

In connection with his *unlabelled shapes*, Harding considered a modified form of equation (1) with the factor 2 in the numerator $2M(r)M(n-r)$ suppressed whenever $r = n - r$. This leads to a very irregular sequence ρ, which he tabulated for $n \leqslant 160$. We have made some fragmentary but very limited progress with this more difficult problem, which in our notation reads as

$$(42) \qquad f(1) = 0, \quad f(n) = \min_{1 \leqslant r \leqslant \frac{1}{2}n} [f(r) + f(n-r) + g(n) + c\delta_{r,n-r}],$$

where

$$(43) \qquad g(n) = \log \tfrac{1}{2}(n-1), \quad c = \log 2,$$

and δ is the Kronecker delta.

In the first place, this new f is a solution of inequality (5) with g replaced by $g(n) + c = \log(n-1)$; and, since $\sum n^{-2} \log(n-1)$ converges, we still have $f(n)/n$ tending to a limit as $n \to \infty$. We have made computer calculations of f for $n \leqslant 8000$; and the limit appears to be approximately 0.587.

Secondly, in place of the previous relation $\rho(n) \geqslant n/3$, we now have

$$(44) \qquad \rho(n) > n/5.$$

To see this, consider the tree (23); and for brevity write

$$(45) \qquad \alpha = \rho(n), \quad \beta = \rho[n - \rho(n)], \quad \gamma = n - \rho(n) - \rho[n - \rho(n)]$$

for the initial splitting of n. The definition of ρ ensures that

$$(46) \qquad \beta \leqslant \gamma;$$

and, of course, we have $\alpha + \beta + \gamma = n$. We consider interchanges of subtrees, as in Section 2; and we say that the *penalty* c is *incurred* wherever a non-zero Kronecker delta occurs.

Suppose that the tree is optimally weighted. Then we cannot have

$$(47) \qquad \alpha < \beta;$$

for in this event an interchange of α and β cannot incur a penalty at the lower level because $\alpha < \gamma$ in view of inequalities (46) and (47), nor at the higher level because $\beta \leqslant \frac{1}{2}[n - \rho(n)] < \frac{1}{2}n$. The change of score must be non-negative, since the tree was optimally weighted. Thus

$$g(\alpha + \gamma) - g(\beta + \gamma) \geqslant 0,$$

contradicting inequality (47). If $\alpha = \beta$, $\gamma = \frac{1}{2}n$, then $\alpha = \frac{1}{4}n$ and inequality (44) is trivial. So we consider

$$(48) \qquad \text{either} \quad \alpha > \beta \quad \text{or} \quad \gamma \neq \tfrac{1}{2}n, \quad \alpha \geqslant \beta.$$

An interchange of α and γ can only incur one penalty at most, perhaps at the upper level in the former case of (48), and perhaps at the lower level in the latter case of (48). The change of score is at most

$$(49) \qquad g(\alpha + \beta) + \log 2 - g(\beta + \gamma) \geqslant 0,$$

whence

$$(50) \qquad \beta + \gamma - 1 \leqslant 2(\alpha + \beta - 1) \leqslant 4\alpha - 2,$$

by (48). Hence $n = \alpha+\beta+\gamma \leqslant 5\alpha-1$; and $\rho(n) = \alpha > n/5$.

Finally, we show that the penalty is never incurred for $n > 2$. Suppose that the penalty is incurred for $n = 2m$, with $m > 1$ and m as small as possible. Then

(51) $$2f(m) = 2f(\tfrac{1}{2}n) \leqslant f(r)+f(n-r)-\log 2 \quad (1 \leqslant r < m).$$

With $r = m-1$, this yields

(52) $$\Delta^2 f(m-1) \geqslant \log 2,$$

where $\Delta^2 h$ is the second forward difference of a function h, and (later) Δh is the first forward difference. For $2 < s \leqslant m+1 < 2m$ the penalty is not incurred, by definition of m, and hence

(53) $$f(s) = f[\rho(s)]+f[s-\rho(s)]+g(s),$$

(54) $$f(s+1) \leqslant f[\rho(s)+1]+f[s-\rho(s)]+g(s+1),$$

whence

(55) $$\Delta f(s) \leqslant \Delta f[\rho(s)]+\Delta g(s).$$

Write

(56) $$r_j = m, \quad r_{j-1} = \rho(r_j), \ldots, \quad r_0 = \rho(r_1),$$

where j is to be chosen presently. Substituting $s = r_i$ $(1 \leqslant i \leqslant j)$ in inequality (55) and adding, we get

(57) $$\Delta f(m) \leqslant \Delta f(r_0) + \sum_{i=1}^{j} \Delta g(r_i).$$

We can write inequality (44) as

(58) $$\tfrac{1}{5}n < \rho(n) \leqslant \tfrac{1}{2}n;$$

and so equations (56) and (58) provide

(59) $$r_i \geqslant 2^i r_0.$$

However

(60) $$\Delta g(n) = \log\left(1+\frac{1}{n-1}\right),$$

which is a decreasing function of n. So

(61) $$\sum_{i=1}^{j} \Delta g(r_i) \leqslant \sum_{i=1}^{j} \log\left(1+\frac{1}{2^i r_0-1}\right) \leqslant \sum_{i=1}^{\infty} \frac{1}{2^i r_0-1}$$

$$\leqslant \frac{2r_0}{2r_0-1} \sum_{i=1}^{\infty} \frac{1}{2^i r_0} = \frac{2}{2r_0-1}.$$

Hence

(62) $$\Delta f(m) \leqslant \Delta f(r_0) + \frac{2}{2r_0 - 1}$$

Similarly, instead of equations (53) and (54), we have

(63) $$f(s+1) = f[\rho(s+1)] + f[s+1-\rho(s+1)] + g(s+1),$$

(64) $$f(s) \leqslant f[\rho(s+1)-1] + f[s+1-\rho(s+1)] + g(s),$$

which yield

(65) $$\Delta f(s) \geqslant \Delta f[\rho(s+1)-1] + \Delta g(s) \geqslant \Delta f[\rho(s+1)-1].$$

This time we write, with J to be chosen presently,

(66) $$R_J = m-1, \quad R_{J-1} = \rho(R_J+1)-1, \ldots, \quad R_0 = \rho(R_1+1)-1.$$

Substituting $s = R_i$ $(1 \leqslant i \leqslant J)$ in inequality (65) and adding, we have

(67) $$\Delta f(m-1) \geqslant \Delta f(R_0).$$

From inequalities (52), (62) and (67), we deduce

(68) $$\log 2 \leqslant \frac{2}{2r_0 - 1} + \Delta f(r_0) - \Delta f(R_0).$$

From equations (56), (58) and (66), we see that

(69) $$\tfrac{1}{5} r_i < r_{i-1} \leqslant \tfrac{1}{2} r_i, \quad \tfrac{1}{5}(R_i+1) < R_{i-1}+1 \leqslant \tfrac{1}{2}(R_i+1).$$

Hence, for $m \geqslant 174$, we can always choose j and J such that

(70) $$6 \leqslant r_0 \leqslant 30, \quad 6 \leqslant R_0 \leqslant 34.$$

Now inequalities (68) and (70) give the contradiction

(71) $$0.693 < \log 2 \leqslant \tfrac{2}{11} + \max_{6 \leqslant r \leqslant 30} \Delta f(r) - \min_{6 \leqslant R \leqslant 34} \Delta f(R) < 0.623.$$

Here we have used the computed values of $\Delta f(n)$ for $6 \leqslant n \leqslant 34$. Thus the first incurrence (after $n = 2$) of a penalty cannot occur for $n = 2m \geqslant 348$. However, the computer studies also show that no penalty is incurred for $2 < n < 348$. This completes the proof that the penalty is never incurred for $n > 2$.

It is possible to have $\rho(n) < \frac{1}{3}n$; and it seems likely from the computer studies that this happens infinitely often. At least is happens for the following values of n:

Table 1. Values of $n \leqslant 8000$ for which $\rho(n) < \frac{1}{3}n$

n	$n-3\rho(n)$	n	$n-3\rho(n)$	n	$n-3\rho(n)$	n	$n-3\rho(n)$
4	1	1369	1	2758	4	5444	14
10	1	1370	2	5392	7	5456	2
22	1	1372	4	5400	3	5461	7
168	3	1374	6	5408	2	5466	12
343	1	1388	2	5414	8	5492	2
345	3	2700	6	5418	6	5507	5
669	6	2702	8	5435	5	5508	6
671	8	2714	2	5437	7	5524	4
682	1	2719	7	5438	8	5525	5
683	2	2722	10	5439	9	5529	9
1346	2	2744	2	5440	10		
1352	8	2757	3	5442	12		

We have attempted to measure the *popularity* $P(n)$ of an integer n by counting the number of integers i for which either $n = \rho(i)$ or $n = i-\rho(i)$ and assigning this number to $P(n)$. Then $P = \{P(n)\}_{n=2,3,\dots}$ is an infinite sequence of integers which we have calculated for $n \leqslant 2666$. The popularity sequence P_1 for solution (2) to equation (1) is

(72) $$P_1 = \{P_1(n)\}_{n=2,3,\dots} = (5, 2, 8, 2, 2, 2, 14, 2, \dots)$$

or

(73) $$P_1(n) = \begin{cases} 2+2^{i-1}+2^i & \text{if } n = 2^i, \\ 2 & \text{otherwise,} \end{cases}$$

and indicates the form of the most popular integers. Continuing this analogy with equation (2), we conjectured that in order to find the 'best' r ('best' by some unknown rule) to minimize the right-hand side of equation (42) it is only necessary to scan the numbers $\{f(r)\}_{r=1,\dots,n-1}$ without consideration of the complementary set $\{f(n-r)\}_{r=1,\dots,n-1}$. (For example equation (2) defines the 'best' r for Harding's *labelled shapes* as a power of 2 lying in $[\frac{1}{3}n, \frac{2}{3}n]$.) Thus, in order to be consistent with our notion of popularity, we desired that for each n the chance choice of $n-r$ should have no effect upon P. So from P we constructed an amended sequence P^* of popularities by taking one away from the lesser of $P[\rho(n)]$ and $P[n-\rho(n)]$ for each $n \leqslant 8000$. (The amended sequence P_1^* for equation (7) is

(74) $$P_1^* = \{P_1^*(n)\}_{n=2,3,\dots} = (4, 0, 7, 0, 0, 0, 13, 0, \dots)$$

or

(75) $$P_1^*(n) = \begin{cases} P_1(2^i)-1 & \text{if } n = 2^i, \\ 0 & \text{otherwise,} \end{cases}$$

which exhibits clearly the nature of equation (2).) Our sequence P^* for unlabelled shapes has non-zero entries as given in Table 2.

Table 2. Values of $n \leqslant 2666$ for which $P^*(n) > 0$

n	$P^*(n)$	n	$P^*(n)$	n	$P^*(n)$	n	$P^*(n)$	n	$P^*(n)$
3	4	219	2	487	1	1133	1	1834	474
4	1	220	3	568	2	1141	6	1838	28
5	2	221	36	664	1	1158	1	1839	28
7	†6	227	112	675	1	1169	2	1840	121
8	†6	229	99	677	3	1207	2	1843	14
12	†2	231	10	685	3	1247	1	1844	77
15	16	232	†1	696	11	1253	2	1845	13
17	†2	233	23	751	2	1265	1	1848	19
20	5	235	†4	783	4	1341	4	1849	55
27	11	236	11	809	2	1360	9	1853	†11
28	3	240	25	885	14	1391	10	1856	4
31	1	245	†4	888	†4	1393	2	1857	19
32	9	247	1	889	†4	1589	2	1858	1
35	14	253	2	891	7	1612	1	1861	9
42	5	255	†4	895	14	1614	8	1865	8
47	1	295	2	898	60	1701	1	1866	31
52	2	323	†2	904	317	1783	8	1944	7
55	26	335	4	906	106	1787	3	2035	1
58	2	341	2	911	65	1789	8	2059	14
59	15	353	2	913	50	1793	7	2076	2
60	4	355	1	914	114	1795	16	2171	14
62	3	408	†1	916	111	1799	29	2258	1
67	16	430	†2	917	1	1802	246	2264	3
77	5	437	8	918	261	1803	11	2266	1
90	8	441	†2	921	5	1804	33	2278	14
97	7	443	1	922	40	1807	23	2322	1
102	3	448	97	925	1	1810	370	2326	1
107	11	450	27	926	40	1815	153	2483	1
113	19	456	331	927	4	1817	39	2485	1
114	77	457	1	930	20	1818	176	2507	2
115	17	458	6	931	26	1819	30	2619	14
117	4	460	29	935	30	1820	39		
119	5	462	62	937	1	1822	26		
121	†4	464	7	942	24	1824	4		
122	3	465	10	944	1	1825	108		
126	12	468	16	948	†11	1827	17		
127	2	469	19	949	1	1829	11		
129	2	473	28	980	3	1830	240		
181	6	475	9	1026	4	1831	24		
216	†4	476	7	1131	1	1832	149		

† P^* was first constructed for values of n such that $P[\rho(n)] \neq P[n-\rho(n)]$. For integers excluded by this condition the comparison process was repeated using P^* as the yardstick. However, for the pairs $[\rho(n), n-\rho(n)] = (7, 8)$, $(12, 17)$, $(121, 216)$, $(232, 408)$, $(235, 255)$, $(245, 255)$, $(323, 441)$, $(430, 441)$, $(888, 889)$, $(948, 1853)$ both $P[\rho(n)] = P[n-\rho(n)]$ and $P^*[\rho(n)] = P^*[n-\rho(n)]$. These values are marked by a dagger.

Table 2 shows that the most popular integers are distributed about the numbers 15, 27, 55, 114, 228, 456, 912, 1824. We do not know why these numbers are chosen, nor have we made any significant progress towards understanding their associated distributions. Indeed, further investigation seems likely to be challenging since any exhaustive quantitative analysis will be very involved. More details of $f(n)$, $\rho(n)$, $P(n)$ ($n \leqslant 8000$) may be obtained from us.

We have just heard of a paper [3], which appears (we have so far only seen the summary and not the paper itself) to deal with closely related, if not identical, material.

REFERENCES

1. E. F. Harding, "The probabilities of rooted tree-shapes generated by random bifurcation", *Adv. Appl. Prob.* **3** (1971), 44–77. (See also 4.1 of this book.)
2. J. M. Hammersley, "Generalization of the fundamental theorem on sub-additive functions", *Proc. Camb. Phil. Soc.* **58** (1962), 235–238.
3. M. Fredman and D. Knuth, "Recurrence relations based on minimization", *Stanford University Computer Science Report No.* 248 (1972).

5

NAVIGATION IN THE PRESENCE OF AN UNCERTAINTY

5.1

A Stochastic Cross-country or Festina Lente

A. W. F. EDWARDS

[Reprinted from *Sailplane and Gliding*, **14** (1963), 12–14]

'*Whatever do you mean by that?*'

'*By what?*'

'*A stochastic cross-country? What does "stochastic" mean?*'

'*It means that there is an element of chance in the flight: you might not reach your goal.*'

'*But all flights are like that.*'

'*Yes.*'

'*Then why bother to call them by a long word when everyone knows this fact?*'

'*Well, it's like this . . .*'

Every cross-country pilot knows that his primary task is to stay up. Only when he is reasonably satisfied about this can he start thinking about the best-speed-to-fly, and why Little Rissington hasn't turned up yet, and other such things. And yet, when he comes to work out his best speed, he will certainly not take into account, mathematically, the possibility of a premature landing, although he will do so in his mind ('Better not fly as fast as that . . . might get too low'). But there is no reason why he shouldn't feed the chance, or stochastic, element into his calculator. Much is known about Stochastic Processes nowadays, and in this article I want to introduce them to gliding in a very simple example: so simple, in fact, as to be rather unrealistic. But one has to start somewhere.

Today there is no wind. Thin cumulus are randomly dotted over the sky, and I have declared Little Rissington. I am determined not to stray from my track, and a cursory glance at the clouds reveals that thermals will be randomly spaced along the route, every d ft on average. My operational height-band will be h ft deep, and—another glance upwards—my rate of climb in thermals will be u ft/s. And, best of all, no down between thermals! Since Little Rissington is nd ft away, I'll need about n thermals to get me there. And I mustn't forget my glider—she sinks at $s = Av^3 + B/v$ ft/s when flown at v ft/s. All ready? Right! Hook on, and let's go.

The distance between adjacent thermals is a random variable, x, which is evidently exponentially distributed with probability density $(1/d)\exp(-x/d)$. (Help! He's in cloud already!) If you don't know about these things, just shut your eyes for the next few minutes. Now, consider the glide from the top of one thermal to the next one, x ft distant, during which the glider is flown at v ft/s. The glide takes x/v seconds, and thus consumes sx/v, or $x(Av^2 + B/v^2)$, feet of height. If this loss exceeds h ft, the glider will land; that is, if x exceeds $h/(Av^2 + B/v^2)$. But the probability of this happening is

$$\frac{1}{d}\int_{h/(Av^2+B/v^2)}^{\infty} e^{-x/d}\,dx,$$

which equals $\exp[-h/d(Av^2 + B/v^2)]$. Thus the probability of still being airborne after n glides between thermals (which, you may remember, will take me to Little Rissington) is

$$\mathbf{P} = \{1 - \exp[-h/d(Av^2 + B/v^2)]\}^n.$$

This is the probability of my reaching the goal. A little thought shows that it has a maximum at $v = (B/A)^{0 \cdot 25}$, which is the speed for best gliding angle, as *The Soaring Pilot* will tell you. This is as it should be, and we deduce that the maximum probability of arrival is

$$[1 - \exp(-h/2d\sqrt{(AB)})]^n.$$

The Soaring Pilot also tells us that the average cross-country speed is

$$w = uv/(u + Av^3 + B/v).$$

In order to maximize this I would have to fly faster than my best-gliding-angle speed, as everyone knows, but the probability of my reaching the goal would then be reduced. By how much? Let's look at an actual example.

Suppose Little Rissington is 100 km away, and the thermals are 4 miles apart on average. d is thus about 21,000 ft, and I will need about $n = 16$ thermals. Suppose the operational height-band, h, is 3000 ft, and the rate of climb in thermals, u, 5 ft/s. If my glider is a Swallow we may guess $A = 4.5 \times 10^{-6}$ and $B = 100$ roughly.

Now the last two equations relate the probability of arrival, **P**, to the average speed, w, by means of the parameter, v. We may therefore draw a graph of **P** against w, keeping an eye on v at the same time. I have done this in Figure 1. We see that if I fly at the best-gliding-angle speed, 40

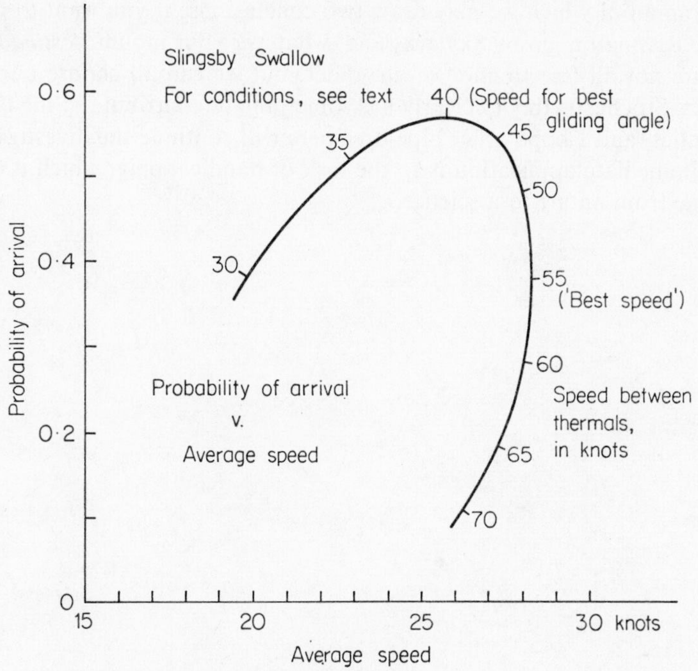

Figure 1

knots, the probability of arrival is 0.57, but if I fly at the 'best-speed-to-fly', 55 knots, the probability is only 0.38, and the average speed has only gone up $2\frac{1}{2}$ knots. An increase of 10% in the average speed costs a reduction of 33% in the probability of arrival. Is it worth it? Well, that depends upon the object of the flight, whether it is a race or not, and, if it is, what marking system is being used. The expectation of points on an any given system can be maximized and the appropriate speed-to-fly found.

A more striking deduction from the graph is what happens around the 'best-speed-to-fly'. It is often said, quite truly, that so long as one 'stuffs the nose down' in between thermals, one will come to within a knot or two of the best possible average speed. Thus, in our example, all speeds between 46 and 65 knots lead to cross-country speeds within 1 knot of the maximum. But look what happens to the probability of arrival: it ranges from 0.54 to 0.18—a factor of three!

It is interesting to compare a Swallow's maximum probability of arrival with that of a Skylark 3, for which AB must be about 2.5×10^{-4}, compared with a Swallow's 4.5×10^{-4}. It turns out to be 0.84, as against 0.57.

From all of which we may draw two conclusions: if you want to get to Little Rissington, go by Skylark; and, whatever your mount, *festina lente*! We are not all free to choose our glider, but we can all choose our own tactics. Stochastic theory clearly has something to contribute to the theory of tactics, and I hope other pipe-dreamers will continue the investigation. One immediate application is to the task of handicapping, which it could change from an art to a science.

5.2

The Fitzwilliam Street Problem

E. M. WILKINSON

5.2.1 INTRODUCTION TO PROBLEM

The quickest way to walk from Trinity to the Chemistry Laboratory is along King's Parade, Trumpington Street, Fitzwilliam Street and Tennis Court Road. When the Statistical Laboratory was at its former site on Lensfield Road, Rollo was faced with this daily walk. He explained to me how he had tried to minimize the distance he had to pace each day. The one problem he had not yet resolved was how best to walk across Fitzwilliam Street.

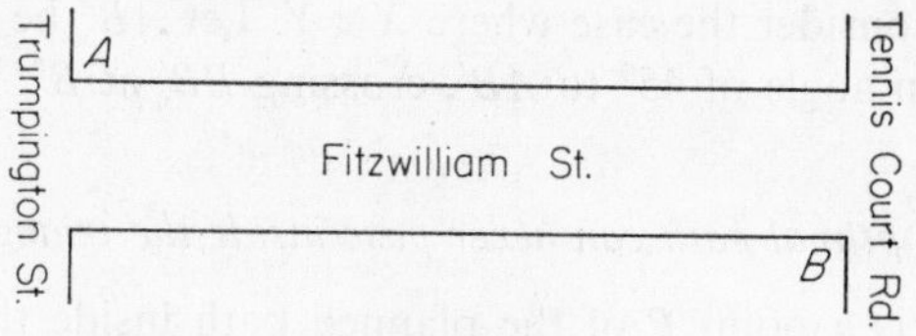

Figure 1

One has to go from *A* to *B* in the minimum expected distance. The catch is, of course, that whenever a car appears in the street, one has to walk directly to the pavement on the *B*-side. We shall assume that cars appear after an exponential holding time.

Rollo felt that the correct answer was the bold strategy of walking in a straight line from *A* to *B*, until a car appeared. I was never able to tell him my solution.

The problem of finding a planned route from *A*, which minimizes the expected distance to walk, is an obvious candidate for the calculus of variations. So far, this approach to the problem has been fruitless. I have, however, used a different method to obtain a solution.

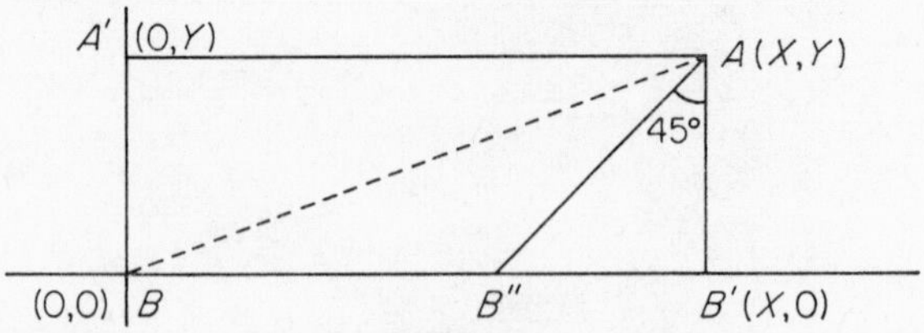

Figure 2

5.2.2 PROPERTIES FOR A PLANNED PATH

Let us suppose that there is an optimal planned path from A. In this section, I shall find certain properties that such a planned path must satisfy.

An important feature of an optimal planned path from A is that if a point $P(x,y)$ is on the path, then the part of the path from P onwards is an optimal planned path from P. So any property that the path has to have at (X, Y) will have to hold at all points (x,y) of the path.

It is clear that sufficiently close to A all points of an optimal curve lie in the rectangle $A'AB'B$. This property will hold for all points of the optimal curve, and so we can deduce that an optimal curve is a path of bounded variation starting at A and finishing at B. At all times it must lie in the rectangle $A'AB'B$.

Let us now consider the case where $X \geqslant Y$. Let AB'' be the straight line through A at an angle of 45° to AB', crossing BB' at B''. I will show that we must have

Lemma 1. *An optimal path can never pass inside the triangle $AB'B''$.*

Proof. If there is a point P of the planned path inside the triangle, then there are points P' and P'' on AB'' such that the arc of the planned path $P'PP''$ lies in the triangle, with P' and P'' being the only points on AB''.

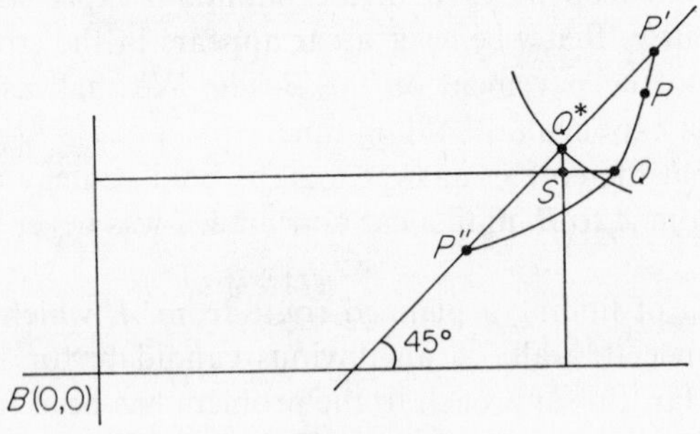

Figure 3

I will show that replacing the arc $P'PP''$ by the straight line $P'P''$ in the planned path produces a new planned path with strictly smaller expected distance. This contradiction will complete the proof.

Firstly the length of the straight line $P'P''$ is strictly shorter than the length of the arc $P'PP''$. Note that all lengths are well defined as the curves must be of bounded variation. Secondly we must consider the case where a car appears while pedestrians are between P' and P''. Let Q and Q^* denote the positions of pedestrians on the two routes when the car appears, so that $P'Q^* = \text{arc } P'Q$. Q lies inside the circle centre P' radius $P'Q^*$. The difference in the distances to walk from Q^* and Q in this case is $Q^*S - QS$. Provided the line $P'P''$ is at an angle greater or equal to 45°, this difference is strictly negative. We shall use this argument again in the next lemma, with the arc $P'PP''$ above the line $P'P''$, and the angle between $P'P''$ and the x-axis less or equal to 45°.

I now want to prove that the curve can never pass inside the triangle $AA'B$. I shall show, in fact, that a stronger result is true.

Lemma 2. *An optimal planned path must be convex from below.*

If this were not true then we could find an arc $P'PP''$ of the supposed optimal planned path such that the path is strictly above the straight line $P'P''$, except at the end points.

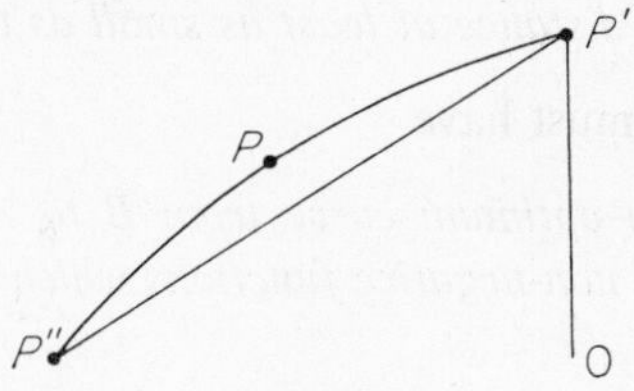

Figure 4

We notice that the angle $P''P'0$ is greater than or equal to 45°. An argument identical to the one above shows that the short cut is advantageous.

Convex curves have a left- and right-hand derivative at all interior points. These derivatives are monotonic increasing from B to A. There can only be a countable number of discontinuities in the derivatives. We have shown that for left-hand derivatives

$$(1) \qquad \frac{Y}{X} \leqslant \left[\frac{dy}{dx}\right]_{(X,Y)} \leqslant 1$$

and hence for the general point (x, y) of an optimal planned path we have the following inequality for the left-hand derivatives

(2) $$\frac{y}{x} \leqslant \left[\frac{dy}{dx}\right]_{(x,y)} \leqslant 1.$$

It can similarly be shown that these inequalities are true for the right-hand derivatives as well.

5.2.3 EXISTENCE OF OPTIMAL PLANNED PATHS

If A is the point $(X, 0)$ or $(0, Y)$, then obviously the unique optimal planned path is the straight line from A to B. If A is the point (X, X), then the inequalities (2) show that there is only one candidate for the optimal path, namely the straight line from A to B. But we will have to prove that there exists an optimal path from (X, X).

In this section I will show firstly that we have

Lemma 3. *Given any path from $A(X, Y)$ to $B(0, 0)$, where $X \geqslant Y$, there is a planned path which is a convex, continuous, increasing, non-negative function on $[B, A]$ and satisfies*

(3) $$\frac{y}{x} \leqslant \left[\frac{dy}{dx}\right]_{(x,y)} \leqslant 1,$$

which has an expected distance at least as small as the given path,

and secondly that we must have

Lemma 4. *There is an optimum curve from B to A in the set of convex, continuous, increasing, non-negative functions which satisfy condition* (3) *on the derivatives.*

These together will prove the existence, but not necessarily the unicity, of an optimum curve.

Proof of Lemma 3. Consider an arbitrary path from A. The first step is to show that there is a path which lies entirely inside the rectangle $AA'BB'$, which is at least as good. This path inside the rectangle can be constructed very simply by 'folding' the plane along the lines $A'A$ and AB'. That is, reflect those parts of the path above the line $y = Y$ to below the line, and then reflect those parts to the right of the line $x = X$ to the left of the line. If this folded curve meets either the x-axis or the y-axis, then replace the remaining part of the curve by the straight line to B. This has clearly produced a planned path inside the rectangle with an expected distance at least as small.

Similarly by folding along the line AB'', we can produce a curve in the quadrangle $AA'BB''$ at least as good as the original curve.

The next stage is to find a path from A to B, given by a monotonic function, that is as good as the path in the quadrangle. If this last curve is given by $(x(s), y(s))$, where s is the distance from A, then construct the curve

$$\left(\min_{0 \leqslant t \leqslant s} x(t), \min_{0 \leqslant t \leqslant s} y(t) \right).$$

If this curve does not reach B, then add on the straight line from B to complete the path. This is then a path defined by a monotonic function, with an expected distance at least as good as the original arbitrary path. Note that the monotonic function may not be continuous.

The lower convex envelope of this monotonic function will be the convex, increasing, non-negative, continuous function that we require. Condition (3) will automatically be satisfied since the path is convex and does not enter the triangle $AB'B''$. So we have only to show that this curve is at least as good as the path given by the monotonic function.

The two paths differ on, at most, a countable number of disjoint intervals along the paths. Consider such an open interval. The lower convex envelope will be the straight line between the two end points. Notice that the angle this line makes with the x-axis is less than or equal to 45°. The construction in Lemma 2 shows that if we include this short cut in a new monotonic function we will have a strictly better planned path. If there is only a finite number of such open intervals, we have the required result.

If there is an infinite number of open intervals, then after the construction of a finite number of short cuts, we can assume that the total arc length of the remaining intervals is less than ε. The differences in total lengths of the monotonic function and its convex envelope will be less than 2ε. Consider points Q^* and Q on the two curves, representing pedestrians' positions after a certain time. There is a point of contact T between the two curves within ε of Q. The difference in arc lengths from B to T is at most 2ε. It follows that the distance between Q and Q^* must be less than 4ε. If a car comes at this point, the difference between the distances to be walked is less than 8ε. So the maximum difference between the lengths of simultaneous journeys along the two paths is at most 8ε. Hence the difference in the expected distances to walk is less than 8ε. This figure can be made less than the strictly positive advantage gained by making the first short cut in the above construction. This completes the proof of Lemma 3.

Let S be the set of convex, continuous, increasing, non-negative functions on $[B, A]$ corresponding to paths from A to B. We shall consider the subset T, consisting of those elements of S that satisfy condition (3), together with a metric given by

$$d(f, g) = \max_{0 \leqslant x \leqslant X} (|f(x) - g(x)|). \tag{4}$$

The limit of a sequence of functions in S is a convex, increasing, non-negative function, so necessarily continuous except perhaps at A. With the added conditions on the difference quotients implied by condition (3), we see that we must also have continuity at A, and that any limit point of a sequence from T is itself a member of T. So we see that T is a closed, bounded subset of $C[0, X]$.

Condition (3) implies that T is an equicontinuous family of functions. The Arzela–Ascoli theorem then tells us that these properties imply the compactness of the set T with respect to the metric topology associated with equation (4).

If we can show that the function which calculates the expected distances for planned paths from T is continuous in the metric topology, we will have proved Lemma 4.

I will use the following result in the proof.

Lemma 5. *If f and g are continuous, convex functions on $[a, b]$, such that $d(f, g) < \varepsilon$, then the difference in lengths of the graphs of the two functions is less than or equal to 4ε.*

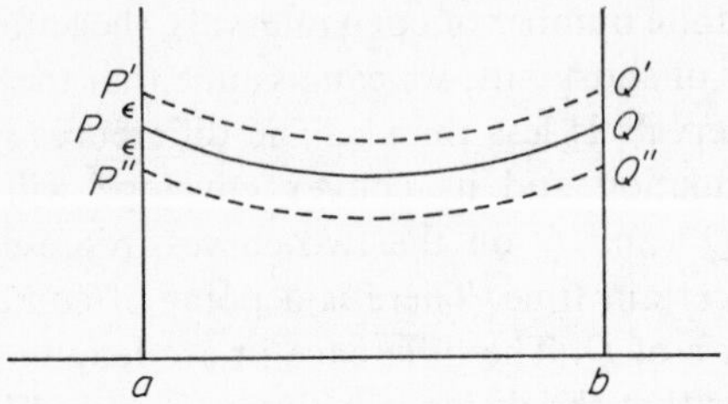

Figure 5

Proof. Consider Figure 5. Let the curve from P to Q be the function f. The dotted lines represent the ε-bounds for the function g. I will show that the shortest convex route from P' to Q' which lies within the ε-bounds is along the dotted line from P' to Q', and that the longest convex route is from P' to P'', along the dotted line to Q'' and then directly to Q'. These distances are L, the length of f, and $L+4\varepsilon$. If L' is the length of the function g then

$$L'+4\varepsilon \geqslant L$$

and

$$L' \leqslant L+4\varepsilon.$$

This will give us the result.

If there is a shorter route from P' to Q', then at some stage the route is not along $P'Q'$. We can show that this leads to a contradiction. We can form an approximating polygonal line to the convex curve from P' to Q', such that the difference in lengths is less than an arbitrary constant ε. We now show that this polygonal line is shorter than any allowed route below it. We do this by induction on n, the number of sides in the polygonal line. For $n = 1$ the result is true.

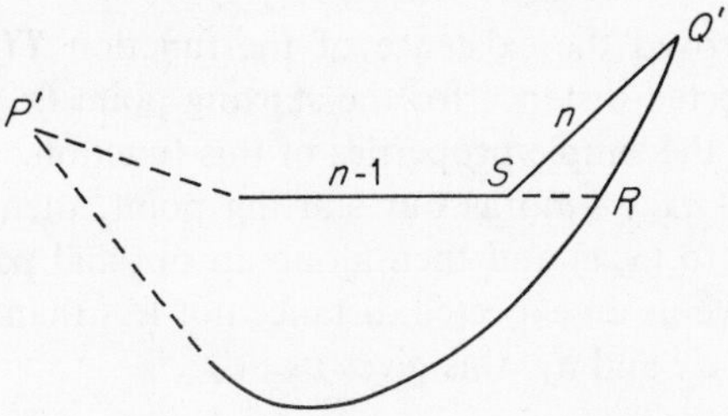

Figure 6

If we have n sides, then we extend the $(n-1)$th side to meet the outside curve at R. By the induction hypothesis we know that

$$\begin{aligned}\text{poly}\, P'S + SQ' &\leqslant \text{poly}\, P'S + SR + \text{arc}\, RQ' \\ &\leqslant \text{arc}\, P'R + \text{arc}\, RQ' = \text{arc}\, P'Q'.\end{aligned}$$

Hence we prove the result for n sides.

Since ε is arbitrary, we have shown that there is no shorter allowed route from P' to Q'.

This type of argument follows through in exactly the same way to prove the result about the longest curve from P' to Q'. This completes the proof of Lemma 5.

It is now easy to show that the function that calculates expected distances is continuous. Let us consider two planned paths f and g, from A to B, with $d(f,g) < \varepsilon$. After a certain time, let Q^* and Q be the position of pedestrians on these two curves. The distance between these two points can be at most 5ε. So if a car comes at this point the difference in distance to be walked is at most 10ε. The difference in total length between the two curves is at most 4ε. The maximum difference in lengths of simultaneous journeys along the two paths is at most 10ε, so the difference in expected distance walked is at most 10ε. Hence the function is continuous.

This completes the proof of Lemma 4.

We know that the best way to walk from $A(X, X)$ to B is the straight line. We have shown existence of a best planned path from any point $A(X, Y)$, where $X \geqslant Y$. But it is clear from the symmetry of the problem that the same result holds for $Y \geqslant X$.

5.2.4 SOLUTION FOR EXPECTED DISTANCES AND PLANNED PATHS

We have now proved the existence of the function $T(x, y)$ which gives the minimal expected distance for the starting point (x, y). I will start by deriving some of the simple properties of this function.

If we take $(x+dx, y+dy)$ as our starting point, then the route which takes us directly to (x, y) and then along an optimal path from (x, y) to the origin will give us an expected distance not less than $T(x+dx, y+dy)$. For infinitesimal dx and dy, this gives us

$$(5) \qquad T(x+dx, y+dy) \leqslant \left(1-\frac{\alpha}{v}ds\right)(T(x,y)+ds)+\frac{\alpha}{v}ds\,(x+y),$$

where $ds^2 = dx^2+dy^2$, α is the exponential parameter and v is the walking velocity. We can rewrite this as

$$(6) \qquad |T(x+dx, y+dy)-T(x,y)| \leqslant ds\left[1+\frac{\alpha}{v}(x+y-T)\right].$$

So $T(x, y)$ satisfies a Lipschitz condition of order one, and so is a continuous function of x and y.

If (x, y) lies on an optimal planned path from $(x+dx, y+dy)$ then we can rewrite inequality (6) for infinitesimal dx and dy as

$$(7) \qquad T(x+dx, y+dy)-T(x,y) = ds\left[1+\frac{\alpha}{v}(x+y-T)\right].$$

So we see that T is differentiable along the optimal curves, and satisfies the following differential equation:

$$(8) \qquad \frac{d}{ds}T[x(s), y(s)] = 1+\frac{\alpha}{v}[(x+y)-T].$$

From equations (6) and (7) we can deduce

Lemma 6. *The optimal planned paths are paths of maximum descent on the surface $z = T(x, y)$.*

I have now to make an assumption about the function T. I assume that the function T is differentiable at all points except the origin. I have been unable to show that this reasonable assumption is true.

We have then from Lemma 6 that

$$\frac{dT}{ds} = \left[\left(\frac{\partial T}{\partial x}\right)^2 + \left(\frac{\partial T}{\partial y}\right)^2\right]^{\frac{1}{2}} \tag{9}$$

and so substituting in equation (8) we get

$$\left[\left(\frac{\partial T}{\partial x}\right)^2 + \left(\frac{\partial T}{\partial y}\right)^2\right]^{\frac{1}{2}} = 1 + \frac{\alpha}{v}[(x+y) - T]. \tag{10}$$

The initial conditions for this partial differential equation are values of T on the x- and y-axes, and also on the line $y = x$. This equation can be reduced to five ordinary differential equations involving T, $\partial T/\partial x$, $\partial T/\partial y$, x and y in terms of a parameter s. For details I refer the reader to Smirnov: *A Course in Higher Mathematics, Integral Equations and Partial Differential Equations*, section 104.

The five equations are

$$\left\{\begin{aligned} \frac{dT}{ds} &= \left[\left(\frac{\partial T}{\partial x}\right)^2 + \left(\frac{\partial T}{\partial y}\right)^2\right]^{\frac{1}{2}}, \\ \frac{dU}{ds} &= \frac{\alpha}{v}(1-U), \\ \frac{dV}{ds} &= \frac{\alpha}{v}(1-V), \\ \frac{dx}{ds} &= \frac{\partial T/\partial x}{[(\partial T/\partial x)^2 + (\partial T/\partial y)^2]^{\frac{1}{2}}}, \\ \frac{dy}{ds} &= \frac{\partial T/\partial y}{[(\partial T/\partial x)^2 + (\partial T/\partial y)^2]^{\frac{1}{2}}}, \end{aligned}\right. \tag{11}$$

where $U = \partial T/\partial x$ and $V = \partial T/\partial y$.

Professor P. Whittle has pointed out that equations (11) can be obtained as a direct consequence of Pontryagin's Maximum Principle.

We see from the last two of these equations that

(12) $$\left(\frac{dx}{ds}\right)^2+\left(\frac{dy}{ds}\right)^2 = 1.$$

So s is a distance parameter along curves of maximum descent, and thus a distance parameter along the optimal planned paths. We shall measure it against the direction of motion.

Solving equations (11) for $\partial T/\partial x$ and $\partial T/\partial y$ we get

(13) $$\begin{cases} \dfrac{\partial T}{\partial x} = P\mathrm{e}^{-\alpha s/v}+1, \\ \dfrac{\partial T}{dy} = Q\mathrm{e}^{-\alpha s/v}+1. \end{cases}$$

If θ is the angle between the tangent to an optimal planned curve and the x-axis, then we have

$$\tan\theta = \frac{Q\mathrm{e}^{-\alpha s/v}+1}{P\mathrm{e}^{-\alpha s/v}+1}.$$

Let us consider curves starting at the origin, $B = [0,0]$. Let s be the distance along the path from the origin. P and Q are constants that satisfy the equation

(15) $$(P+1)^2+(Q+1)^2 = 1.$$

The gradient of a particular path at the origin is given by

(16) $$\tan\theta = \frac{Q+1}{P+1}.$$

The curve that touches the x-axis at B has $Q = -1$, and $P = 0$. In this special case the equation is

(17) $$\tan\theta = 1-\mathrm{e}^{-\alpha s/v}.$$

It is clear that points in the area shaded in Figure 7, where the lower boundary is the function (17), can be reached by reversed paths starting

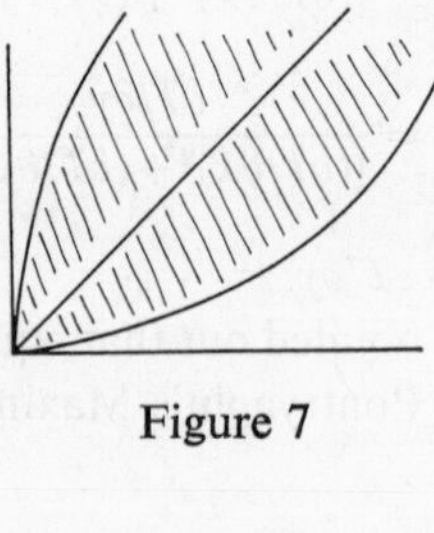

Figure 7

at the origin with initial gradient between 0° and 90°. We now have to investigate reversed paths to points outside the shaded area.

Figure 8. Examples of planned paths of unit length with $\alpha = 1$ and $v = 1$

The differential equations (11) hold only in the interior of the positive quadrant. The breakdown on the axes is due essentially to the term $(x+y)$ in equation (10), which should have been written $(|x|+|y|)$. The partial derivatives are not defined on the axes.

A path from a point outside the shaded area must reach one of the axes and then proceed along it to the origin. Because each axis is itself an optimal path, the direction of maximum descent at points on the axes is along the axes. So a path must touch the axis. If it touches the x-axis, then the curve will be parallel to the curve given by equation (17). We do not in fact have to use the differentiability of T on the x-axis to show that

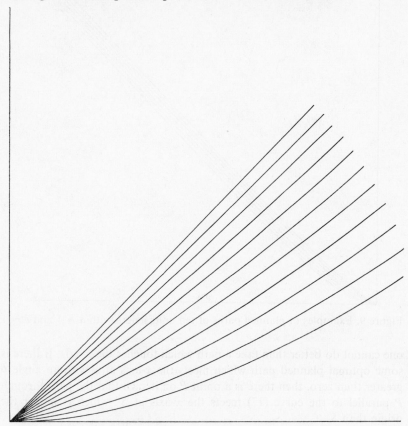

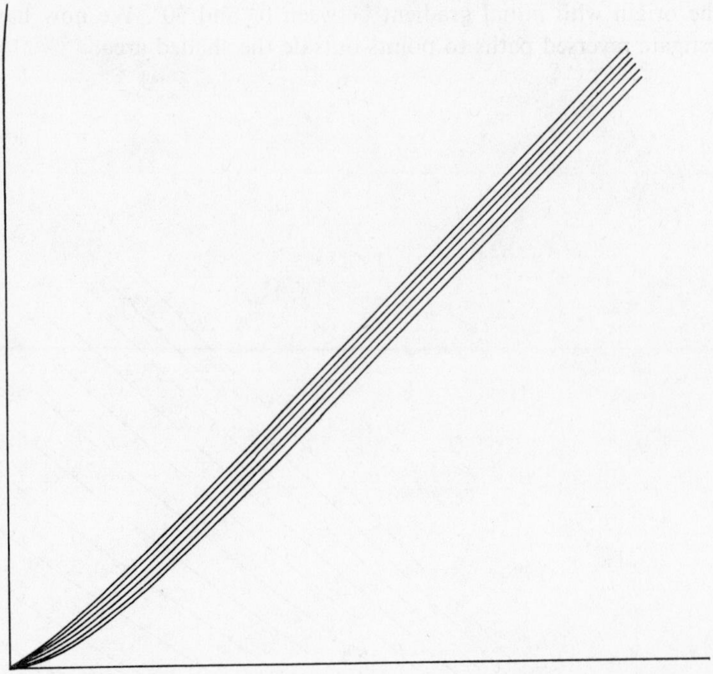

Figure 9. Examples of planned paths of ten units length with $\alpha = 1$ and $v = 1$

one cannot do better than take a path which touches the x-axis. If there is some optimal planned path which meets the x-axis at S with an angle θ greater than zero, then there is a point P on it such that the path through P parallel to the curve (17) meets the x-axis at S' which is nearer the origin than S.

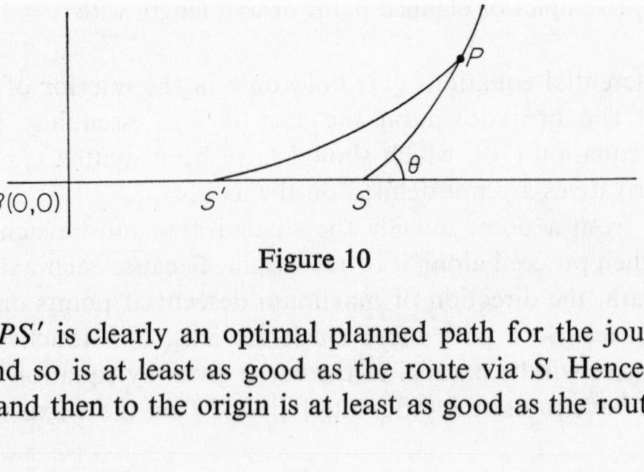

Figure 10

The arc PS' is clearly an optimal planned path for the journey from P to S', and so is at least as good as the route via S. Hence the route along PS' and then to the origin is at least as good as the route via S.

This proves a most interesting feature of the Fitzwilliam Street problem. If one wishes to cross a road to reach some shops quite a way up the road, the path one follows will be independent of which shop one is actually making for.

Figures 8 and 9 give plots of some of the optimal curves for the case $\alpha = 1$ and $v = 1$.

The solution for T is found by integrating along these paths. This solution is differentiable at all points except the origin, since the partial derivatives are continuous. The routes of maximum descent given in intrinsic form by equation (14) are the solutions to the differential equation

$$\frac{dy}{dx} = \frac{\partial T/\partial y}{\partial T/\partial x}. \tag{18}$$

We can prove by using a standard existence theorem for differential equations that the solutions to this equation with initial conditions $Y = y(X)$ are uniquely determined in a neighbourhood of (X, Y) provided Y and X are positive. However the Lipschitz condition on $(\partial T/\partial y)/(\partial T/\partial x)$ which is required by this standard theorem breaks down on both axes, which explains why the curves of maximum ascent which start from A touching the axes are not unique. But clearly optimal planned paths from a fixed starting point are uniquely determined.

For small α/v a good approximation to the best planned path is the straight line from A to B. If however α/v is large, then a good approximation is a path at an angle of 45° across the road, and then along the pavement.

5.2.5 GENERALIZED PROBLEM

There are two main ways in which to generalize this problem. Firstly one can consider the case where, when a car comes, one has a choice as to which pavement to rush to, and secondly the situation where Fitzwilliam Street has been modernized to a multi-lane highway, where one only needs to run to the next set of lane markings when a car comes. Both these problems remain to be solved.

ACKNOWLEDGMENT

I should like to thank Professor D. G. Kendall and Professor P. Whittle for helpful discussions on this work, and the Science Research Council for financial support.

6

RANDOM MEASURES AND RANDOM SETS

6.1

An Elementary Approach to the Spectral Theory of Stationary Random Measures

D. VERE-JONES

6.1.1 INTRODUCTION

In recent years, spectral theory has become an important tool in the theory and statistical analysis of point processes and other random measures. In the statistical setting, the essential ideas of this theory were set out by Bartlett [1] and [2], and have been further developed by Cox and Lewis [5]. More recently, it has been pointed out by Daley [6], [7] that a rigorous approach to this theory can be based either on the theory of processes with stationary increments (as in Chapter 11 of Doob [9]), or on the theory of generalized random functions (as in the monograph by Gel'fand and Vilenkin [10]). Brillinger [4] has outlined a third approach, based on the theory of random interval functions developed by Bochner in [3].

The aim of the present paper is to put forward a direct approach to the spectral theory of random measures, based on the idea that smoothing a random measure leads to a continuous random process to which the standard spectral theory can be applied. This technique is, of course, one of the standard tools of generalized function theory; the point is that in the special case of measures the techniques become very simple and allow the main results of the theory to be built up with only elementary mathematical machinery. Bartlett suggested a similar approach in the discussion to [16], and attributed the idea to Moyal.

The main new results relate to the elementary properties of positive, positive-definite measures (p.p.d. measures) developed in Section 6.1.3, and already quoted in [8]. A number of open questions remain in this area, and some of these are mentioned in the text with conjectures as to their solutions. Of the others, we may mention here the problem (raised by L. Shepp) of determining the extreme rays of the convex cone of p.p.d. measures.

6.1.2 THE COVARIANCE STRUCTURE OF STATIONARY RANDOM MEASURES

A random measure $N(\cdot)$ may be defined by specifying the finite-dimensional distributions of the random variables $N(A)$, where, for different choices of the bounded Borel set A on the real line, $N(A)$ represents the random mass allocated to A. In talking of a random measure we assume $N(A) \geqslant 0$; otherwise we shall talk of a random signed measure.

The moment structure of a random measure is specified by the set functions

$$M(A) = \mathbf{E}[N(A)],$$

$$M_2(A \times B) = \mathbf{E}[N(A)\,N(B)].$$

We assume from now on that the set functions $M(\cdot), M_2(\cdot)$ are countably additive and take finite values when A, B are bounded Borel sets. For discussions of the basic properties of random measures and their associated moment measures we refer to [11], [12], [15]. A summary of the theory is given in [8].

For the rest of the paper we shall also assume that the random measure $N(\cdot)$ is *second-order stationary*, by which we mean that the values of $M(A), M_2(A \times B)$ are unchanged by simultaneous shifts of the sets A, B through any time interval τ. This invariance has an important effect on the structure of the moment measures, for it implies that they can be written out in terms of the product of Lebesgue measure down a 'diagonal' and a further measure on the orthogonal diagonal. For the particular measures $M(\cdot), M_2(\cdot)$ this reduction takes the forms (using differential notation for brevity)

$$dM(t) = m\,dt,$$

$$M_2(ds \times dt) = d\gamma(s-t)\,dt,$$

where m is a constant we shall refer to as the *mean rate* of the process, and γ is a symmetric measure (the symmetry deriving from the symmetry $M_2(A \times B) = M_2(B \times A)$) that we shall refer to as the *reduced second-moment measure*.

Diagonal reductions of this type are familiar in the discussion of measures on groups, and since elementary proofs are given in [4] and the appendix to [7] it seems unnecessary to provide one here. It should be noted that if the random measure is absolutely continuous with density $n(t)$, then these reductions correspond to the reductions

$$\mathbf{E}(n(t)) = m; \quad \mathbf{E}[n(s)\,n(t)] = c(s-t),$$

where $c(\cdot)$ is the density of $\gamma(\cdot)$. However, the most important examples of random measures are point processes, and labelled point processes, in which case $N(\cdot)$ is purely atomic, and $\gamma(u)$ has an atom of mass λb at $u = 0$, where λ is the rate of occurrence of atoms of N and b is the mean square value of the mass of one of these random atoms.

It is also possible to introduce a *covariance measure* by means of the equation

$$C_2(A \times B) = M_2(A \times B) - M(A)\,M(B) = \operatorname{cov}(N(A), N(B)).$$

In the stationary case, there is a corresponding reduced measure $C(\cdot)$, which is again a symmetric measure, and given by

$$dC(u) = d\gamma(u) - m^2\,du. \tag{1}$$

The reduced second-moment measure possesses two obvious but important properties, which hold the key to the development in the next section. Firstly, it is *positive* ($\gamma(A) \geqslant 0, \gamma(\cdot)$ not identically zero), and secondly it is *positive definite*: for bounded Borel functions $\varphi(\cdot)$ of bounded support,

$$\int d\gamma(x) \left[\int \varphi(u)\,\bar{\varphi}(x+u)\,du \right] \geqslant 0. \tag{2}$$

The reason for choosing a Borel function of bounded support is that (a.s.)

$$I(\varphi) = \int \varphi(t)\,dN(t)$$

can be defined in a straightforward manner, as a Lebesgue–Stieltjes integral with respect to the individual realizations; then inequality (2) is simply the inequality $\mathbf{E}[|I(\varphi)|^2] > 0$.

Any measure satisfying the above two properties will be called, as in the introduction, a *p.p.d. measure*. It follows from what we have just said that the reduced second-moment measure of any second-order stationary random measure is a p.p.d. measure; also, it follows from equation (1) that the reduced covariance measure of such a random measure differs from a p.p.d. measure only by a multiple of Lebesgue measure.

An important open question is to determine whether the converse of these assertions are true and, in particular, whether any p.p.d. measure can be represented as the reduced second-moment measure for some stationary random measure. The usual construction appealing to Gaussian processes is not available here, since we are restricted to non-negative sample realizations. Similarly it is an open question to characterize the class of measures which can be reduced covariance measures, or, when the atom at the origin is removed and the covariance measure is otherwise absolutely continuous, to determine the class of functions which can be *covariance densities*. The best result available to date appears to be that of Milne and Westcott, in [14], who use the example of a Gauss–Poisson process to demonstrate that any non-negative integrable function can be a covariance density.

6.1.3 PROPERTIES OF P.P.D. MEASURES

In this section we develop some properties of p.p.d. measures which will apply, in view of the results just discussed, to the reduced second-moment measure of a (second-order) stationary random measure, and, after the subtraction of a multiple of Lebesgue measure, to its reduced covariance measure. On the other hand, all the results we know for second-moment measures are also true for all p.p.d. measures, which is one reason for conjecturing that the two classes coincide.

The first properties we derive may be motivated by a consideration of the mean-square continuous process $V(t)=N(t,t+1]$, which has covariance function

$$g_v(u)=\int_{-1}^{1}(1-|y|)\,d\gamma_y(u+y).$$

Of course this function may be considered whether or not γ derives from an underlying random measure, and in any case it represents a continuous positive-definite function. It is therefore bounded and such that the limit of

$$\frac{1}{2T}\int_{-T}^{T}g_v(x)\,dx$$

exists as $T\to\infty$. These properties can be immediately transferred to $\gamma(\cdot)$, making use of non-negativity to assert the inequalities

$$\gamma(x,x+1)\leqslant\int_{x-1}^{x+2}g_v(y)\,dy,$$

$$\int_{-T+1}^{T-1}g_v(x)\,dx\leqslant\gamma(-T,T)\leqslant\int_{-T-1}^{T+1}g_v(x)\,dx.$$

We therefore obtain the following result, in which we use $\mathscr{P}_+$ to denote the class of p.p.d. measures, and R to denote the real line.

Theorem 1. *For any $\gamma \in \mathscr{P}_+$, the function $\gamma(x, x+1)$ is uniformly bounded for $x \in R$, and the limit of $\gamma(-T, T)/(2T)$ as $T \to \infty$ exists and is finite.*

Although the results extend immediately to the covariance measure of a random measure, the proof we have given depends essentially on the non-negativity of $\gamma(\cdot)$, and we conjecture that the corresponding results for the covariance measure of a random *signed* measure are false in general.

A similar device may be used to derive a Fourier representation for $\gamma(\cdot)$, but here it is slightly more convenient to work in terms of the function

$$g_x(u) = \frac{1}{2}\int e^{-|x-y|}\, d\gamma(y), \tag{3}$$

corresponding, when $\gamma(\cdot)$ is a moment measure, to the process

$$X(t) = \int_{-\infty}^{t} e^{-(t-s)}\, dN(s). \tag{4}$$

Making use of the boundedness property of the previous theorem, it is easy to verify that $g_x(u)$ is again a bounded continuous function. Moreover, $g_x(\cdot)$ is positive definite, and it may therefore be represented in the form

$$g_x(u) = \int e^{i\theta u}\, d\mu_x(\theta), \tag{5}$$

where $\mu_x(\cdot)$ is a positive, symmetric (since $g_x(u)$ is real and symmetric) totally finite measure on R (Bochner's Theorem).

It will be useful to write equation (5) also in the form of a Parseval relation

$$\int g_x(u)\,\varphi(u)\, du = \int \tilde{\varphi}(\theta)\, d\mu_x(\theta), \tag{6}$$

where

$$\tilde{\varphi}(\theta) = \int e^{i\theta x}\,\varphi(x)\, dx.$$

Equation (6) is valid whenever $\varphi(\cdot)$ is integrable (with respect to Lebesgue measure) and hence in particular whenever $\varphi \in S$, where S is the class of all functions which are infinitely differentiable, and, together with all their derivatives, converge to zero at infinity faster than any inverse power of x. This class of functions is used in the theory of distributions, and has the property that it is mapped onto itself by the Fourier transformation

$\varphi \to \tilde{\varphi}$. Another important property is that, if equation (6) is given for all $\varphi \in S$, it uniquely defines $\mu_x(\cdot)$ in terms of g_x, and vice versa.

Now any spectral measure $\mu(\cdot)$, which can be associated with $\gamma(\cdot)$, should be related to $\mu_x(\cdot)$ by the frequency domain analogue of equation (3) namely

$$d\mu_x(\theta) = \frac{1}{1+\theta^2}\, d\mu(\theta).$$

We therefore *define* the spectral measure of $\gamma(\cdot)$ to be

$$d\mu(\theta) = (1+\theta^2)\, d\mu_x(\theta)$$

and then show that γ is the Fourier transform of μ, in the sense that the Parseval equation

$$\int \psi(u)\, d\gamma(u) = \int \tilde{\psi}(\theta)\, d\mu(\theta) \tag{7}$$

holds for all $\psi \in S$. But this is just a matter of rewriting equation (6) in terms of $\gamma(\cdot)$ and $\mu(\cdot)$, setting

$$\psi(x) = \frac{1}{2}\int \mathrm{e}^{-|x-y|}\, \varphi(y)\, dy, \quad \tilde{\psi}(\theta) = \frac{1}{1+\theta^2}\tilde{\phi}(\theta) = \int \mathrm{e}^{i\theta x}\, \psi(x)\, dx,$$

and noting that $\varphi \to \psi$ defines a 1 : 1 mapping of S onto itself.

This result is sufficient to imply that $\gamma(\cdot)$ may be regarded as the Fourier transform of μ, and that $\gamma(\cdot)$ and $\mu(\cdot)$ determine each other uniquely. However equation (7) can be extended to a class of functions much larger than S. For example, it follows from equation (6) that equation (7) is valid whenever ψ is the smoothed version of an integrable function.

To obtain a more convenient condition, note that

$$\int \psi(x)\, \varphi(x+y)\, dx$$

is such a function whenever $\psi \in S$, and φ is integrable, so that we have

$$\int \psi(x) \left[\int \varphi(x+y)\, d\gamma(y)\right] dx = \int \tilde{\psi}(\theta)\, \tilde{\varphi}(\theta)\, d\mu(\theta). \tag{8}$$

If in addition $\tilde{\varphi}(\theta)$ is μ-integrable, we may appeal to Fubini's theorem to rewrite the right-hand side of this expression in the form

$$\iint \mathrm{e}^{i\theta x}\, \psi(x)\, \tilde{\varphi}(\theta)\, d\mu(\theta)\, dx = \int \psi(x) \left[\int \mathrm{e}^{i\theta x}\, \tilde{\varphi}(\theta)\, d\mu(\theta)\right].$$

Thus for all $\psi \in S$ we have

$$\int \psi(x) \left[\int \varphi(x+y)\, d\gamma(y)\right] du = \int \psi(x) \left[\int \mathrm{e}^{i\theta x}\, \tilde{\varphi}(\theta)\, d\mu(\theta)\right] dx.$$

But this result implies that the two expressions in square brackets are equal a.e., so that for φ integrable and $\tilde{\varphi}$ μ-integrable

$$\int \varphi(x+y)\, d\gamma(y) = \int e^{i\theta x}\, \tilde{\varphi}(\theta)\, d\mu(\theta) \quad \text{a.e.} \tag{9}$$

Now the right-hand side of equation (9) is a continuous function of x whenever φ is μ-integrable, and so equation (9) will hold identically whenever the left-hand side is continuous; more particularly the relation (7) will hold whenever the left-hand side is continuous at $x = 0$. The a.e. condition cannot be dropped in general, for it is always possible to alter $\varphi(\cdot)$ at a finite number of points, thereby leaving the right-hand side of equation (9) unchanged, but changing the left-hand side whenever one of the anomalous points coincides with an atom of $\gamma(\cdot)$.

Theorem 2. *For each $\gamma \in \mathscr{P}_+$ there exists a unique $\mu \in \mathscr{P}_+$ such that the Parseval equation* (7) *holds for $\psi \in S$. More generally, equation* (9) *is valid a.e. whenever ϕ is integrable and $\tilde{\phi}$ is μ-integrable; then equation* (7) *holds when the left-hand side of equation* (9) *is continuous at $x = 0$.*

Let us stress that Theorem 2 implies that the class $\mathscr{P}_+$ of p.p.d. measures is invariant under the operation of taking Fourier transforms. For the positive-definiteness of γ implies the positivity of μ, and the positivity of γ implies the positive-definiteness of μ.

Several further corollaries may be drawn from the Parseval equations (7) and (9). In the first place, it is now easy to identify the limit of

$$\gamma(-T, T) = \int_{-\infty}^{\infty} \frac{\sin \theta T}{\theta T}\, d\mu(\theta)$$

in Theorem 1 with $\mu\{0\}$. Indeed it requires only a slightly more elaborate argument to show that for any real θ,

$$\lim_{T\to\infty} \frac{1}{2T} \int_{-T}^{T} e^{-i\theta u}\, d\gamma(u) = \mu\{\theta\}$$

($\mu\{\theta\}$ denotes the mass of μ at the point θ). Since $\mu(\cdot)$ is also a p.p.d. measure, there is a corresponding result with the roles of γ and μ interchanged.

An explicit inversion theorem can be obtained from equation (7) if we take $\psi(x)$ to be the indicator function $\chi_I(x)$ of an interval $I = (a, b)$ of continuity for $\gamma(\cdot)$, convolved with a suitable smoothing function. In this way we can obtain, for example,

$$\gamma(a, b) = \lim_{\lambda\to\infty} \int_{-\infty}^{\infty} \frac{e^{-ib\theta} - e^{ia\theta}}{i\theta} \frac{\lambda^2}{\lambda^2 + \theta^2}\, d\mu(\theta).$$

Again, a similar result holds with the roles of γ and μ interchanged. On the other hand it does not seem possible to retain the inversion theorem in its classical form for general p.p.d. measures.

As a third result of interest, let us take $\psi(x)$ in equation (7) to be the triangular function $1-|x|/T\ (-T\leqslant x\leqslant T)$. In the case that $\gamma(\cdot)$ is the second-moment measure of a random measure, this yields the representation

$$\mathbf{E}[N(0,T)]^2 = T^2\int\left(1-\frac{|x|}{T}\right)d\gamma(x) = \int\left(\frac{\sin\frac{1}{2}\theta T}{\frac{1}{2}\theta}\right)^2 d\mu(\theta).$$

If we restate this equation in terms of the covariance measure $dC(u)$ and its Fourier transform, say $dF(\theta) = d\mu(\theta) - m^2\,\delta(\theta)\,d\theta$ ($\delta(\theta)$ is a Dirac delta), we obtain the representation

$$V(T) = \operatorname{var}[N(0,T)] = \int\left(\frac{\sin\frac{1}{2}\theta T}{\frac{1}{2}\theta}\right)^2 dF(\theta),$$

which is essentially the form given by Daley in [6] and [7].

6.1.4 THE CRAMÉR REPRESENTATION OF A STATIONARY RANDOM MEASURE

We next show how these techniques can be extended to yield Cramér-type representations for the random measure itself. Throughout this section, $N(\cdot)$ will denote a second-order stationary random measure with reduced second-moment measure γ and corresponding spectral measure μ.

We first establish a relation between the Hilbert spaces generated by the random measure $N(\cdot)$ and its smoothed version

$$X(t) = \int_{-\infty}^{t} e^{-(t-u)}\,dN(u).$$

Let $L_2(N), L_2(N;\,a)$ be the Hilbert spaces spanned by the random variables $N(s,t)$ for $-\infty<s<t<\infty$ and for $-\infty<s<t\leqslant a$ respectively; and $L_2(X), L_2(X;\,a)$ the Hilbert spaces spanned by $X(t)$ for $-\infty<t<\infty$ and for $-\infty<t\leqslant a$ respectively. Then we have the following result.

Lemma. *$L_2(N) = L_2(X)$, and for each a, $-\infty<a<\infty$, $L_2(N;\,a) = L_2(X;\,a)$.*

Proof. We have to show that $N(a,b)$ can be approximated in mean square by linear combinations of $X(t)$, $t\leqslant b$, and that $X(t)$ can be approximated in mean square by linear combinations of $N(a,b)$ for $a<b\leqslant t$.

The first of these results is readily achieved by making use of the inequalities

$$X(t+h) - e^{-h}\,X(t) \leqslant N(t,t+h) \leqslant e^{h}\,X(t+h) - X(t),$$

forming a finite sum of the terms so as to cover (a,b) with intervals of length h, and letting $h \to 0$.

Similarly, if we set

$$S_h = \sum_1^{\infty} \mathrm{e}^{-(nh)} N(t-nh, t-\overline{n-1}h)$$

then

$$S_h \leqslant I \leqslant \mathrm{e}^h S_h \quad (I = X(t))$$

and hence

$$\mathbf{E}|I - S_h|^2 \leqslant (\mathrm{e}^h - 1)^2 \mathbf{E}|S_h|^2 \leqslant (\mathrm{e}^h - 1)^2 \mathbf{E}|I|^2,$$

so that I is approximated in mean square by S_h as $h \to 0$. This completes the proof of the lemma.

Now $X(t)$ is mean square continuous, and it therefore admits the Cramér representation

$$X(t) = \int \mathrm{e}^{it\theta} \, dZ_X(\theta) \quad \text{(a.s.)}, \tag{10}$$

where $Z_X(\theta)$ is a process of orthogonal increments satisfying

$$\mathbf{E}|dZ_X(\theta)|^2 = d\mu_X(\theta) = \frac{1}{1+\theta^2} d\mu(\theta).$$

More generally, equation (10) may be thought of as illustrating an isometric isomorphism between $L_2(X)$ and the Hilbert space $L_2(\mu_X)$ of functions square integrable with respect to μ_X. The general form of the isomorphism is represented by the mean square integral

$$Y_X(\varphi) = \int \varphi(\theta) \, dZ_X(\theta), \tag{11}$$

where $\varphi \in L_2(\mu_X)$ and $Y_X(\varphi)$ is the element in $L_2(X)$ corresponding to φ; for the definition of the mean square integral with respect to a process of orthogonal increments see Doob [9].

To turn these results into relations between N and μ we introduce the process of orthogonal increments

$$dZ(\theta) = (1 + i\theta) \, dZ_X(\theta)$$

which satisfies

$$\mathbf{E}|dZ(\theta)|^2 = d\mu(\theta).$$

If we also write $\varphi_1(\theta) = \varphi(\theta)/(1+i\theta)$, then $\varphi \to \varphi_1$ defines a 1 : 1 isometric isomorphism of $L_2(\mu_X)$ onto $L_2(\mu)$, and setting $Y_X(\varphi) = Y_N(\varphi_1)$ we can rewrite equation (11) in the form

$$Y_N(\varphi_1) = \int \varphi_1(\theta)\, dZ(\theta). \tag{12}$$

It is clear that equation (12) now defines an isometric isomorphism between $L_2(N)$ and $L_2(\mu)$. In particular, $N(a,b)$ and $(e^{ib\theta} - e^{ia\theta})/i\theta$ are corresponding elements in this isomorphism, so we obtain the representation

$$N(a,b) = \int \frac{e^{ib\theta} - e^{ia\theta}}{i\theta}\, dZ(\theta) \quad \text{(a.s.)}. \tag{13}$$

The inverse relationship giving $Z(\cdot)$ in terms of $N(\cdot)$ requires more care, since the function $(e^{-it\beta} - e^{-it\alpha})/(-it)$ is not integrable as a function of t, and it is not immediately clear how its integral with respect to $dN(\cdot)$ should be defined. However, the truncated version

$$\int_{-T}^{T} \frac{e^{-it\beta} - e^{-it\alpha}}{-it}\, dN(t) = \int S_T(\theta)\, dZ(\theta), \tag{14}$$

where

$$S_T(\theta) = 2\int_{\theta-\beta}^{\theta-\alpha} \frac{\sin \omega T}{\omega}\, d\omega,$$

presents no difficulties, as we can approximate the integral on the left by linear combinations of step functions and substitute from equation (13). As $T \to \infty$, the function $S_T(\theta) \to S(\theta)$, where $S(\theta) = 2\pi$ $(\alpha < \theta < \beta)$, $S(\alpha) = S(\beta) = \pi$, and $S(\theta) = 0$ if $\theta < \alpha$ or $\theta > \beta$. Moreover, the functions $S_T(\theta)$ are locally uniformly bounded, and near infinity uniformly $O(\theta^{-1})$. Since μ integrates θ^{-2} at infinity, it follows that $S_T(\theta) \to S(\theta)$ in $L_2(\mu)$, which proves the existence of the mean square limit on the right of equation (14) as $T \to \infty$, and establishes the inversion formula

$$Z(\alpha,\beta) + \tfrac{1}{2}Z\{\alpha\} + \tfrac{1}{2}Z\{\beta\} = \underset{T\to\infty}{\text{l.i.m.}} \frac{1}{2\pi} \int_{-T}^{T} \frac{e^{-it\beta} - e^{-it\alpha}}{-it}\, dN(t). \tag{15}$$

Theorem 3. *Let the random measure N be second-order stationary with spectral measure μ. Then there exists a process of orthogonal increments $dZ(\theta)$ such that*

$$\mathbf{E}|dZ(\theta)|^2 = d\mu(\theta)$$

and the random integral

$$I(\varphi) = \int \varphi(\theta)\, dZ(\theta)$$

sets up a 1 : 1 *isometric isomorphism between* $L_2(N)$ *and* $L_2(\mu)$ *in which the processes* $N(\cdot)$ *and* $Z(\cdot)$ *are related by the reciprocal formulae*

$$N(a, b) = \int \frac{e^{i\theta b} - e^{i\theta a}}{i\theta}\, dZ(\theta) \quad \text{(a.s)},$$

$$Z(\alpha, \beta) + \tfrac{1}{2}Z\{\alpha\} + \tfrac{1}{2}Z\{\beta\} = \underset{T\to\infty}{\text{l.i.m.}} \int_{-T}^{T} \frac{e^{-it\beta} - e^{-it\alpha}}{-it}\, dN(t),$$

where the first integral is a mean square integral with respect to Z, and the second is a Lebesgue–Stieltjes integral existing a.s. with respect to the measures $dN(t)$.

We conclude this section by briefly discussing the validity of the general Parseval formula

$$\int \varphi(t)\, dN(t) = \int \tilde{\varphi}(\theta)\, dZ(\theta). \tag{16}$$

We shall assume that $\varphi(t)$ is in L_1 (integrable with respect to Lebesgue measure) and that $\tilde{\varphi}(\theta)$ is in $L_2(\mu)$. Since functions $(e^{i\theta b} - e^{i\theta a})/i\theta$ form a basis for $L_2(\mu)$ (this is already implicit in the isomorphism theorem), the right-hand side can be approximated in mean square by integrals of functions of this form, and hence (using equation (13)) by a sequence of integrals of the form

$$\int s_n(t)\, dN(t),$$

where $s_n(t)$ is a step function. If, therefore, we are prepared to regard the integral on the left-hand side as a mean square limit of step functions, equation (16) is essentially a tautology and requires no further justification.

However, this is not a satisfactory discussion of the problem, as the integral on the left-hand side of equation (16) has a more natural definition as a Lebesgue–Stieltjes integral, which will exist a.s. whenever $\varphi \in L_1$, with respect to the individual random measures $dN(t)$. Thus the question arises as to whether these two definitions of the random integral on the left-hand side of equation (16) are at least a.s. equal.

Although this is clearly the case for a large class of functions, including all functions $\varphi \in S$, and all bounded Borel functions of bounded support, I do not know whether or not it remains true for all $\varphi \in L_1$ with $\tilde{\varphi} \in L_2(\mu)$, which seem to be the natural conditions in this context.

6.1.5 THE WOLD DECOMPOSITION AND BACKWARD MOVING AVERAGE REPRESENTATIONS

In this section it will be convenient to work with the zero-mean processes

$$d\Lambda(t) = dN(t) - m\,dt,$$
$$\xi(t) = X(t) - m.$$

We shall say that Λ (or N) is *deterministic* if

$$\bigcap_a L_2(\Lambda;\, a) = L_2(\Lambda)$$

and *purely non-deterministic* if

$$\bigcap_a L_2(\Lambda;\, a) = \{0\}.$$

The lemma of the preceding section implies that the $\Lambda(\cdot)$ and $\xi(\cdot)$ processes are deterministic or purely non-deterministic together. In the general case, ξ has a decomposition

$$\xi(t) = \xi_1(t) + \xi_2(t)$$

into two orthogonal processes, where ξ_1 is deterministic and ξ_2 is purely non-deterministic. We now reconstruct random variables $\Lambda_1(a, b)$, $\Lambda_2(a, b)$ from the ξ_1, ξ_2 processes to obtain a similar decomposition of Λ. Indeed, $\Lambda_1(a, b)$ can be built up as the mean square limit of the same linear combination of $\xi_1(t)$ as $\Lambda(a, b)$ is of $\xi(t)$, convergence in mean square of the ξ_1 combinations following from the orthogonality properties. From the countable additivity of $\Lambda(\cdot)$,

$$\Lambda\left(\bigcup_1^\infty I_n\right) = \sum_1^\infty \Lambda(I_n) \quad \text{(a.s.)},$$

when the I_n are disjoint intervals and

$$\bigcup_1^\infty I_n$$

is bounded; substituting in terms of Λ_1 and Λ_2, we obtain the relation

$$\Lambda_1\left(\bigcup_1^\infty I_n\right) - \sum_1^\infty \Lambda_1(I_n) = \Lambda_2\left(\bigcup_1^\infty I_n\right) - \sum_1^\infty \Lambda_2(I_n).$$

Since the two sides here are orthogonal, they must both coincide a.s. with the zero element; so

$$\Lambda_1\left(\bigcup_1^\infty I_n\right) = \sum_1^\infty \Lambda_1(I_n) \quad \text{(a.s.)},$$
$$\Lambda_2\left(\bigcup_1^\infty I_n\right) = \sum_1^\infty \Lambda_2(I_n) \quad \text{(a.s.)}.$$

Thus Λ_1 and Λ_2 are both countably additive random interval functions. (However we cannot deduce from this argument that Λ_1 and Λ_2 have the stronger property that their realizations are a.s. signed measures.) Finally, a further application of the lemma shows that Λ_1 is deterministic and Λ_2 purely non-deterministic.

The condition for the deterministic component to be absent can be phrased in terms of the spectrum of the Λ-process, taking over the corresponding condition for the ξ-process. We shall use the notation

$$dF(\theta) = d\mu(\theta) - m^2\,\delta(\theta)\,d\theta$$

for the spectrum of the Λ-process, the measure F being the Fourier transform of the covariance measure $dC(u)$ of the random measure $N(\cdot)$.

Now, ξ is purely non-deterministic if and only if its spectrum is absolutely continuous, with spectral density, say $f_\xi(\theta)$, satisfying

(17) $$\int \frac{\log f_\xi(\theta)}{1+\theta^2}\,d\theta > -\infty.$$

Since $dF(\theta) = (1+\theta^2)\,dF_\xi(\theta)$, Λ has a.c. spectrum if and only if ξ has a.c. spectrum, so that Λ is purely non-deterministic if and only if its spectrum is a.c. with density satisfying

$$\int \frac{\log\{f(\theta)(1+\theta^2)\}}{1+\theta^2}\,d\theta > -\infty.$$

Since

$$\int \frac{\log(1+\theta^2)}{1+\theta^2}\,d\theta < \infty,$$

this is equivalent to the usual condition

$$\int \frac{\log f(\theta)}{1+\theta^2}\,d\theta > -\infty.$$

Theorem 4. *Any second-order stationary random measure may be written (after removal of the mean) uniquely in the form*

$$\Lambda = \Lambda_1 + \Lambda_2,$$

where Λ_1, Λ_2 are mutually orthogonal, countably additive interval functions, Λ_1 is deterministic and Λ_2 is purely non-deterministic. The deterministic component is absent if and only if the zero-mean spectrum $dF(\theta)$ is a.c., and its density $f(\theta)$ satisfies

(18) $$\int \frac{\log f(\theta)}{1+\theta^2}\,d\theta > -\infty.$$

The condition (17) is equivalent to the existence of a canonical factorization for $f_\xi(\theta)$

$$f_\xi(\theta) = |g_\xi(\theta)|^2,$$

where $g_\xi(\theta)$ is an L_2-function which can be represented as the Fourier transform of an L_2-function with support in $(-\infty, 0]$. The canonical factorization is realized if $g_\xi(\theta)$ is zero-free as well as analytic in the lower half-plane $\text{Im}(\theta) < 0$, and with this extra condition the factorization is unique up to a constant of unit modulus. The corresponding factorization of the spectral density of Λ is then given by

$$f(\theta) = |g(\theta)|^2,$$

where

$$g(\theta) = (1 + i\theta)\, g_\xi(\theta).$$

This corresponds exactly to the canonical factorization that would be obtained by treating the process as a random distribution (cf. Jowett and Vere-Jones [13]). It is equivalent to a backward moving average representation for Λ which may be written formally

$$\frac{d\Lambda(t)}{dt} = \int_{-\infty}^{0} G(x)\, dB(t-x),$$

where B is a process of orthogonal stationary increments, and $G(x)$ must in general be interpreted as a distribution. In fact, in all the examples known to us, $G(\cdot)$ reduces to a measure, and under further conditions (for example, if the density $f(\theta)$ is a rational function) this measure is absolutely continuous except perhaps for an atom at $x = 0$. An interesting conjecture is that the boundedness conditions which hold for p.p.d. measures are sufficient to imply that G is always a measure, but this we are able neither to prove nor to disprove. For random measures with rational density, these results lead to simple prediction formulae for the conditional mean rate of the process at the time $t + \tau$, given the realization up to time t. These results are surveyed, and illustrated with some simple examples, in [13].

ACKNOWLEDGMENTS

I am grateful to Dr. D. J. Daley for many valuable discussions on the topics of this paper, and for encouraging my intrusions on an area where he had made the pioneering contribution. I have also benefited from many discussions with Mr. J. E. Jowett.

In a different vein I should like to thank Professor Kendall and Mr. Harding for the opportunity of contributing to this volume, and hence of adding my voice to others in expressing sorrow and regret at the untimely death of a fine talent and respected colleague.

REFERENCES

1. M. S. Bartlett, "The spectral analysis of point processes", *J. R. statist. Soc. B* **25** (1963), 264–296.
2. M. S. Bartlett, *An Introduction to Stochastic Processes*, 2nd ed., Cambridge University Press, Cambridge (1966).
3. S. Bochner, *Harmonic Analysis and the Theory of Probability*, University of California Press, Berkeley (1955).
4. D. R. Brillinger, "The spectral analysis of stationary interval functions", *Proc. 6th Berkeley Symposium*, Vol. 1, University of California Press, Berkeley (1972), 483–513.
5. D. R. Cox and P. A. W. Lewis, *The Statistical Analysis of Series of Events*, Methuen, London (1966).
6. D. J. Daley, "Spectral properties of weakly stationary point processes", *Bull. Int. Stat. Inst.* **37** (1969), No. 2, 344–346.
7. ——, "Weakly stationary point processes and random measures", *J. R. Statist. Soc. B* to appear.
8. D. J. Daley and D. Vere-Jones, "A summary of the theory of point processes", *Proceedings of the Stochastic Point Process Conference* (1971), Wiley, New York (1972), 299–303.
9. J. L. Doob, *Stochastic Processes*, Wiley, New York (1953).
10. I. M. Gel'fand and N. Ya. Vilenkin, *Generalized Functions*, Vol. 4 (translation), Academic Press, New York (1964).
11. T. E. Harris, *The Theory of Branching Processes*, Springer, Berlin (1963).
12. M. Jirina, "Asymptotic Behaviour of Measure-valued Branching Processes", *Rozpravy Ceskoslovenska Akad. Ved., Rada. Mat. Privod. Ved.* **76** (1966), No. 3.
13. J. Jowett and D. Vere-Jones, "The prediction of stationary point processes", *Proceedings of the Stochastic Point Process Conference* (1971), Wiley, New York (1972), 305–335.
14. R. K. Milne and M. Westcott, "Futher results for Gauss–Poisson processes", *Adv. Appl. Prob.* **4** (1972), 151–176.
15. J. E. Moyal, "The general theory of population processes", *Acta Math.* **108** (1962), 1–31.
16. D. Vere-Jones, "Stochastic models for earthquake occurrence", *J. R. Statist. Soc. B* **32** (1970), 1–62.

6.2

Foundations of a Theory of Random Sets

D. G. KENDALL

6.2.1 INTRODUCTION

In this paper† I want to describe the foundations of a general theory of random sets; its further developments will only be mentioned very briefly. As noted at the end of the introductory chapter [10] to this book, other approaches to random-set theory were developed at about the same time. Most of the present work was in fact carried out five years ago, and it was presented in lecture-courses in Cambridge and Heidelberg as well as in an informal seminar at Oberwolfach ('Point-processes as seen from a Hunting-lodge'). A consolidated review of these parallel developments would be very valuable, but will not be attempted here. I do however want to make clear that the key idea, very simple and yet extremely powerful, on which the whole of the present approach is based is due to Rollo Davidson and can be found in his Smith's Prize Essay [4] which was in fact (so far as this matter is concerned) an account of his first *term* of research.

I should also like to acknowledge that many of the motivating ideas can in fact be found in a seminal paper of Choquet [3]; I did not at first appreciate the connection, but was ultimately much influenced by it, as all who are familiar with that paper will at once realize.

Everyone knows how to set up a (uniform) Poisson point-process on the line. We take an arbitrary origin, and then mark off points on the line, to left and right of this origin, in distance-steps which are independent random variables distributed with the exponential density $\lambda e^{-\lambda x}$, where

† This article is closely linked to my chapters 1.1, 3.7 and 5.4 in the sister volume *Stochastic Analysis* (= *SA*) (Ed. Kendall and Harding), John Wiley and Sons, London (1973), and the reader will need to refer to these from time to time. For this reason the four linked articles in *S.A.* and *S.G.* have been written in a uniform style with several notational conventions in common.

x is the coordinate on the line and λ is the 'intensity' of the process. The 'origin' is *not* a point of the process; the apparent paradox here is a familiar one and we need not dwell on it.

The construction of a uniform Poisson process in k dimensions (say $k = 2$) is a little more complicated. One can, for example, cut up the plane into half-open unit squares and associate with these squares independent Poisson random variables of expectation λ. If X, a non-negative integer, is the random variable so associated with a generic square Q, we then scatter X points in Q so that they are independently uniformly distributed therein, the insertion operations for the several squares being mutually independent. (Of course an analogous procedure can be followed also when $k = 1$.) We are left, however, with a feeling of clumsiness; the Cartesian structure is there all right, but was it quite 'natural' to use it? How are we to generalize this procedure to carrier-spaces which are locally compact groups not admitting such a simple Cartesian product structure? And so on.

When we turn to non-uniform Poisson processes our unease becomes yet more acute; even on the line the problem has lost its simplicity. Finally we note that, once uniformity has been dropped, there is no reason why the carrier-space should be a group; why not an ellipsoid, or other manifold? Why not Lebesgue's 'crumpled piece of paper'?

This nagging worry having been with me for several years, I began looking about for a natural way of describing *a random set on a more or less arbitrary carrier-space*, the latter being endowed with just enough structure for the job in hand. Specific examples which one might have in mind, in addition to the Poisson process, are the set of zeros of Brownian motion, the set of times at which a Markov process is in a given state, and so on. As these examples indicate, methods (cf. Ryll-Nardzewski [22]) which depend on 'counting' (or taking the Lebesgue measure of) the intersection of the random set Z with a 'test set' T will be inappropriate even in some instances where they are meaningful; thus if Z is the set of zeros of Brownian motion, and T is an open interval, then the number of points in $Z \cap T$ is a.s. zero or infinite, and the Lebesgue measure of $Z \cap T$ is a.s. zero (cf. Taylor [23]).

Davidson's idea, which is the basic one in the present work, is that instead of working with some version of the 'size' of $Z \cap T$, *we should merely note whether $Z \cap T$ is empty or not*. If the system of 'test sets' T is rich enough in some suitable sense, then the information acquired in this way (just one 'bit' per 'test set') should suffice to identity Z. Obviously this is true in the extreme case when all singletons are available as 'test sets', but of course we shall want to manage with a smaller system of 'test sets' than that, and we may have to pay some modest price for this. What follows is merely the working out of this basic idea.

6.2.2 TRAPPING SYSTEMS, AND INCIDENCE FUNCTIONS

We shall think of the random set Z as a wild animal in a forest; as the huntsman in charge of operations we want to locate it, and to this end we have a system of *traps* throughout the forest, electrically connected to the hunting-lodge. If the animal Z is in contact with the trap T, then a corresponding light on the control panel will be illuminated, and otherwise not. The huntsman inspects the control panel, notes which lights are shining, and so deduces the position of the animal in the forest (see Figure 1).

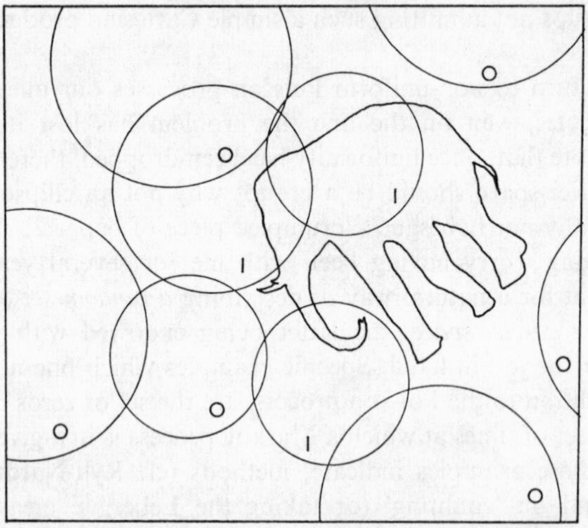

Figure 1. A typical 'sample function' for the stochastic elephant process. (Some of the 'traps' T are shown as circles, and the values of $fz(T)$ are indicated.)

We now make this informal description precise. Let C be the *carrier-space* (the forest), so that the random set Z is some subset of C. We suppose that C is equipped with a *trapping-system* $\mathscr{T}$ consisting of subsets T of C such that

TS(i): *every T is non-empty;*

TS(ii): *the traps T cover C.*

Later (in Section 6.2.5) we shall add two further axioms.

Now let Z be an *arbitrary* subset of C, and let us use it to define the function $f_Z: \mathscr{T} \to \{0, 1\}$ as follows;

$$f_Z(T) = 0, \quad \textit{when } Z \cap T \textit{ is empty},$$
$$f_Z(T) = 1, \quad \textit{otherwise}.$$

The function f_Z contains 'all the information available to the huntsman'.

Two fundamental questions now pose themselves. Firstly, does f_Z determine Z uniquely, and if not, what is the equivalence relation between subsets Z of C associated with the non-unicity? Secondly, given f (a zero–one function defined over $\mathscr{T}$), how can we tell whether $f = f_Z$ for some subset Z of C, and how can we construct this Z (or the equivalence class of such Zs)? In the case of non-uniqueness there will also be a third fundamental question: is there a member of the class of solutions Z to the equation

$$f = f_Z$$

which enjoys some sort of canonical status? We shall now proceed to answer all these questions concerning the 'incidence functions' f_Z.

We have described the function f_Z uniquely determined by the subset Z of C as an 'incidence function', and so in fact it always will be, but we now want to introduce incidence functions as objects in their own right, and so we make the following definitions.

SIF: *a zero–one function f over $\mathscr{T}$ is called a strong incidence function when $T \subset \bigcup T_\alpha$ and $f(T) = 1$ imply $f(T_\alpha) = 1$ for* some α, *for every covering of the trap T by traps T_α.*

WIF: *a zero–one function f over $\mathscr{T}$ is called a weak incidence function when condition* SIF *holds in the weakened form in which only* finite *coverings $T \subset \bigcup T_i$ $(1 \leqslant i \leqslant n)$ are considered.*

Notice that in SIF the traps T_α of a covering need not, and in general will not, form a countable collection. Obviously every strong incidence function (sif) is a weak incidence function (wif), and here is an example to show that the converse statement would be false.

Example 1. *Let C be the real line. Let $\mathscr{T}$ consist of all intervals of the form $T = [\alpha, \beta)$, where $-\infty < \alpha < \beta < \infty$, and define f as follows:*

$$f(T) = 1, \quad \textit{if } \alpha \leqslant 0 \leqslant \beta,$$
$$f(T) = 0, \quad \textit{otherwise}.$$

This is a wif, but it is not a sif. The verification is trivial and is left to the reader.

It is obvious that f_Z is a sif for every subset Z of C, and so the function defined in Example 1 cannot be expressed in the form f_Z. We naturally hope that the condition SIF may prove to be necessary and sufficient for a zero–one function f to be expressible in the form $f = f_Z$, and this is true, as the following construction shows.

Let f be a sif, and define

$$Z = \left(\bigcup_{f(T)=0} T \right)^* = \bigcap_{f(T)=0} T^*, \tag{1}$$

where here and always we use an asterisk to denote the formation of the set-theoretic complement. Now either $f(S) = 0$, or $f(S) = 1$, for an arbitrary trap S. If $f(S) = 0$, then S is one of the Ts occurring in the first formula for Z, and so S does not meet Z. If $f(S) = 1$, then S cannot be covered by the union of Ts occurring in that formula (because f is a sif), and thus S must meet Z. Therefore $f = f_Z$. This construction will be of very great importance in our theory, and we shall return to it in a moment. Let us first show that when f is a sif, the equation $f = f_Z$ can have more than one solution.

Example 2. *Let C be the real line and let $\mathscr{T}$ consist of the non-empty finite open intervals. Let $Z = (0,1)$, and let $Z' = [0,1]$. Now a trap T will meet Z if and only if it meets Z', and so $f_Z = f_{Z'}$.*

We therefore have the anticipated non-uniqueness, and must do something to take the sting out of it. To this end we introduce the concept of the *$\mathscr{T}$-closure* of a set Z with respect to a trapping-system $\mathscr{T}$; we define this to be

$$\mathrm{Cl}(Z; \mathscr{T}) = \left(\bigcup_{T \cap Z = \varnothing} T \right)^* = \bigcap_{T \cap Z = \varnothing} T^*. \tag{2}$$

We prove as our first result

Theorem 1. (*a*) *Every set Z is covered by its $\mathscr{T}$-closure.*

(*b*) *Inclusion relations are preserved by $\mathscr{T}$-closure.*

(*c*) *T hits Z if and only if T hits* $\mathrm{Cl}(Z; \mathscr{T})$.

(*d*) *$\mathscr{T}$-closure is an idempotent operation.*

(*e*) $\bigcup \mathrm{Cl}(Z_\alpha; \mathscr{T}) \subset \mathrm{Cl}(\bigcup Z_\alpha; \mathscr{T})$, *and* $\mathrm{Cl}(\bigcap Z_\alpha; \mathscr{T}) \subset \bigcap \mathrm{Cl}(Z_\alpha; \mathscr{T})$.

Proof. (*a*) The union in equation (2) is covered by Z^*. (*b*) If $Z \subset Z'$, then any trap which avoids Z' must avoid Z, so the union for Z' is covered by the union for Z. (*c*) Because of (*a*), we need only show that if T avoids Z, then it must avoid the $\mathscr{T}$-closure of Z. But if T avoids Z then it figures in, and so is covered by, the union in equation (2), and so is covered by the complement of $\mathrm{Cl}(Z; \mathscr{T})$. (*d*) Let $W = \mathrm{Cl}(Z; \mathscr{T})$, and let us call a trap T *clean* if it avoids Z, and *very clean* if it avoids the larger set W. By (*c*), clean$\equiv$very clean, for traps, and now equation (2) shows at once that

$$\mathrm{Cl}(\mathrm{Cl}(Z; \mathscr{T}); \mathscr{T}) = \mathrm{Cl}(W; \mathscr{T}) = \mathrm{Cl}(Z; \mathscr{T}).$$

(*e*) Each Z_α is covered by $\bigcup Z_\beta$, and so (from (*b*)),

$$\mathrm{Cl}(Z_\alpha; \mathscr{T}) \subset \mathrm{Cl}(\bigcup Z_\beta; \mathscr{T}), \quad \text{for all } \alpha,$$

and the result then follows on forming the union over α; the inclusion relation involving intersections is proved similarly.

Theorem 1 tells us that $\mathscr{T}$-closure is 'almost' a topological closure, but the following examples show that this statement cannot be strengthened.

Examples 3. *Let C be the real line, and let $\mathscr{T}$ consist of all the half-infinite open intervals of either species. Let $Z = (-\infty, -1]$, and let $Z' = [1, \infty)$. It is easy to verify that each of these sets is its own $\mathscr{T}$-closure. Now consider $Z \cup Z'$; this set is hit by every trap, and so its $\mathscr{T}$-closure is the whole of R. Thus*

$$\mathrm{Cl}(Z \cup Z'; \mathscr{T}) \neq \mathrm{Cl}(Z; \mathscr{T}) \cup \mathrm{Cl}(Z'; \mathscr{T}),$$

although of course the inclusion implied by (*e*) *of the theorem holds. The point of this example is that the first part of Theorem* 1(*e*) *cannot be improved. The second part cannot be improved, either. To see this, take $C = \{x, y, z\}$, let $\mathscr{T}$ consist of $\{x, z\}$ and $\{y, z\}$, and take*

$$Z_1 = \{x, y\}, \quad Z_2 = \{z\}.$$

Each of Z_1 and Z_2 has C as its $\mathscr{T}$-closure, yet $Z_1 \cap Z_2 = \emptyset$.

We now define $\mathscr{T}$-closed sets in the obvious way; a set Z will be called $\mathscr{T}$*-closed* if and only if it is its own $\mathscr{T}$-closure. We then have

Theorem 2. (*a*) *$\emptyset$ and C are $\mathscr{T}$-closed, and so is each T^*.*

(*b*) *$\mathrm{Cl}(Z; \mathscr{T})$ is $\mathscr{T}$-closed, for every Z.*

(*c*) *If each Z_α is $\mathscr{T}$-closed, then so is $\bigcap Z_\alpha$.*

(*d*) *If Z and Z' are $\mathscr{T}$-closed, then $Z \cup Z'$ need not be so.*

(*e*) *The $\mathscr{T}$-closure of Z is the smallest $\mathscr{T}$-closed set covering Z.*

(*f*) *A necessary and sufficient condition for Z to be $\mathscr{T}$-closed is that*

$$Z^* \subset \bigcup_{T \cap Z = \emptyset} T. \tag{3}$$

Proof. We first establish the useful criterion at (f). It is obvious from equation (2) that it is necessary, so let us suppose that the inclusion (3) holds. Then from equation (2) we see that $Z^* \subset W^*$, where $W = \mathrm{Cl}(Z; \mathscr{T})$, and so $W \subset Z$. But from Theorem 1(a) we know that $Z \subset W$, so $Z = W$ and Z is $\mathscr{T}$-closed.

(a) This follows from the inclusion (3) and the facts that $\mathscr{T}$ is a covering, and that T is one of the traps which avoid T^*. (b) This is just Theorem 1(d). (c) Let $Z = \bigcap Z_\alpha$; then

$$Z^* = \bigcup Z_\alpha^* = \bigcup_\alpha \left(\bigcup_{T \cap Z_\alpha = \emptyset} T \right) \subset \bigcup_{T \cap Z = \emptyset} T,$$

and the conclusion follows from the criterion at (3). (d) See Example 4 below. (e) Let $W = \mathrm{Cl}(Z; \mathscr{T})$. We know that W is a $\mathscr{T}$-closed set which covers Z, while if V is a $\mathscr{T}$-closed set covering Z then

$$V = \mathrm{Cl}(V; \mathscr{T}) \supset \mathrm{Cl}(Z; \mathscr{T})$$

by Theorem 1(b), so $W = \mathrm{Cl}(Z; \mathscr{T})$ is the smallest such V (indeed the intersection of all such Vs).

Once again we notice that we 'almost' have the axioms for a system of closed sets in a topology, but *we cannot close the gap*, as Theorem 2(d) and the following example show.

Example 4. *In Examples* 3, Z *and* Z' *are each* $\mathscr{T}$*-closed. Their union, however, is not (for it is not its own* $\mathscr{T}$*-closure). In fact it is easily seen from equation* (2) *that in this example the* $\mathscr{T}$*-closed sets are precisely the closed intervals (infinite, half-infinite, finite, singleton or empty).*

As a corollary to Theorem 2 we have the important

Remark. *If condition* (3) *is satisfied, then* (3) *holds with equality. Thus (using* (a) *and* (c) *of Theorem* 2) *a set is* $\mathscr{T}$*-closed if and only if its complement is a union of traps.*

We now prove the fundamental

Theorem 3. *Let* f *be a zero–one function over a trapping-system* $\mathscr{T}$. *Then* f *can be written in the form* f_Z *if and only if it is a* strong *incidence function, and in that case one solution is obtained by taking* Z *to be the* $\mathscr{T}$*-closed set*

$$Z = \left(\bigcup_{f(T)=0} T \right)^* = \bigcap_{f(T)=0} T^*.$$

This is the unique $\mathscr{T}$*-closed solution, and the other solutions are precisely those subsets of the carrier-space* C *which have* Z *as their* $\mathscr{T}$*-closure.*

Proof. The statement contained in the first sentence has already been established, save that we must show Z to be $\mathscr{T}$-closed. This, however, is an immediate consequence of Theorem 2(*a*) and (*c*). Now let f be a sif, so that $f = f_Z$, and Z is $\mathscr{T}$-closed. First we note that a trap T hits a set W if and only if it hits $\mathrm{Cl}(W; \mathscr{T})$ (Theorem 1(*c*)), and so

$$(4) \qquad f_W(T) = 1 \quad \text{if and only if } T \text{ hits } \mathrm{Cl}(W; \mathscr{T}).$$

On the other hand we know that

$$f_Z(T) = 1 \quad \text{if and only if } T \text{ hits } Z.$$

Now any two zero–one functions are identical if and only if they agree on the sets to which they assign the value 1, while on the other hand, from equation (2) and the definitions, any two $\mathscr{T}$-closed sets are identical if and only if they avoid the same traps, i.e. if and only if they hit the same traps. Thus $f_W = f_Z = f$ if and only if $\mathrm{Cl}(W; \mathscr{T}) = Z$, and so the solutions to $f = f_W$ are just those sets W whose $\mathscr{T}$-closure is Z, and of these Z alone is $\mathscr{T}$-closed. This completes the proof.

It follows from the last theorem that there is a *natural identification* of strong incidence functions (relative to a given trapping system $\mathscr{T}$) with $\mathscr{T}$-closed sets. Thus *the theory of random $\mathscr{T}$-closed sets is coextensive with the theory of random strong incidence functions.* If, therefore, we wish to discuss random sets of a certain type on a certain carrier-space C, and if we can choose the trapping-system $\mathscr{T}$ in such a way that the sets of this type are exactly the $\mathscr{T}$-closed sets, then the way ahead is clear.

It may however be the case that the sets of interest do not admit a representation as the system of $\mathscr{T}$-closed sets relative to a suitable trapping-system. In that case we must choose $\mathscr{T}$ (and there may be options open to us here) in such a way that all the sets of interest (and some others) are $\mathscr{T}$-closed, and then ultimately we will have to ensure that the probability is carried by those particular sifs (i.e. by those particular $\mathscr{T}$-closed sets) which were of primary interest to us.

From the last paragraph it is clear that we need to be able to recognize a system of $\mathscr{T}$-closed sets when we meet one, and this problem is solved by

Theorem 4. *A system $\mathscr{E}$ of subsets E of the carrier-space C is identifiable with the system of $\mathscr{T}$-closed sets relative to some trapping-system $\mathscr{T}$ if and only if $\mathscr{E}$ contains $\varnothing$, C, and all intersections of $\mathscr{E}$-sets. The representation can then be realized by taking $\mathscr{T}$ to consist of all the sets E^*, where $E \in \mathscr{E}$ and $E \neq C$.*

This trapping-system will often be uneconomically large, and it will suffice to take $\mathscr{T}$ to consist of all the sets F^*, for $F \in \mathscr{F}$, where $\mathscr{F}$ is a subclass of $\mathscr{E}$ not containing C, such that any E $(\neq C)$ in $\mathscr{E}$ can be expressed as an intersection of Fs in $\mathscr{F}$; this is the most general possible choice of $\mathscr{T}$, if $\mathscr{E}$ is to be the class of $\mathscr{T}$-closed sets.

Proof. It is clear from Theorem 2 that the nsc stated in the first sentence is indeed necessary; suppose then that it is satisfied and let $\mathscr{T}$ be set up in accordance with the first prescription. The $\mathscr{T}$ so obtained is a trapping-system, because $E^* = \emptyset$ is excluded and $E^* = C$ is included. Any $\mathscr{T}$-closed set V can (by equation (2)) be expressed in the form

$$\bigcap_{V \subset E \in \mathscr{E}} E,$$

and so will belong to $\mathscr{E}$. On the other hand, if E_0 is a particular $\mathscr{E}$-set then plainly

$$E_0 = \bigcap_{\substack{E_0 \subset E \\ C \neq E \in \mathscr{E}}} E = \bigcap_{\substack{E_0 \cap E^* = \emptyset \\ C \neq E \in \mathscr{E}}} (E^*)^* \tag{5}$$

(the intersection being over vacuously many factors and so yielding the value C in case E_0 itself is C), so that in all cases E_0 is a $\mathscr{T}$-closed set (being either C or an intersection of T^*s).

We now turn to the second part of the theorem, where a more economic choice of $\mathscr{T}$ is suggested. The proof that the new $\mathscr{T}$ is a trapping-system follows from the facts that no $F = C$ (so no $F^* = \emptyset$), and that $\emptyset \in \mathscr{E}$, whence $\emptyset = \bigcap F_\alpha$, whence $C = \bigcup F_\alpha^*$. The proof that each $\mathscr{T}$-closed set (with the new $\mathscr{T}$) is in $\mathscr{E}$ goes as before, while the proof that every set in $\mathscr{E}$ is a $\mathscr{T}$-closed set follows if we extend the earlier argument by the remark that each factor-set E in equation (5) can be written in the form

$$E = \bigcap F_\alpha = \bigcap (F_\alpha^*)^*.$$

No other type of solution is possible, since (*i*) by Theorem 2(*a*) each T^* must be in $\mathscr{E}$, whence each T must be the complement of an $\mathscr{E}$-set; (*ii*) no $T = \emptyset$; and (*iii*) (from equation (2)) each element of $\mathscr{E}$ must be an intersection of T^*s.

When the nsc in Theorem 4 is not satisfied then we have to choose $\mathscr{T}$ so as to yield a class of $\mathscr{T}$-closed sets which is strictly bigger than $\mathscr{E}$. Here the following theorem is helpful.

Theorem 5. *Let $\mathscr{E}$ be any non-vacuous class of subsets of the carrier-space C. Let $\mathscr{F}$ be any class of subsets of C satisfying the three conditions,*

(*i*) *C is not an element of $\mathscr{F}$,*

(*ii*) *the intersection of all the elements of* $\mathscr{F}$ *is* Ø,

(*iii*) *each element of* $\mathscr{E}(\neq C)$ *can be expressed as an intersection of* $\mathscr{F}$*-sets.*

Then $\mathscr{T} = \{F^*: F \in \mathscr{F}\}$ *is a trapping-system relative to which each* $\mathscr{E}$*-set is* $\mathscr{T}$*-closed. Moreover, the conditions* (*i*)–(*iii*) *are necessary as well as sufficient for this to be so.*

Proof. Conditions (*i*) and (*ii*) are just the nsc's for $\mathscr{T}$ to be a trapping-system. If (*iii*) holds then each $E\,(\neq C)$ can be written in the form $\bigcap F_\alpha = \bigcap (F_\alpha^*)^*$, and so is $\mathscr{T}$-closed, and C is always $\mathscr{T}$-closed. The proof of the converse is just the observation that if E is $\mathscr{T}$-closed then by equation (2) it is an intersection of complements of traps, whence an intersection of Fs, as required for (*iii*).

We conclude this section with a few general remarks, and some further examples. First, we have remarked that even when the nsc of Theorem 4 is satisfied, the trapping-system constructed there may be much too large, and that it will usually be possible to take a smaller one. We shall see later that a certain amount of importance might be attached to the case in which the trapping-system $\mathscr{T}$ can be chosen to be countable; we shall speak of this as 'the countable case', and it is clear that we shall be in that situation if the class $\mathscr{F}$ occurring in Theorem 4 (or Theorem 5) can be chosen to be countable.

Example 5. *Let* C *be the real line, and suppose that our aim is to discuss random closed intervals. We may proceed as in Examples* 3 *and* 4, *but then* $\mathscr{T}$ *will not be countable. However, Theorem* 4 *shows that we can place ourselves in the countable case by taking* $\mathscr{F}$ *to consist of all half-infinite closed intervals with rational end-points;* $\mathscr{T}$ *will then consist of all half-infinite open intervals with rational end-points.*

Example 6. *The classical theory of random sets is almost wholly restricted to random* locally finite *sets on the line* R (*'locally finite'* = *'has a finite intersection with every compact interval'* = *'has no cluster points'*). *Obviously these can never by themselves form a system of* $\mathscr{T}$*-closed sets because* $C = R$ *is not locally finite, and it does not help to consider the complements of locally finite sets, instead, because this would exclude* Ø. *We can however take* $\mathscr{T}$ *to consist of all locally finite sets together with one extra set* R, *and then the nsc of Theorem* 4 *will trivially be satisfied. Of course we are here a long way from being in the countable case, and the question arises, can we make use of the second part of Theorem* 4 *so as to arrive at a countable class* $\mathscr{T}$?

To carry out this programme we must choose a countable class of locally finite sets F_n such that every locally finite set can be represented as the intersection of some subsequence of $\{F_n: n=1,2,\ldots\}$. This, however, is impossible. *Among the locally finite sets which would have to be so represented there will be $F(t)=\{t\}$, where $-\infty<t<\infty$, and therefore (for every such t) there must exist $n(t)$ such that $t \in F_n$. But each F_n is countable and so $\bigcup F_n$ is countable and cannot possibly cover R, so the assumption that there is a countable $\mathscr{F}$ leads to a contradiction.*

This example is especially interesting because it brings out well the importance of Theorem 5. If we are prepared to have the locally finite sets (and C) *and some others* $\mathscr{T}$-closed, then a countable solution is possible. Indeed we have only to note that locally finite sets are closed in the usual topology for the line, and that when $\mathscr{T}$ consists of all non-empty bounded open intervals with rational end-points then the $\mathscr{T}$-closed sets are exactly the topologically closed sets (use the remark following Theorem 2); in this way we obtain a solution of the type desired. Thus we can discuss random locally finite sets and stay within the countable case by considering them as instances of random closed sets.

***Example* 7.** *Another interesting example is that in which $\mathscr{E}$ consists of all closed convex sets in the plane (C is now R^2). From Theorem* 4 *and known results in the theory of convexity we see that the traps may be taken to be all the open half-planes. There are of course uncountably many of these, and a reduction to the countable case is again difficult (it will not do to take open half-planes with rationally located boundaries, although as Professor Kingman has pointed out to me this will be adequate if we are concerned with* compact *convex sets). Once again, however, we can regard the theory of random closed convex sets as a part of the theory of random closed sets in the plane, and then we can operate within the countable case.*

Other more practically motivated examples can also be found—problems of blood testing, problems in the 'theory of search', and so forth; we shall not, however, pursue these further here.

As a final general comment on Theorems 1–5, we note (as the reader will have observed) that the proofs are very elementary and strongly reminiscent of the first ten pages of any book on general topology. They are, however, perhaps deceptively simple, for not all the topological structure is present, and in order to construct them one must, as it were, climb a well-known rock face with one hand tied behind one's back. It is natural to ask if similarly weakened sets of axioms, generating part but not all of general topology, have been considered before, and in fact this is so; the reader will find a variety of 'semi-topologies' examined in the interesting book [18] by Mamuzić. Using theorems to be found there, one can show that $\mathscr{T}$-closures can be characterized by the four properties: $\mathrm{Cl}(\emptyset) = \emptyset$; $Z \subset \mathrm{Cl}(Z)$; $\mathrm{Cl}(\cdot)$ preserves inclusion relations; $\mathrm{Cl}(\cdot)$ is idempotent. This characterization can also be established *directly from our own Theorems* 1, 2 *and* 4. We leave the proof to the reader as an instructive and amusing exercise.

6.2.3 FURTHER RESULTS CONCERNING STRONG AND WEAK INCIDENCE FUNCTIONS

We have seen that sifs are in a natural one-to-one correspondence with $\mathscr{T}$-closed sets, and we ought now to see to what extent this identification is an isomorphism. First it is clear that the empty set $\emptyset$ corresponds to the sif which is identically zero, while the whole carrier space C corresponds to the sif which is identically equal to unity.

Next we observe that if Z_1 and Z_2 are $\mathscr{T}$-closed sets, and if f_1 and f_2 are the corresponding sifs, then $Z_1 \subset Z_2$ if and only if $f_1 \leqslant f_2$. ('Only if' is obvious from the definitions, while 'if' is an immediate consequence of equation (1).) We therefore have

***Theorem* 6.** *The relation between strong incidence functions and $\mathscr{T}$-closed sets is an order-isomorphism. On the sif scale the order bounds are $f \equiv 0$ and $f \equiv 1$. If (Z_α) and (f_α) are families of $\mathscr{T}$-closed sets and (corresponding) sifs then* $\sup f_\alpha$ *is a sif which corresponds to the $\mathscr{T}$-closure of $\bigcup Z_\alpha$, while $\bigcap Z_\alpha$ is a $\mathscr{T}$-closed set which corresponds to the largest sif lying below every f_α. Thus $\bigcap Z_\alpha$ corresponds to the sif* $\mathbf{S}\inf f_\alpha$, *where the operator* $\mathbf{S}$ *is that introduced in Theorem* 7 *below.*

Completion of proof. It is easily verified from the definition that the supremum of any family of sifs is itself a sif; as it is the smallest sif lying above all members of the family, it must correspond to the smallest $\mathscr{T}$-closed set covering $\bigcup Z_\alpha$; i.e. to the $\mathscr{T}$-closure of this union. The second part follows in a similar way, using the fact that $\bigcap Z_\alpha$ is $\mathscr{T}$-closed, so is the largest $\mathscr{T}$-closed set lying below all members of the family, and so corresponds to the largest sif lying below all the f_αs. The argument shows that this 'largest sif' must *exist*. To *identify* it, we need the operator $\mathbf{S}$ introduced below.

We now introduce the vitally important operator $\mathbf{S}$. Let g be any zero–one function over $\mathscr{T}$, and define another such function $\mathbf{S}g$ by

$$(\mathbf{S}g)(T) = 0, \quad \text{if } g(T) = 0;$$

$(\mathrm{S}g)(T) = 0, \quad$ if $g(T) = 1$, but $T \subset \bigcup T_\alpha$ where each $g(T_\alpha) = 0$;

$(\mathrm{S}g)(T) = 1, \quad$ otherwise.

Then we have

Theorem 7. *For any zero–one function* g *over* $\mathscr{T}$, Sg *is the largest sif lying below* g.

Proof. It is obvious that $\mathrm{S}g \leqslant g$, and it is easily verified that Sg is a sif, for suppose that $T \subset \bigcup T_\alpha$ where $(\mathrm{S}g)(T_\alpha) = 0$ for all α; then each $T_\alpha \subset \bigcup T_{\alpha\beta}$ where every $g(T_{\alpha\beta}) = 0$, and so $(\mathrm{S}g)(T) = 0$. Finally let $f \leqslant g$, where f is a sif. We want to show that $f \leqslant \mathrm{S}g$, and so for the sake of a contradiction we suppose that there exists a trap T such that

$$(\mathrm{S}g)(T) = 0, \quad f(T) = g(T) = 1.$$

Because Sg vanishes on T we know that $T \subset \bigcup T_\alpha$, where every $g(T_\alpha) = 0$. But this implies that every $f(T_\alpha) = 0$, and so we contradict the sif character of f.

Two other properties of the operator S are worth noting, and are trivial to prove:

(*i*) S$g = g$ *if and only if* g *is a sif;*

(*ii*) $\mathrm{S}^2 = \mathrm{S}$.

More substantial is

Theorem 8. *Let* g *be any zero–one function over* $\mathscr{T}$; *then the* $\mathscr{T}$*-closed set associated with the sif* Sg *is the complement of the union of all those traps on which* g *vanishes.*

Proof. This follows from Theorem 3 and the identity,

$$\bigcup_{(\mathrm{S}g)(T)=0} T = \bigcup_{g(T)=0} T.$$

While, as Theorem 6 shows, the operator S arises quite naturally in the lattice-theoretic study of the objects before us, our primary purpose in introducing it lies in the fact that, if g is a *weak* incidence function, then there exists a largest *strong* incidence function Sg lying below g. Our basic strategy will be to formalize the concept of a stochastic *weak* incidence function g, then to pass to a stochastic *strong* incidence function Sg, and so finally to arrive at a stochastic $\mathscr{T}$-closed *set* Z by putting $\mathrm{S}g = f_Z$.

Two related theorems will be of value to us in carrying out this programme; here they are.

Theorem 9. *Let g be a weak incidence function. Then:*

(*a*) *if* $T_1 \subset T_2$, *we must have* $g(T_1) \leqslant g(T_2)$;

(*b*) *if* $T = T_1 \cup T_2 \cup \ldots \cup T_n$, *then* $g(T) = 0$ *if and only if each*

$$g(T_j) = 0 \quad (j = 1, 2, \ldots, n);$$

(*c*) *if* $T_1 \cup \ldots \cup T_m = T'_1 \cup \ldots \cup T'_n$, *then* $g(T_j) = 0$ (*all* j) *implies that* $g(T'_k) = 0$ (*all* k).

Proof. (*a*) We need only consider the case when $g(T_2) = 0$, and then the result follows at once from condition WIF. (*b*) Obvious from WIF. (*c*) Likewise.

This theorem has the interesting consequence that we can extend the domain of definition of any weak incidence function g from $\mathscr{T}$ to $\mathscr{T}_+$, where the latter consists of the empty set and of all *finite* unions of traps; we have only to define

$$g(\varnothing) = 0,$$

and (if $\theta = T_1 \cup \ldots \cup T_n$)

$$\begin{aligned} g(\theta) &= 0, \quad \text{if } g(T_1) = \ldots = g(T_n) = 0, \\ &= 1, \quad \text{otherwise.} \end{aligned}$$

It is easy to verify that g, when so extended, satisfies the condition WIF relative to the extended trapping system $\mathscr{T}_+ \backslash \{\varnothing\}$. Theorem 9(*c*) guarantees that the extended g is well defined. At later stages of this work we shall frequently need to extend a wif g to $\mathscr{T}_+$, and will do so without further comment.

The other theorem shows that there is a natural way of splitting a wif g into two components, one of which is the associated sif, $\mathbf{S}g$.

Theorem 10. *Let g be* any *zero–one function over* $\mathscr{T}$, *and define the operator* $\mathbf{R}$ *by*

$$\mathbf{R}g = g - \mathbf{S}g,$$

so that $\mathbf{R}g$ *is also a zero–one function over* $\mathscr{T}$. *Then* $h = \mathbf{R}g$ *is the unique solution to the equations,*

$$g = h \vee (\mathbf{S}g), \quad 0 = h \wedge (\mathbf{S}g),$$

where the lattice operations are the standard ones for function-spaces, and

$$\mathbf{R}^2 = \mathbf{R}, \quad \mathbf{S}^2 = \mathbf{S}, \quad \mathbf{RS} = \mathbf{SR} = 0.$$

If g is a weak *incidence function, then* $\mathbf{R}g$ *has the property,*

$$(\mathbf{R}g)(T_1 \cup \ldots \cup T_n) = 1 \quad \textit{implies} \quad (\mathbf{R}g)(T_j) = 1 \quad \textit{for}\text{ some } j.$$

Proof. The first part of the theorem is easily proved by reference to the following table, which shows all possible combinations of values of $g(T)$, $(\mathbf{S}g)(T)$ and $(\mathbf{R}g)(T)$.

$$g(T) = 0, \quad 1, \quad 1;$$
$$(\mathbf{S}g)(T) = 0, \quad 0, \quad 1;$$
$$(\mathbf{R}g)(T) = 0, \quad 1, \quad 0.$$

We already know that $\mathbf{S}^2 = \mathbf{S}$. If $(\mathbf{R}g)(T) = 0$, then the table shows that $(\mathbf{R}^2 g)(T) = 0$, while if $(\mathbf{R}g)(T) = 1$ then T is not g-null but can be covered by g-null sets, so is not $\mathbf{R}g$-null, but can be covered by $\mathbf{R}g$-null sets, whence $(\mathbf{R}^2 g)(T) = 1$. Thus $\mathbf{R}^2 = \mathbf{R}$.

Next we note that $\mathbf{RS}g = \mathbf{S}g - \mathbf{S}^2 g = \mathbf{S}g - \mathbf{S}g = 0$, so that $\mathbf{RS} = 0$, while $\mathbf{SR}g$ is a sif lying underneath both $\mathbf{R}g$ and g, so underneath both $\mathbf{R}g$ and $\mathbf{S}g$, so must vanish; that is $\mathbf{SR} = 0$.

For the last part, g is now a wif, and $\mathbf{R}g$ can have its domain extended to every θ in $\mathscr{T}_+$ by extending the domains of both g and $\mathbf{S}g$ in the manner already explained. Remember that a sif is a wif! If $\mathbf{R}g$ takes the value 1 on $\theta = T_1 \cup \ldots \cup T_n$, then $g(\theta) = 1$ and $(\mathbf{S}g)(\theta) = 0$, so $\mathbf{S}g$ vanishes on each T_j, but g takes the value unity on one T_j at least, and then $\mathbf{R}g$ takes the value unity on this T_j, too.

The importance of this theorem is that, just as $\mathbf{S}$ associates a sif in a canonical way with every wif, so $\mathbf{R}$ tells us when a wif g is in fact a sif (namely, when $\mathbf{R}$ annihilates it), and also tells us which are the traps (or θs) on which a non-sif wif behaves anomalously (namely, exactly those for which $\mathbf{R}g$ takes on a non-zero value).

6.2.4 RANDOM WEAK INCIDENCE FUNCTIONS

We now suppose that the carrier-space C and the trapping-system $\mathscr{T}$ are fixed, and we let W denote *the set of all weak incidence functions.* We propose to investigate in complete generality the conversion of W into a probability-space. A random weak incidence function will not be, for us, an object of great intrinsic interest, but it will be a stepping-stone towards a theory of random strong incidence functions and so of random $\mathscr{T}$-closed sets.

We can think of a random wif g as a stochastic process $\{g(T): T \in \mathscr{T}\}$, where $\mathscr{T}$ is the parameter-set (i.e. the analogue of 'time') and each $g(T)$ is a random variable. To this rather novel variant of the more familiar situation in which $\mathscr{T}$ is the real line, or part thereof, we adapt as appropriate several standard concepts and procedures from the theory of stochastic processes.

For our 'stochastic process' the parameter-set is a very complicated one, but on the other hand the state-space is quite trivial; it is just the 2-point space $\{0, 1\}$. We can therefore employ the theory of stochastic processes over arbitrary parameter-sets appropriate to the case when the state-space is compact, Hausdorff and second-countable, by simply giving to $\{0, 1\}$ the discrete topology. If we write $2 = \{0, 1\}$, then we see that W is a subset of $2^{\mathscr{T}}$, and it is that subset which is determined by the satisfaction of all the conditions comprised in WIF. Now there is a vast infinity of such conditions, but each one restricts the generic element of $2^{\mathscr{T}}$ at only a finite number of T-values, so that if we give to $2^{\mathscr{T}}$ the product topology (in which case it becomes a compact Hausdorff space, possibly *not* second countable), then W will be a closed, so *compact*, subset thereof.

The indispensable measurable sets in $2^{\mathscr{T}}$ are the Baire sets. We recall that these can be defined either as the elements of the smallest σ-algebra $\mathscr{B}_0$ which contains all the compact $\mathscr{G}_\delta$s, or (perhaps more elegantly) as the elements of the smallest σ-algebra with respect to which every real-valued continuous function over $2^{\mathscr{T}}$ is a measurable function. What is more relevant probabilistically is the fact that the Baire sets in $2^{\mathscr{T}}$ are exactly the elements of the σ-algebra $\mathscr{B}_0$ generated by the finite-dimensional cylinder-sets which play a famous role in the Daniell–Kolmogorov existence theorem. (Note that, in $2^{\mathscr{T}}$, *all* finite-dimensional cylinder-sets of whatever 'base' are both closed and open, and so are Baire sets.) That theorem tells us, in fact, that subject to the usual natural consistency conditions on the finite-dimensional distributions, there exists a unique probability measure on the Baire sets of $2^{\mathscr{T}}$, of which the said distributions are the finite-dimensional 'projections'. (All this was explained at some length in [10].)

We are of course by now quite accustomed to the fact that the Baire sets, though indispensable (for they and they alone correspond to events whose occurrence or non-occurrence one might observe), are inadequate as a σ-algebra in which to work with ease and comfort. The present instance is no exception; we shall want our probability measure to be supported by the compact set W, but W will in general not be a Baire set. We can overcome this particular difficulty by a device whose value was first emphasized by Kakutani (see Doob [5]), although recent studies (e.g. Dudley [7]) have shown that we cannot regard it as a universal panacea.

The Kakutani device depends on the fact that if the Baire sets are inadequate, then we can turn to another and larger σ-algebra, $\mathscr{B}$, that comprising the Borel sets. This is defined as the smallest σ-algebra with respect to which all of the open/closed/compact sets are measurable (the three definitions are here equivalent). This is a definitely larger σ-algebra when (as will be usual for us) $\mathscr{T}$ is not countable, for example because all compact sets are Borel sets, while the Baire σ-algebra contains just those very special compact sets which happen to be $\mathscr{G}_\delta$s. It is known that every probability measure over the Baire σ-algebra has at least one extension to a probability measure over the Borel σ-algebra, but this by itself would not be of much comfort to us because the extension can fail to be unique. The essence of the Kakutani argument is that is picks out one uniquely natural extension, and uses that.

It is known that every probability measure pr on Baire sets is *regular* in the sense that

$$\text{(6)} \qquad \mathrm{pr}\,(B_0) = \sup \mathrm{pr}\,(K_0) = \inf \mathrm{pr}\,(G_0),$$

where B_0 denotes a generic Baire set, K_0 a generic Baire compact set covered by B_0, and G_0 a generic Baire open set covering B_0. The relevant theorem is that, among the various probability measures which extend pr from the Baire sets to the Borel sets, there is *one and one only* which has a corresponding 'regularity' property:

$$\text{(7)} \qquad \mathrm{pr}\,(B) = \sup \mathrm{pr}\,(K) = \inf \mathrm{pr}\,(G),$$

where B is a generic Borel set, K is a generic compact set covered by B, and G is a generic open set covering B. We therefore construct the Daniell–Kolmogorov measure on Baire sets, and then extend it in this unique way to a regular measure on Borel sets; we shall use the same notation pr for the extended measure, as there can be no ambiguity here. Notice that as W is compact, the aim of our construction, namely the requirement that

$$\text{(8)} \qquad \mathrm{pr}\,(W) = 1,$$

has now become at least meaningful.

We call the Kakutani extension 'natural' and *not* 'canonical' because in our introductory chapters [10] we agreed to use the word 'canonical' in a very special sense. This conflicts, therefore, with the terminology of Meyer [20] who calls the Daniell–Kolmogorov model 'the First Canonical Process', and its Kakutani extension 'the Second Canonical Process'. In his terminology 'canonical' means 'uniquely constructible from the name of the process' (see [10], and also Nelson [21]).

All the above will be quite familiar to most readers, but now I want to stress one point concerning the Kakutani extension of the Daniell–Kolmogorov theorem which seems to me of great practical importance, and which has received very little attention until now. It is splendid that we can attach a meaning to pr(B), for any Borel set B, in a natural way, but *how can we calculate the number* pr(B) *in practice?* One might say, it can be calculated by the formula

$$\text{(9)} \qquad \mathrm{pr}(B) = \inf_{B \subset G} \sup_{K_0 \subset G} \mathrm{pr}(K_0),$$

where G denotes a generic open set covering the Borel set B, and K_0 denotes a generic *Baire* compact set (i.e. a compact $\mathscr{G}_\delta$) covered by G, but this is hardly practical politics because the infimum will have to be formed over such a vast collection.

It is natural to ask whether any solution is possible which evades this difficulty at least in special cases, and indeed this is so; here are some prescriptions for calculating pr(B), of which (*e*) and (*f*) are thought to be new.

We should mention that Nelson [21] impinges tangentially on these questions, but there is a different objective in his analysis; contrast, for example, his Theorem 2.3 with (*e*) below.

We first note that we may take as accessible to us not merely the measures pr(B_0) of Baire sets B_0, but also the outer and inner *Baire* measures of any set E, these being defined by

$$\mathrm{pr}_0^*(E) = \inf_{E \subset B_0} \mathrm{pr}(B_0)$$

and

$$\mathrm{pr}_{0*}(E) = \sup_{B_0 \subset E} \mathrm{pr}(B_0),$$

B_0 denoting a generic Baire set. Notice that while we do not need to distinguish between pr on Baire sets and pr on Borel sets, we *must* distinguish (by a suffix $|_0$) between inner and outer measures formed using Baire sets (which we can assume calculable), and those formed using Borel sets (which are not calculable unless we beg the whole question). We also note that any probability measure over a σ-algebra can be extended in the usual way to the 'completion' of that σ-algebra (the completion depending on the measure in use); we therefore write pr$^+$ for the extension of pr from the Borel sets to the completed Borel σ-algebra $\mathscr{B}^+$. (It is relevant to notice that in forming pr_0^* and pr_{0*} it does not matter whether we use pr_0, or pr_0^+). We now have the following results.

(*a*) *Let* $\{K_\alpha\}$ *be an* (*arbitrarily large*) *assemblage of compact* (*whence Borel*) *sets each of* pr-*measure one. Then* $\bigcap K_\alpha$ *is compact* (*so Borel*), *and it too is of* pr-*measure one.* (*This is useful when each* K_α *is a Baire set.*)

(*b*) *Let* $\{G_\alpha\}$ *be an (arbitrarily large) assemblage of open (whence Borel) sets each of* pr-*measure zero. Then* $\bigcup G_\alpha$ *is open (so Borel), and it too is of* pr-*measure zero. (This is useful when each* G_α *is a Baire set.)*

(*c*) *Let the Borel set B be compact, or a countable union of compact sets, or a countable intersection of countable unions of compact sets, or ...; then*

$$\mathrm{pr}(B) = \mathrm{pr}_0^*(B).$$

(*d*) *Let the Borel set B be open, or a countable intersection of open sets, or a countable union of countable intersections of open sets, or ...; then*

$$\mathrm{pr}(B) = \mathrm{pr}_{0*}(B).$$

For the next two results we need to recall the following definition. Let $\mathscr{A}$ denote an arbitrary family of sets; then Souslin ($\mathscr{A}$) denotes the class of sets E which can be expressed in the form

$$E = \bigcup_{\mu} \bigcap_{k \geq 1} A_{m_1, m_2, \ldots, m_k},$$

where each suffixed set A is in $\mathscr{A}$ and the union is to be extended over all infinite sequences $\mu = (m_1, m_2, \ldots)$ of positive integers. It is not difficult to show that $\mathscr{A} \subset$ Souslin ($\mathscr{A}$), and that all intersections and unions of countable sequences of Souslin ($\mathscr{A}$) sets are in Souslin ($\mathscr{A}$), and from these facts it follows that the results (*e*) and (*f*) below contain as special cases the results (*c*) and (*d*) above. Let $\mathscr{K}$ ($\mathscr{G}$) denote the classes of compact (open) sets, respectively.

(*e*) *Let E belong to Souslin* ($\mathscr{K}$)*; then E lies in the* pr-*completion* $\mathscr{B}^+$ *of the Borel* σ-*algebra, and* $\mathrm{pr}^+(E) = \mathrm{pr}_0^*(E)$.

(*f*) *Let E belong to Souslin* ($\mathscr{G}$)*; then E lies in the* pr-*completion* $\mathscr{B}^+$ *of the Borel* σ-*algebra, and* $\mathrm{pr}^+(E) = \mathrm{pr}_{0*}(E)$.

The proof of (*e*) and (*f*) (for which see Fremlin and Kendall [9]) is rather long, but we shall not need to use these two results in what follows. The last two prescriptions (*g*) and (*h*) generalize (*a*) and (*b*).

(*g*) *Let* $\{K_\alpha\}$ *be a system of compact sets, and let K (compact) be their intersection. Then*

$$\mathrm{pr}(K) = \inf \mathrm{pr}\left(\bigcap_f K_\alpha\right),$$

where the intersection is to be of finitely many factors, and all such intersections contribute to the infimum. (This is useful when each K_α *is a Baire set.)*

(*h*) *Let* $\{G_\alpha\}$ *be a system of open sets, and let* G *(open) be their union. Then*

$$\mathrm{pr}(G) = \sup \mathrm{pr}\left(\bigcup_f G_\alpha\right),$$

using a notation dual to that of (*g*). (*This is useful when each* G_α *is a Baire set.*)

It is only prescriptions (*a*) (and later, (*c*)) that we really need; we note that each one of the infinitely many conditions comprised in WIF restricts us to a *Baire* compact set. Thus our requirement (8) reduces to the equivalent requirement that each one of these *Baire* sets is to have probability one. These conditions, together with the Daniell–Kolmogorov consistency conditions, are necessary and sufficient for the 'finite dimensional distributions' to be such that they determine a unique regular probability measure on the Borel sets which gives measure one to W. Thus they present us with necessary and sufficient conditions for our procedure to yield a 'random weak incidence function'.

We now proceed to set out these conditions in a manageable form. We shall write φ for a (perhaps vacuous) collection of finitely many different traps (of cardinal $|\varphi|$), so that typically $\varphi = \{T_1, \ldots, T_n\}$ with $|\varphi| = n$, and we shall write $\theta = \bigcup \varphi = T_1 \cup \ldots \cup T_n$. Each finite-dimensional distribution will be associated with some φ, and will specify the joint distribution of the random variables

$$g(T_1), g(T_2), \ldots, g(T_n).$$

We write pr_φ for this finite-dimensional distribution (a distribution over a space of $2^{|\varphi|}$ points), and we note that the consistency conditions are simply those which express the fact that if $\varphi_1 \subset \varphi_2$, then pr_{φ_2} is to project into pr_{φ_1}, together with the usual permutational invariance. The φs form a directed system, so that the pr_φs form a net, and we have just described what we mean by a *consistent* net. We shall say that a consistent net is *proper* whenever

$$\mathrm{pr}_\varphi(g(T) = 1, \text{ and } g(T_j) = 0 \text{ for } 1 \leqslant j \leqslant n) = 0$$

for all $\varphi = \{T, T_1, \ldots, T_n\}$ *such that* $T \subset T_1 \cup \ldots \cup T_n$. What we have proved so far can then be summarized in

Theorem 11. *There is a one-to-one correspondence between the proper consistent nets* $\{\mathrm{pr}_\varphi\}$ *of finite-dimensional distributions and the regular probability measures on the Borel sets in* $2^{\mathcal{T}}$ *which give measure one to the compact set* W, *and project into the net* $\{\mathrm{pr}_\varphi\}$.

We can greatly improve Theorem 11 by noting that the probabilities assigned by the pr_φs can all be expressed in terms of a much simpler object, which we shall call the *avoidance function*, defined by

$$(10)\qquad \begin{cases} (i) & A: \mathcal{T}_+ \to R, \\ (ii) & A(\varnothing) = 1, \\ (iii) & A(\bigcup \varphi) = \mathrm{pr}_\varphi \ (g \text{ vanishes on each trap} \in \varphi). \end{cases}$$

It is important to notice here that if φ_1 and φ_2 satisfy $\bigcup \varphi_1 = \bigcup \varphi_2$, and if the net is proper and consistent, the g is almost surely a wif and so (by Theorem 9(*c*)) the Baire sets

$$\{g \text{ vanishes on each trap} \in \varphi_1\}$$

and

$$\{g \text{ vanishes on each trap} \in \varphi_2\}$$

have the same probability (for in fact they differ only on a set of pr-measure zero—necessarily also a Baire set). Thus the avoidance function is well defined over what we have called $\mathcal{T}_+$, provided that the net $\{\mathrm{pr}_\varphi\}$ is both proper and consistent.

Two proper consistent nets which share the same avoidance function must be identical, for there is no difficulty in reconstructing pr_φ for each φ when the avoidance function is given. To see this, let φ denote any finite collection of different traps, let v denote any assignment of zeros and ones to the traps in φ, and let φ_v denote the subcollection of those traps in φ which are assigned the value zero by v. It will be clear that the event $\{g \equiv v \text{ on } \varphi\}$ is identical with that part of the event

$$\{g \equiv 0 \text{ on } \varphi_v\}$$

which is disjoint from each one of the k events

$$\{g \equiv 0 \text{ on } \varphi_v \cup \{T_j\}\},$$

where $T_1, T_2, \ldots, T_k$ is an enumeration of $\varphi \backslash \varphi_v$. On performing an inclusion–exclusion calculation identical with that which establishes the classical identity of Poincaré in elementary probability theory, we find that

$$(11)\qquad \mathrm{pr}_\varphi(g \equiv v \text{ on } \varphi) = \sum_{\varphi_v \subset \psi \subset \varphi} (-1)^{|\psi| - |\varphi_v|} A(\bigcup \psi),$$

so that *the whole set* $\{\mathrm{pr}_\varphi\}$ *can be constructed when the avoidance function is known.* (Cf. Kingman [16].)

This last result tells us that the avoidance function associated with a proper consistent net must at least have the following two properties:

(*i*) *it must be normalized:*

(12) $$A(\varnothing) = 1;$$

(*ii*) *it must be completely monotonic:* that is, if $\theta \in \mathscr{T}_+$ and if

$$\varphi = \{T_1, T_2, \ldots, T_n\},$$

then

(13)
$$A(\theta) - \sum_i A(\theta \cup T_i) + \sum_{i<j}\sum A(\theta \cup T_i \cup T_j) - \ldots + (-1)^n A(\theta \cup T_1 \cup \ldots \cup T_n) \geqslant 0.$$

This definition of complete monotony is a natural generalization of that which is familiar in the theory of real-valued functions of a real variable (where it occurs in the characterization of Laplace–Stieltjes transforms of positive measures), and in its present form it plays an important role in capacity theory. Note that we have proved inequality (13) only when the Ts are distinct, and do not occur as members of some one φ for which $\bigcup \varphi = \theta$; these restrictions are unnecessary, however, for when they are relaxed one obtains an inequality which can be reduced via degeneracies either to inequality (13) in its restricted form or to the tautology $0 \geqslant 0$.

Notice also that (*i*) and (*ii*) (in its derestricted form) imply that $0 \leqslant A(\theta) \leqslant 1$ for all θ in $\mathscr{T}_+$, and that $A(\theta_1) \geqslant A(\theta_2)$ whenever $\theta_1 \subset \theta_2$. We now conclude the present discussion by proving

Theorem 12 (First Fundamental Theorem). *The formulae* (10) *and* (11) *express a natural one-to-one correspondence between* (*i*) *proper consistent nets* $\{\mathrm{pr}_\varphi\}$ *and* (*ii*) *normalized completely monotonic avoidance functions A. Thus our model for a random weak incidence function is completely characterized by the associated normalized completely monotonic function A, and every such function can arise in this way. What is more, our model is 'universal'.*

Proof. All we have to do (apart from the matter of 'universality') is to show that every normalized c.m. function A arises from the application of formula (10) to some (necessarily unique!) proper consistent net. That is, we must establish the truth of the following six assertions. Let φ and v have the same significance as before, and let $F(\varphi; v)$ denote the right-hand side of equation (11).

(1) Each $F(\varphi; v) \geqslant 0$.

(2) $\sum F(\varphi; v) = 1$, where the summation is over all the $2^{|\varphi|}$ vs associated with φ.

(3) $F(\varphi; v) = \sum F(\varphi \cup \psi; w)$, where ψ is any finite collection of distinct traps having no member in common with φ, and the summation is to be extended over all those zero–one functions w on $\varphi \cup \psi$ which reduce to v when their domain is restricted to φ.

(4) $F(\varphi; v) = F(\varphi'; v')$ whenever φ' is a permutation of φ, and v' is the corresponding rearrangement of v.

(5) $F(\varphi; v) = 0$ whenever $v = 1$ at an element (trap!) of φ which can be covered by the v-null elements (traps!) of φ.

(6) If E in $\mathscr{T}_+$ is equal to $\bigcup \varphi$, and if v vanishes everywhere on φ, then $F(\varphi; v) = A(E)$.

Here (1) and (2) say that F determines a probability distribution on the cylinder-sets in $2^{\mathscr{T}}$; (3) and (4) say that the net $\{\mathrm{pr}_\varphi\}$ determined in this way is consistent; (5) says that it is proper; and finally (6) says that this proper consistent net has A as its avoidance function. It will be convenient to prove (1)–(6) in reverse order.

First we note that (6) follows immediately from the definition of F, because if v vanishes over φ, then $\varphi_v = \varphi$ and the right-hand side of equation (11) reduces to the single term $A(\bigcup \varphi)$. To prove (5), suppose that $v(T) = 1$, where T belongs to φ, and suppose that T is covered by the union of the traps in φ_v. Then every ψ having T as an element contributes to the right-hand side of equation (11) a term equal in magnitude but opposite in sign to the term associated with $\psi \backslash \{T\}$, and so the whole expression reduces to zero.

Next we look at (4) and (3). The truth of (4) is obvious from inspection of the definition of F. To establish (3), we observe that

$$\sum_{w=v \text{ on } \varphi} F(\varphi \cup \psi; w) = \sum_{w=v \text{ on } \varphi} \sum_{(\varphi\cup\psi)_w \subset \chi \subset \varphi\cup\psi} (-1)^{|\chi|-|(\varphi\cup\psi)_w|} A(\bigcup \chi).$$

Now if ψ is vacuous, it is obvious that the right-hand side reduces to $F(\varphi; v)$, so consider next the case in which ψ consists of a single trap T (not an element of φ!). Because w is wholly specified over φ, only the value of $w(T)$ is free, and so the first summation is over the two possibilities, $w(T) = 1$ and $w(T) = 0$. If $w(T) = 0$, then the collection of traps χ is required to have T as an element, whereas if $w(T) = 1$ then this is optional. Inspection of the right-hand side of the last identity shows that the terms involving collections χ containing T as a member can be arranged in pairs of equal magnitude and opposite sign, so that they cancel out, while the terms involving collections χ not containing T as a member constitute the expanded form of $F(\varphi; v)$, as required. The general case of (3) can now be established by a trivial induction in which ψ is built up by adjoining one new trap at a time.

We then consider (2) and (1). In order to prove (2), we must show that we obtain the value 1 when the right-hand side of equation (11) is summed over all possible vs. We thus have to deal with a double summation over v and ψ; let us sum over v first, holding ψ fixed. If $|\psi| = k$, then we obtain a value zero except when $k = 0$, because

$$\sum_j (-1)^j \, {}^kC_j = 0,$$

but when $k = 0$ (i.e. when ψ is vacuous), we obtain the value 1. This proves (2). As for (1), this follows from the definition of complete monotony on setting $\theta = \bigcup \varphi_v$ and enumerating the traps in $\varphi \backslash \varphi_v$. The First Fundamental Theorem will thus have been proved as soon as we establish the 'universal' character of our model, and to this last and very important step we now turn.

By *universal* we mean that *every* stochastic process representing a random weak incidence function contains *in its name-class* (for which, see [10]) a model which is an instance of the First Fundamental Theorem. Thus 'universal' is weaker than 'canonical'. We shall examine the position as regards canonicity after we have completed the proof of Theorem 12.

Let $(\Omega, \mathscr{F}, P; \{0,1\}, \mathscr{T}; Y)$ be a stochastic process defined over a probability space $(\Omega, \mathscr{F}, P)$ with 'state-space' $\{0, 1\}$ and the trapping-system $\mathscr{T}$ as 'parameter-set', Y denoting a family of zero–one random variables Y_T $(T \in \mathscr{T})$. Thus for fixed ω in Ω, $Y_{.}(\omega)$ is a zero–one function over $\mathscr{T}$, and for fixed T, $Y_T(.)$ is a random variable. Suppose further that for ω outside an $\mathscr{F}$-set N of P-measure zero, $Y_{.}(\omega)$ *is a wif*. In other words, *consider an arbitrary version of a random wif*.

Now if we compare this stochastic process with the canonical (Daniell–Kolmogorov) process in the same name-class, which will be that associated with the probability space $(2^{\mathscr{T}}, \mathscr{B}_0, \text{pr})$ carrying $\text{pr} = PY^{-1}$, and having the coordinate random variables $(.)_T$ $(T \in \mathscr{T})$ defined over it, then the canonical mapping from the given process to this canonical one with the same name will be

$$Y: \Omega \to 2^{\mathscr{T}},$$

where

$$Y(\omega) = (\ldots, Y_T(\omega), \ldots),$$

and outside the P-null $\mathscr{F}$-set N we shall have $Y(\omega) \in W$. In the terminology used in the discussion of canonicity given in the introductory chapters [10] of *SA* and *SG*, we can say that *the Ω-process possesses W as a 'nice property'*, and thus W must be $(\mathscr{B}_0, \text{pr})$-thick; that is we must have

$$\text{pr}_0^*(W) = 1.$$

But W is compact! From (*c*) it follows immediately that if we pass from the canonical (Daniell–Kolmogorov) model to the Kakutani model, then we shall have $\text{pr}(W) = 1$. Thus in the sense defined above our model of a random $\mathscr{T}$-closed set *is* universal; this detail completes the proof of Theorem 12.

We now complement this proof by one or two additional remarks, some of which are of practical importance, while others help to round off the earlier discussion [10] of canonicity in this and also in the general case.

First notice that the above Ω-process may enjoy other 'properties' which in the context of a given application or a given methodology may be seen as 'nice'. If these, together with W, are labelled and identified as subsets Γ_λ of $2^{\mathscr{T}}$ $(\lambda \in \Lambda)$, then the canonical (Λ)-extension of the canonical process will make them *all* (W included) measurable and of measure 1, and will do so in a minimal way. This will often be a useful procedure, but it is most especially so when W is the only 'nice' property in view.

For then, if we retract $2^{\mathscr{T}}$ to W, we obtain a probability measure over W itself, the σ-algebra being the trace of $\mathscr{B}_0$ on W, and as W is compact *this is exactly the Baire σ-algebra in W*. (The proof of this follows almost immediately from Tietze's extension theorem.)

When the state-space $Z = \{0, 1\}$ and the parameter-set $A = \mathscr{T}$ is a trapping-system over C, we shall write $\mathscr{B}_0(W)$ for the minimal extension of $\mathscr{B}_0$ to a σ-algebra containing W as an element. The model

(14) $$(\{0,1\}^{\mathscr{T}}, \mathscr{B}_0(W), \text{pr}_W; \{0,1\}, \mathscr{T}; (.)_T\ (T \in \mathscr{T})),$$

where pr_W is the unique extension of pr from $\mathscr{B}_0$ to $\mathscr{B}_0(W)$ which gives measure 1 to W, will be called *the canonical model of a random weak incidence function*.

It fully deserves this name, because if (as in the proof of universality) we are given *any* version of a stochastic process representing a weak incidence function, then by the canonicity theory [10] we know that we can set up a canonical extension of the D–K process in which the new σ-algebra is the smallest one extending $\mathscr{B}_0$ and containing as elements all the 'nice' properties (including the 'wif property' W) associated with the Ω-process, and in which all these 'nice' properties hold almost surely. *This canonical extension of the canonical* (Daniell–Kolmogorov) *model will be an extension of the model at* (14) *above*, whence our 'canonization' of the latter.

We call the Kakutani-based model 'universal' rather than 'canonical' because if there *are* other 'nice' properties of the Ω-process which we want to carry over to the canonical extension of the canonical model, then we must expect to have to face the awkward fact that on making these other properties 'almost sure', we may be putting ourselves into a situation in which we cannot simultaneously have the Kakutani probability-one sets 'almost sure'. As we have stressed in [10], this is a situation which in the nature of things must occur from time to time; we have a fairly free choice of cakes, but the eating of some will in general exclude the consumption of some others.

When W is the only 'nice' property discernible in the Ω-process we will often opt for the Kakutani model because the topological apparatus therein is frequently very useful. It is of some interest that the Kakutani model on $(Z^A, \mathscr{B})$ (where $\mathscr{B}$ denotes the σ-algebra of Borel sets) is always a canonical extension of the canonical model on $(Z^A, \mathscr{B}_0)$ (the Daniell–Kolmogorov model), even in the very general situation when Z is an arbitrary second-countable compact Hausdorff space and A is arbitrary.

To see that this is so, recall the 'interpolation theorem' for Z^A; if B is any Borel set, and if K is a compact kernel and G an open cover for B, so that $K \subset B \subset G$, then there exist a compact $\mathscr{G}_\delta$-set K_0 and an open $\mathscr{K}_\sigma$-set G_0 such that

$$K \subset K_0 \subset G_0 \subset G.$$

(This is an immediate consequence of the Urysohn theorem.) Now each of K_0 and G_0 is a Baire set; also from the regularity of the Kakutani model we know that each of $\mathrm{pr}(G \backslash B)$ and $\mathrm{pr}(B \backslash K)$ can be made as small as we please. Thus we can find Baire sets B_n^0 such that

$$\mathrm{pr}(B_n^0 \triangle B) < \frac{1}{2^n},$$

for each $n \geqslant 1$, and then if $B_0 = \limsup B_n^0$ we shall have $\mathrm{pr}(B_0 \triangle B) = 0$; that is,

$$B = B_0 \triangle N \quad \text{and} \quad \mathrm{pr}(B_0) = \mathrm{pr}(B),$$

where B_0 is Baire, and N is Borel and null. Now consider the canonical extension of the canonical model in which the sets 'proclaimed' to be measurable and full (i.e. of measure one) are exactly the Borel sets of Kakutani probability 1. This is possible, for they are all $(\mathscr{B}_0, \mathrm{pr})$-thick, and the family is countable-intersection closed. Obviously, if these full sets become measurable and full, then also the Kakutani null sets will become measurable and null. Thus every Borel set B must become measurable, and will have its Kakutani measure; that is, this canonical extension *is* the Kakutani model.

This provides an *a posteriori* justification for the terminology employed by Nelson and Meyer, but notice that we have only shown that the Kakutani model is *a* canonical extension; there will be others, incompatible with it, which we might prefer on occasion.

To close this discussion we remind the reader that random weak incidence functions are only of interest to us here as a resting place on our route towards random strong incidence functions and so random $\mathscr{T}$-closed sets. But the study of weak incidence functions in their own right is a mathematically tempting project, and I hope to present an account of it in another place [14].

6.2.5 RANDOM STRONG INCIDENCE FUNCTIONS, AND RANDOM $\mathscr{T}$-CLOSED SETS

In Section 6.2.2 we devoted a little time to the discussion of what we called 'the countable case', but the advantages to be gained in this situation are surprisingly slight; the reason for this is that, even when the trapping-system is countably infinite, there will in general be an uncountable number of ways of covering a given trap with an infinite sequence of traps, and so the advantage of countability in the sif case is somewhat illusory. There is a minute gain in the wif case, for then the conditions which have to be satisfied by a zero–one function over $\mathscr{T}$, if it is to be a wif, will also be countable, and this means that the subset W of $2^{\mathscr{T}}$ which consists of all the wifs will now be a Baire compact set, so that $\mathscr{B}_0(W)$ can be replaced by $\mathscr{B}_0$. It seems likely that the use of a countable trapping-system, when available, will be convenient in practical problems involving random wif's, because of the fact already noted that the *value* of a Kakutani probability pr (B) may be difficult to *calculate* when B is a Borel set but not a Baire set. From the general theoretical point of view, however, it seems desirable to avoid using the crutch of countability if we can manage without it, and this will be the view adopted here (without prejudice to the conduct of later more practically oriented stages of the investigation).

If we recall the details of the measurability crises encountered in more familiar stochastic process situations, when the parameter-set A is the real line, it will be noticed that what is used to evade them is not the countability of the parameter-set itself, but rather the existence of some countable 'separating basis' for the parameter-set. This separating basis need not be unique, and there will not normally be any way of choosing one such basis rather than another, other than on grounds of expediency. All that is required is that such a basis, unique or not, should exist, and then we can prove the desired general theorems in which, however, the basis itself will not appear. For purposes of calculation we customarily exploit the non-unicity of the basis by choosing one (e.g. the binary rationals) especially suited to computation.

The above remarks form a good introduction to the course we propose to follow. We shall be driven to strengthen our axiomatic structure, and this is hardly surprising, for the structure used until now, which merely demands the existence of a covering of C by non-empty sets T, is slight in the extreme, and it is perhaps rather extraordinary that it has sufficed to support the development up to this point. We shall now enrich it by (I speak loosely for the moment) a kind of local countability and a sort of local compactness. The new axioms have in fact been constructed by taking a very special solution which works in a well-behaved topological case (see Example 8 below), and then discarding everything but what seems to be absolutely necessary for the abstract framework which we propose to set up.

Axioms TS(i) and TS(ii) will be retained, so that we shall continue to enjoy all that follows from our earlier weaker structure, and in terms of these it will now be convenient to make the following definition:

a trap T will be called a *c-trap*, when every covering *by traps* of $\mathrm{Cl}(T; \mathscr{T})$ can be reduced to a finite subcovering.

We now strengthen the axiomatic basis as follows.

TS(i): *every trap T in $\mathscr{T}$ is non-empty.*

TS(ii): *the traps T cover C.*

TS(iii): *with every trap T we can associate a countable system $\mathscr{S}(T)$ of subtraps of T in such a way that we can construct all possible subtraps of T by forming suitable unions of subtraps in the system $\mathscr{S}(T)$.*

TS(iv): *if x belongs to T, then x belongs to a c-trap whose $\mathscr{T}$-closure is covered by T.*

When the axioms are satisfied we shall call $(C, \mathscr{T})$ a *trapping-space*. Here the third axiom provides what we called local countability, while the fourth provides a weak version of local compactness. We illustrate the situation by the following important example of a trapping-space.

Example 8. *We take C to be any second-countable locally compact Hausdorff space, and identify the traps T with the elements of any (not necessarily countable) basis of non-empty open sets for the topology of C.*

By hypothesis there does exist a countable basis, but it is not *assumed that this is the trapping system* $\mathscr{T}$. *Now every subspace of a second countable space is second countable, and so is Lindelöf; in particular, this is true for each open set* G_r $(r = 1, 2, \ldots)$ *comprising the countable basis for the topology. So each* G_r *is a (perhaps very large) union of traps and is Lindelöf, so is a union of* countably *many traps. If we bring together all the countable collections of traps needed to represent* $G_1, G_2, \ldots$ *in this way, we obviously obtain a countable basis for the topology consisting entirely of traps, and incidentally we show that* TS(*iii*) *is satisfied. (Of course it is obvious that* TS(*i*) *and* TS(*ii*) *are satisfied.) Now we use the local compactness. If x belongs to T (an open set) then there exists an open G with compact closure such that*

$$x \in G \subset \bar{G} \subset T.$$

But G is a union of traps, and one of these must contain x; say T_0 *does so. Then* $T_0 \subset G$, *and thus the closure of* T_0 *will be covered by the* (*compact*) *closure of G, and so will itself be compact, and in fact we shall have*

$$x \in T_0 \subset \bar{T}_0 \subset \bar{G} \subset T, \quad \textit{where } \bar{T}_0 \textit{ is compact.}$$

But $\bar{T}_0$ *is here the same as* $\mathrm{Cl}(T_0; \mathscr{T})$, *and its compactness implies that* T_0 *is a c-trap, so* TS(*iv*) *is satisfied.*

We are now going to set up a universal model for a random $\mathscr{T}$-closed set in an *arbitrary* trapping-space C. We begin by proving

Theorem 13. *With each trap T in* $\mathscr{T}$ *we can associate* (*in perhaps more than one way*) *a sequence* $\{{}_jT : j = 1, 2, \ldots\}$ *of c-traps such that*

$$T = \bigcup_j {}_jT = \bigcup_j \mathrm{Cl}({}_jT; \mathscr{T}). \tag{15}$$

Proof. We start with $\mathscr{S}(T)$; this is a countable system of subtraps of T, and it acts within T rather like a basis, in that all other subtraps of T can be built out of it by forming unions, but the elements of $\mathscr{S}(T)$ may not be c-traps. Now let $x \in T$, and apply axiom TS(*iv*). Then there exists a c-trap T_0 such that

$$x \in T_0 \subset \mathrm{Cl}(T_0; \mathscr{T}) \subset T,$$

where $\mathrm{Cl}(T_0; \mathscr{T})$ is such that *any* covering *by traps* can be reduced to a finite subcovering. Now T_0 is a subtrap of T, and so is a union of some of the 'special' traps from $\mathscr{S}(T)$, and one of these must therefore contain x; let it be, say, T_{00}. We shall have

$$x \in T_{00} \subset T_0 \subset T,$$

and also (from Theorem 1(*b*)).

$$\mathrm{Cl}(T_{00}; \mathscr{T}) \subset \mathrm{Cl}(T_0; \mathscr{T}) \subset T.$$

We now show that T_{00} is a c-trap. For suppose we have a covering by traps of $\mathrm{Cl}(T_{00}; \mathscr{T})$. This may not cover $\mathrm{Cl}(T_0; \mathscr{T})$, but it will do so if augmented by all the traps disjoint with T_{00}, for the union of these last is exactly (by definition) the complement of $\mathrm{Cl}(T_{00}; \mathscr{T})$. However, the traps disjoint with T_{00} are *exactly* the traps disjoint with $\mathrm{Cl}(T_{00}; \mathscr{T})$, from Theorem 1(*c*), so we have the following situation: a trap-covering of $\mathrm{Cl}(T_{00}; \mathscr{T})$ together with a system of traps disjoint with $\mathrm{Cl}(T_{00}; \mathscr{T})$ provides a covering of $\mathrm{Cl}(T_0; \mathscr{T})$. As T_0 is a *c*-trap, a finite subcovering of $\mathrm{Cl}(T_0; \mathscr{T})$ will exist, and this will also be a finite subcovering of $\mathrm{Cl}(T_{00}; \mathscr{T})$, *from which all the traps disjoint with* $\mathrm{Cl}(T_{00}; \mathscr{T})$ *can obviously be dropped.* Thus we obtain a finite refinement of the original covering of $\mathrm{Cl}(T_{00}; \mathscr{T})$, and this completes the proof that T_{00} is a *c*-trap.

We now let x vary throughout T, and so obtain a system of *c*-traps T_{00} all satisfying $\mathrm{Cl}(T_{00}; \mathscr{T}) \subset T$. But these are all 'special' traps in $\mathscr{S}(T)$, and so *there is at most a countable infinity of them.* Enumerate them as ${}_1T, {}_2T, \ldots$, and we obtain the desired sequence of *c*-traps satisfying equation (15).

We now recall that a wif g, if it is not a sif, displays its bad behaviour on just those traps T with $g(T) = 1$ which can be covered by g-null traps. In the notation of Section 6.2.3 these are precisely the traps T for which $\mathbf{R}g(T) = 1$. We combine this fact with the information contained in Theorem 13 to get

Theorem 14. *Let g be any weak incidence function and T any trap. Then* $(\mathbf{R}g)(T) = 1$ *if and only if*

$$g(T) = 1 \quad \textit{and} \quad g({}_jT) = 0 \quad \textit{for } j = 1, 2, \ldots.$$

Thus if W denotes the set of weak (and S the set of strong) incidence functions, we shall have

$$W \backslash S = \bigcup_{T} \bigcap_{n \geq 1} \{g \in W, g(T) = 1, g({}_jT) = 0 \ (1 \leq j \leq n)\}.$$

Proof. If $g(T) = 1$ and if each $g({}_jT) = 0$, then $\mathbf{R}g$ takes on the value 1 at T because $\{{}_jT\}$ is a covering of T. On the other hand, if $g(T) = 1$ and if T has some g-null covering, then so has every $\mathrm{Cl}({}_jT; \mathscr{T})$, whence every $\mathrm{Cl}({}_jT; \mathscr{T})$ has a *finite* g-null covering. But then each ${}_jT$ has a finite g-null covering and g is a wif, so $g({}_jT)$ must be zero for every j. This proves the first part of the theorem, and the second part follows on noting that $g \in W$ will be in S if and only if $\mathbf{R}g$ vanishes on every T.

Theorem 14 as it stands lacks anything even approaching canonical status, because the traps $\{_jT\}$ are not uniquely determined by T, but this does not matter because Theorem 14 will only be used as a technical device in order to prove a much more important result, and this latter will be free of all mention of the $\{_jT\}$.

Let us first note that the identity at the end of Theorem 14 is equivalent to

$$S = W \setminus \bigcup_T \{g(T) = 1, g(_jT) = 0\ (j \geqslant 1)\},$$

and so S is obtained from the compact set W by deleting its overlap with a perhaps very large union of compact Baire sets (compact $\mathscr{G}_\delta$s) (note that there are as many terms figuring in this union as there are traps in $\mathscr{T}$). In the 'countable case' ($\mathscr{T}$ itself countable) this will of course tell us that S is a Baire set, *but in general S will not even be Borel.*

We have thus reached a point in our work at which the Kakutani device is no longer sufficient to resolve our measurability difficulties, and we therefore turn for help to the 'standard modifications' of J. L. Doob [6]. The general properties of these have been treated somewhat cursorily in the literature, and in order to repair this state of affairs, and to complement our discussion of 'standard extensions' (here and in [10]), we shall now break off from our main argument for a while. The parenthesis which follows is essential for our later work, but it can be read (together with the relevant parts of [10]) as a self-contained preliminary account of a 'theory of versions' for stochastic processes.

6.2.6 ADMISSIBLE PROPERTIES, STANDARD MODIFICATIONS, AND CANONICAL EXTENSIONS OF THE CANONICAL MODEL

Let A be a fixed parameter-set and Z a fixed second-countable compact Hausdorff state-space. As usual $\mathscr{B}_0$ will denote the Baire σ-algebra in Z^A, and $(\Omega, \mathscr{F}, \mathrm{pr}; Z, A; Y_\alpha\ (\alpha \in A))$ will denote a generic stochastic process having Z as state-space and A as parameter-set. We have (in [10]) explained that any two such processes are to be regarded as equivalent if they determine the same probability measure $\mathrm{pr}\, Y^{-1}$ over $\mathscr{B}_0$. The corresponding equivalence classes have been called *name-classes*, the *name* of a name-class (or of any one of its members) being this common probability measure on $\mathscr{B}_0$. From the Daniell–Kolmogorov theorem we know that each probability measure on $\mathscr{B}_0$ corresponds to some name-class, so that there is a natural one-to-one correspondence between such measures and the name-classes associated with Z and A. That theorem also tells us that if μ is any probability measure on $\mathscr{B}_0$, then the model

$$(Z^A, \mathscr{B}_0, \mu; Z, A; X_\alpha\,(\alpha \in A))$$

belongs to the name-class with name μ, and we have called this the *canonical model* in the name-class. (Here X_α is the αth coordinate function, so that X_α is the natural mapping from Z^A to Z 'at α'.

If Γ is any subset of Z^A, we shall say that Γ is an *admissible property* (of the 'sample paths') for the name-class when there is a stochastic process in the class for which $Y_{\cdot}(\omega) \in \Gamma$ whenever ω lies outside an $\mathscr{F}$-set N of pr-measure zero. Similarly we shall say that an indexed family $\{\Gamma_\lambda\colon \lambda \in \Lambda\}$ of subsets of Z^A is an *admissible family of properties* for the name-class when there is a single stochastic process in the class for which, for each $\lambda \in \Lambda$, $Y_{\cdot}(\omega) \in \Gamma_\lambda$ whenever ω lies outside an $\mathscr{F}$-set N_λ of pr-measure zero. We now rephrase some of the results of [10] in this new terminology.

(*a*) *A subset* Γ *of* Z^A *is an admissible property for the name-class with name-measure* μ *on* $\mathscr{B}_0$ *if and only if* $\mu^*(\Gamma) = 1$, *this outer measure being formed with respect to* $\mathscr{B}_0$ *or its completion (it does not matter which). When this condition is satisfied then there is a unique minimal* σ*-algebra* $\mathscr{B}_0(\Gamma)$ *containing* Γ *and all Baire sets, and there is a unique extension* μ_Γ *of* μ *to this* σ*-algebra which makes* Γ *full. The extension can be uniquely continued to the unique minimal* σ*-algebra* $\mathscr{B}_0^+(\Gamma)$ *which is complete under* μ_Γ. *The model*

$$(Z^A, \mathscr{B}_0(\Gamma), \mu_\Gamma; Z, A; X_\alpha\,(\alpha \in A))$$

is called the canonical Γ*-extension of the canonical model; for it, sample paths have the property* Γ *with probability one.*

(*b*) *A family* $(\Lambda) = \{\Gamma_\lambda : \lambda \in \Lambda\}$ *of subsets* Γ_λ *of* Z^A *is an admissible family of properties for the name-class if and only if* (*i*) *each* Γ_λ *is separately admissible and* (*ii*) *the family is closed under the formation of countable intersections. When these conditions are satisfied then there is a unique minimal* σ*-algebra* $\mathscr{B}_0[\Lambda]$ *containing all the* Γ_λ*s and all Baire sets, and there is a unique extension* $\mu_{(\Lambda)}$ *of* μ *to this* σ*-algebra which makes every* Γ_λ *full. The extension can be uniquely continued to the unique minimal* σ*-algebra* $\mathscr{B}_0^+[\Lambda]$ *which is complete under* $\mu_{(\Lambda)}$. *The model*

$$(Z^A, \mathscr{B}_0[\Lambda], \mu_{(\Lambda)}; Z, A; X_\alpha\,(\alpha \in A))$$

is called the canonical (Λ)*-extension of the canonical model; for it, for each* λ, *the sample paths have property* Γ_λ *with probability one.*

We recall that it is vain to hope for a 'final' canonical extension of the canonical model which extends all the others. On the other hand the partly-ordered system of all (Λ)-extensions is 'universally adequate' in the following sense.

Let $(\Omega, \mathscr{F}, \mathrm{pr}; Z, A; Y_\alpha\ (\alpha \in A))$ be any stochastic process in the name-class which is defined on a *complete* probability-space (we lose no real generality here because we could always first complete $\mathscr{F}$ under pr), and let (Λ) consist of all the $\mathscr{F}$-sets having pr-measure one. Then it is not difficult to show that the mapping

$$\varphi: \Omega \to Z^A,$$

$$\varphi(\omega) = \{Y_\alpha(\omega): \alpha \in A\},$$

is a measurable and measure-preserving mapping of $(\Omega, \mathscr{F}, \mathrm{pr})$ into *each one* of the probability-spaces

$$(Z^A, \mathscr{B}_0[\Lambda], \mu_{(\Lambda)}), \quad (Z^A, \mathscr{B}_0^+[\Lambda], \mu_{(\Lambda)}),$$

and that it maps the given stochastic process into the canonical (Λ)-extension of the canonical model (or its completion, respectively) in the sense that

$$X_\alpha(\varphi(\omega)) = Y_\alpha(\omega)$$

for all α and ω. Thus whatever stochastic process in the name-class we are presented with, there is always a canonical extension of the canonical model, completed if we so wish, which does just as good a job. Moreover *all* of the $\mathscr{B}_0^+[\Lambda]$-measurable Z-valued random variables over Z^A can be carried back (by the use of φ) so that they become $\mathscr{F}$-measurable Z-valued random variables over Ω, without change of joint distribution.

There are, I think two reasons why (*a*) and (*b*) do not quite suffice to deal with all possible measurability troubles of the '$\omega \in \Gamma$' type.

In the first place, the most common *practical* method of finding out what properties or families of properties are admitted by a name-class is to start with some convenient version and then to trim it until it behaves as well as we should like it to do; it is of course necessary to check that the 'trimming' process follows the rules laid down for such '*standard modifications*', since only by respecting this rule can we ensure (*i*) that we do not leave the name-class and (*ii*) that we keep in contact with the initial process.

Having constructed a trimmed version in this way, we can then invoke (*b*) above, to find a canonical extension of the canonical model whose 'trim' is just as good. The initial process can then be thrown away.

It may, however, be the case that the initial process carries some *other* stochastic structure to which we wish to continue to refer (as we may, because of (*ii*) above). We can do so by keeping the original version, together with the standard modification which we have 'built on top of it', and this will normally be desirable *unless* all foreseeably interesting stochastic structures on $(\Omega, \mathscr{F}, \text{pr})$ are comprised within the stochastic process we are modifying. As the parameter-set A is unrestricted, it may be felt that the 'unless' clause could always in principle be invoked, but it is clear that this would sometimes be inconvenient.

It appears, therefore, that both of the arguments advanced for the study of standard modifications are at bottom based on expediency, and I think that probably this is a correct view of the matter. It does not, however, impair their utility.

The trapping-space problem provides an excellent illustration of this situation. We have used the Kakutani model to make W, the set of wifs, measurable (in virtue of being compact) and of measure one, and W is then an admissible property for the name-class in the terminology used here. We shall shortly have to discover (using standard modifications) when the smaller set S of sifs is also an admissible property for the same name-class, and obviously we can then have both S and W admissible because W covers S, and so is always 'at least as thick' as S. We can *not*, however, expect all the other full Borel sets to be simultaneously admissible, and so on transferring to the canonical S-extension of the canonical model we lose touch with the possibly valuable Borelian structure. By working instead with a standard modification built on top of the Kakutani model we retain contact with the Borelian structure, and this might at a later date 'come in useful'.

We now formally introduce standard modifications, first in an abstract form which may seem rather superficial, and then (by specializing this) in the form appropriate to stochastic-process theory. Let $(\Omega, \mathscr{F}, \text{pr})$ be any probability space; normally it will be complete, but this is not essential. We suppose that $\mathscr{F}$ is some extension of the basic σ-algebra $\mathscr{F}_0$ 'natural to the problem'. We shall say that a mapping

$$\mathbf{M}: \Omega \to \Omega$$

determines a standard modification of the *probability-space* when the following two conditions are satisfied.

SM(i): **M** *is measurable when we give to its domain-space the* σ*-algebra* $\mathscr{F}$, *and to its range-space the* σ*-algebra* $\mathscr{F}_0$.

SM(ii): *each* $\mathscr{F}_0$*-set* F_0 *is 'almost invariant'; that is,*

(16) $$\text{pr}(F_0 \triangle \mathbf{M}^{-1} F_0) = 0.$$

Notice that SM(i) merely asks that $\mathbf{M}(.)$ be an 'Ω-valued random variable', and the σ-algebras are assigned in such a way as to make the fulfilment of this requirement as easy as possible. SM(*ii*) has the important consequence that

$$\mathrm{pr}(\{\omega: \mathbf{M}\omega \in F_0\}) = \mathrm{pr}(\mathbf{M}^{-1}F_0) = \mathrm{pr}(F_0) = \mathrm{pr}(\{\omega: \omega \in F_0\}),$$

and so for all practical purposes (because of the special role of the σ-algebra $\mathscr{F}_0$), the 'random variable' (.) can be replaced by the 'random variable' $\mathbf{M}$(.). What must be stressed, however, is that this is just a rather weak consequence of SM(*ii*); SM(*ii*) is in fact equivalent to the much stronger assertion that

(17) $\mathrm{pr}(\{\omega: \omega \in F \text{ and } \mathbf{M}\omega \in F_0\}) = \mathrm{pr}(\{\omega: \omega \in F \cap F_0\})\ (F \in \mathscr{F}, F_0 \in \mathscr{F}_0).$

It is because of equation (17) that we are able to make use of the standard modification $\mathbf{M}$ and still when necessary make further use of the original structure associated with $\mathscr{F}$.

We now say that the standard modification $\mathbf{M}$ *admits* the 'nice' properties $\{\Gamma_\lambda: \lambda \in \Lambda\}$ when

SM(*iii*): $\mathbf{M}\omega \in \Gamma_\lambda$ *for all* $\omega \in \Omega \backslash N_\lambda$, *where* N_λ *is a null* $\mathscr{F}$*-set.*

The family of 'nice' properties will, as usual, be referred to as (Λ) (a convenient abuse of notation), save in the important special case when it consists of just one member Γ.

We can now make a number of fairly obvious remarks which are similar in content to and supplement our remarks made earlier about standard extensions.

First, we can always without loss of generality suppose that the family (Λ) is countable-intersection closed. Next, each member of the family is necessarily ($\mathscr{F}_0$, pr)-thick. Thirdly, if $\mathscr{F}_0$ is given, and if (Λ) is any countable-intersection closed family of ($\mathscr{F}_0$, pr)-thick sets, then it is possible to choose $\mathscr{F}$ so that the identity-mapping is a standard modification admitting the family. Fourthly, if L is a fixed countable subset of the index-set Λ, then

$$N = \bigcup_{\lambda \in L} N_\lambda \quad \text{and} \quad \bigcap_{\lambda \in L} \Gamma_\lambda$$

will be respectively a null $\mathscr{F}$-set and an ($\mathscr{F}_0$, pr)-thick set. The thick set cannot be vacuous, so let ω_1 be some point in it. If we then define

$$\mathbf{M}'\omega = \omega_1 \quad \text{for } \omega \in N,$$
$$\mathbf{M}'\omega = \mathbf{M}\omega \quad \text{for } \omega \in \Omega \backslash N,$$

then we obtain another standard modification admitting the family, but now $\mathbf{M}'\omega \in \Gamma_\lambda$ *surely*, for all λ in L. Of course the countability restriction on L is essential here.

Another rather interesting point is that if for each countable $L \subset \Lambda$ there exists a version in the name-class having the 'nice' properties $\{\Gamma_\lambda: \lambda \in L\}$, then there exists a single version (for example the canonical (Λ)-extension of the canonical model) which has the 'nice' property Γ_λ for *all* $\lambda \in \Lambda$. In other words, a family $\{\Gamma_\lambda: \lambda \in \Lambda\}$ is admissible for the name-class if and only if each countable subfamily is admissible for the name-class. This fact can be reformulated in various ways in terms of the existence of standard modifications when the σ-algebra $\mathscr{F}$ is at our choice. *Normally, however, it is not;* it is more usual for $\mathscr{F}$ to be prescribed in some specially convenient way, perhaps (*but not necessarily*) in connection with some canonical construction. This will become clearer after we have proved Theorem 15.

We now turn to standard modifications built 'on top of' models, or more generally on top of versions belonging to some given name-class. We prove

Theorem 15. *Let* $\mathbf{M}$ *be a standard modification of* $(Z^A, \mathscr{F}, \mathrm{pr})$, *where* A *is a parameter-set and* Z *is a second-countable compact Hausdorff space, and* $\mathscr{F}_0$ *of the definition has been replaced by the* σ*-algebra* $\mathscr{B}_0$ *of Baire sets in* Z^A. *Let* $\mathbf{M}$ *admit the countable-intersection closed family* $\{\Gamma_\lambda: \lambda \in \Lambda\}$. *Then, if* X *now denotes a generic point of* Z^A, *with* α*th coordinate* X_α,

(*i*) $\{X: (\mathbf{M}X)_\alpha \in D\}$ *is* $\mathscr{F}$*-measurable for each* $\alpha \in A$ *and for each Baire set* D *in* Z;

(*ii*) $\mathrm{pr}(\{X: (\mathbf{M}X)_\alpha = X_\alpha\}) = 1$ *for each* $\alpha \in A$;

(*iii*) *for each* λ *in* Λ, $\mathbf{M}X \in \Gamma_\lambda$ *when* X *lies outside an* $\mathscr{F}$*-set* N_λ *of* pr*-measure zero.*

Conversely (*i*)–(*iii*) *are also sufficient for* $\mathbf{M}$ (*relative to* $\mathscr{F}$) *to yield a standard modification built on top of the coordinate-process which admits the family* $\{\Gamma_\lambda: \lambda \in \Lambda\}$.

Proof. The equivalence of (*i*) and SM(*i*) is just an application of the monotone class theorem (or of Dynkin's π/λ theorem [8]). Note that (*i*) just says that each $(\mathbf{M}X)_\alpha$ is a Z-valued random variable. Given (*i*), we now prove that (*ii*) is equivalent to SM(*ii*). Let $\{G_n: n \geqslant 1\}$ be a countable basis of open sets for the Z-topology. Let us put

$$B_n^\alpha = \{X: X_\alpha \in G_n\};$$

this will be a Baire set in Z^A, and we shall have

$$\{X: (\mathbf{M}X)_\alpha \neq X_\alpha\} = \bigcup_n (B_n^\alpha \triangle \mathbf{M}^{-1} B_n^\alpha), \tag{18}$$

whence SM(*ii*) certainly implies (*ii*). On the other hand if (*ii*) holds, let $\mathscr{E}$ be the class of Baire sets B_0 in Z^A for which $\mathrm{pr}\,(B_0 \triangle \mathbf{M}^{-1} B_0) = 0$; this is a σ-algebra and contains every B_n^α. But the B_n^αs generate $\mathscr{B}_0$, so $\mathscr{E} = \mathscr{B}_0$, and SM(*ii*) follows. Finally (*iii*) is just a restatement of SM(*iii*).

This theorem shows that our definition of standard modifications *of probability-spaces* is consistent with Doob's definition of standard modifications *of stochastic processes* in the only situation in which the two overlap, namely when the probability-space is a Cartesian product, and the stochastic process consists initially of the coordinate-projections. The mapping **M** converts

$$(Z^A, \mathscr{F}, \mathrm{pr};\, Z, A;\, X_\alpha\ (\alpha \in A))$$

into

$$(Z^A, \mathscr{F}, \mathrm{pr};\, Z, A;\, (\mathbf{M}X)_\alpha\ (\alpha \in A)),$$

and these two are versions of one and the same name-class, namely that whose 'name' is 'pr on $\mathscr{B}_0$'.

It is always desirable to remember, however, that the new process is more properly to be described as

$$(Z^A, \mathscr{F}, \mathrm{pr};\, Z, A;\, X_\alpha\ (\alpha \in A), \text{ and } (\mathbf{M}X)_\beta\ (\beta \in A));$$

this could be called a probability-space simultaneously carrying two separate versions of the same name-class. One of these (the **M**-version) has the 'nice' properties in the family (Λ), while the other (the original version) may not have these properties, but will usually have other properties which make it still of value to us. If the original version was of no particular interest in itself and was just 'as good a place to start from as any other', then we may wish to scrap it completely. In that case we can profitably pass immediately to the canonical (Λ)-extension of the canonical model,

$$(Z^A, \mathscr{B}_0[\Lambda], \mathrm{pr}_{(\Lambda)};\, Z, A;\, X_\alpha\ (\alpha \in A)),$$

which we know to exist from the existence and properties of **M**. **M** will then have served its purpose, and can be allowed to retire from the scene.

Notice that if the original version

$$(Z^A, \mathscr{F}, \mathrm{pr};\, Z, A;\, X_\alpha\ (\alpha \in A))$$

was itself some canonical (Λ')-extension of the canonical model, we can *not* then immediately infer from the existence of **M** that the systems (Λ) and (Λ') of 'nice' properties can simultaneously be adjoined as measurable and full, although sometimes it might happen that this is so.

An interesting point is that if $\mathscr{F}_1 = \mathscr{B}_0[\Lambda_1]$ and $\mathscr{F}_2 = \mathscr{B}_0[\Lambda_2]$, where the system of 'nice' properties (Λ_2) contains as members all the 'nice' properties in the system (Λ_1), then any standard modification built on top of the canonical (Λ_1)-extension of the canonical model can be carried back and built on top of the canonical (Λ_2)-extension of the canonical model. In particular, a standard modification constructed with $\mathscr{F} = \mathscr{B}_0^+$ can be carried back and rebuilt on top of *any* completed canonical extension of the canonical model, *or indeed on top of any complete version whatsoever in the name-class.*

We can now begin to see something of the contrast between the two assertions:

(*I*) the completed Kakutani model (or the Kakutani model itself) has the 'nice' property Γ as a measurable set of measure 1;

(*II*) the completed canonical (Daniell–Kolmogorov) model

$$(Z^A, \mathscr{B}_0^+, \text{pr}; \ldots)$$

permits the construction on top of this model of a standard modification admitting the 'nice' property Γ.

In general it would seem that neither (*I*) nor (*II*) implies the other. They *both*, of course, imply that Γ is $(\mathscr{B}_0, \text{pr})$-thick, and so further imply that there exists a canonical extension of the canonical model

$$(Z^A, \mathscr{B}_0, \text{pr}; \ldots)$$

for which Γ is measurable and of measure 1, and which is otherwise minimal. Assertion (*I*) adds to this that the (completed) canonical Γ-extension can in this case be consistently further extended to the (completed) Kakutani model ('completed' being deleted here if it does not appear in (*I*)), so that the two are comparable in the partial ordering for canonical extensions. Assertion (*II*), on the other hand, permits the valuable extra assertion to be made, that *any complete version whatsoever* (in the name-class) permits the construction on top of that version of a standard modification admitting the 'nice' property Γ.

At least one class of Borel measurable sets Γ is known for which (*I*) is equivalent to (*II*). This is the class of subsets Γ_H of Z^A, where H is a countable subset of A, and Γ_H consists of those mappings f of A into Z for which

$$\text{Graph}(f|H) \textit{ is dense in } \text{Graph}(f), \textit{ in the topology of } A \times Z.$$

Here it is assumed that A is topologized, and for the result just mentioned (for which, in the general case, see [11]) we shall need to assume that A

is Hausdorff and second countable. (By $f|H$ we here mean the mapping f with its domain restricted to H.) *Such a mapping f will here be called H-separable.* Notice that this is a topological definition, and that *it has nothing to do with probability.*

Doob in [6] introduced the profoundly important probabilistic concept of the H-separability of a stochastic process, in the special case when Z is the (compactified) real line and A is the real line or some interval thereof. In this 'classical' case, and also in the abstract case of interest here, we can say that a version in the name-class being studied is an *H-separable version* if it is such that, outside a measurable null set, *each individual sample path is H-separable*; i.e. if it has Γ_H as a 'nice' property.

Among the many important theorems proved by Doob is one (extended to the abstract case in [11]) which asserts that *every* complete version in the name-class admits an H-separable standard modification if

$$\text{pr}\,(\{\omega\colon X_\alpha(\omega) \in \overline{X_{H\cap I}(\omega)}) = 1 \tag{19}$$

for each α in A and for each open interval I of A containing α. Here $X_E(\omega)$ denotes the range of $X_{\cdot}(\omega)$ over the restricted domain E, and the bar denotes topological closure in $\bar{R}$.

Notice that condition (19) is a condition on the name-class alone, for the set whose probability occurs in equation (19) is a Baire set.

Now it is easy to show (and all this is true, for Hausdorff second countable A, in the abstract case also) that Γ_H is a Borel set; in fact it is a $\mathscr{K}_{\sigma\delta}$. It therefore must always have a probability with respect to the Kakutani model (completion is not needed). Adaptation of an argument of Meyer [20] shows that $\text{pr}\,(\Gamma_H) = 1$ *if and only if* equation (19) is satisfied. Thus, for Γ_H, *(I) implies (II)*. However, *(II)* implies $(\mathscr{B}_0, \text{pr})$-thickness, and so because Γ_H is a $\mathscr{K}_{\sigma\delta}$, *(II) implies (I)* (on using the result (*c*) of Section 6.2.4).

In practice one supplements these results (and again a generalization to the abstract case is possible) by Doob's further theorem that equation (19) holds for *every* countable dense set H in A, provided that the process is 'continuous in probability'. (Once again it is relevant to note that this condition, which requires rather careful formulation in the abstract case, is a condition on the name-class only.)

We therefore wind up this discussion by proposing the following problem: *for which Borel sets* Γ *is (I) equivalent to (II)*?

We shall not discuss this question further here, as it is not immediately relevant to our present task, but we note (from the result (*e*) of Section 6.2.4) that *(II)* always implies *(I)* (the 'completed' version) when Γ happens to be Souslin over the compacts.

6.2.7 RANDOM STRONG INCIDENCE FUNCTIONS AND RANDOM $\mathscr{T}$-CLOSED SETS (CONCLUDED)

We now take up our main argument again at the point at which we left it (the end of Section 6.2.5). We suppose that we have before us an instance of our canonical model for a random *weak* incidence function, the minimal appropriate probability-space having been completed under the measure. That is, we have the model

$$(2^{\mathscr{T}}, \mathscr{B}_0^+(W), \mathrm{pr}_W; 2, \mathscr{T}; g_T\ (T \in \mathscr{T})),$$

where the probability measure pr_W is uniquely characterized by a normalized completely monotonic avoidance function determining a proper consistent net of finite-dimensional distributions and so a measure pr (with $\mathrm{pr}^*(W) = 1$) on $(2^{\mathscr{T}}, \mathscr{B}_0)$, this being extended to pr_W on $\mathscr{B}_0^+(W)$ in the unique manner already explained. Notice that we have switched our notation so that it is now $g_{\mathrm{T}} = g(T)$ (instead of X_{T}) which denotes the Tth coordinate random variable, while g now denotes a generic point in $2^{\mathscr{T}}$; this is to bring our notation into line with that used when previously working with incidence functions of either type. (The use of $(\cdot)_T = (\cdot)(T)$ would be more logical but is too clumsy.)

Our strategy will be to apply Theorem 15 to the situation just described, so that the σ-algebra $\mathscr{B}_0^+(W)$ plays the role of $\mathscr{F}$, *and to identify the mapping* $\mathbf{M}$ *with the mapping* $\mathbf{S}$ *introduced in Section* 6.2.3. We know at once that $\mathbf{S}g$ will then *always* be a strong incidence function; that is, $\mathbf{S}g \in S$ for *all* g. If we can show, using Theorem 15, that $\mathbf{S}$ is a standard modification, then we shall have in

$$(2^{\mathscr{T}}, \mathscr{B}_0^+(W), \mathrm{pr}_W; 2, \mathscr{T}; (\mathbf{S}g)_T\ (T \in \mathscr{T}))$$

a model for a random *strong* incidence function, but of course we must not expect to be in a position to do this unless the avoidance function $A(.)$ satisfies some restrictive condition. Identifying a necessary and sufficient form of this condition will be our main objective. Of course $\mathscr{F}_0$ will be $\mathscr{B}_0$.

We look now at the first of the conditions in Theorem 15, and note that no generality will be lost if we take the Baire set D to be the singleton $\{0\}$. From the table in Section 6.2.3 we see that $(\mathbf{S}g)(T) = 0$ if and only if either

(*a*) $g(T) = 0$

or

(*b*) $(\mathbf{R}g)(T) = 1$,

these two possibilities being mutually exclusive. Now (*a*) determines a

Baire set in $2^{\mathcal{T}}$, while (*b*) determines a subset of $2^{\mathcal{T}}$ which meets W (the set of all wif's) where the following Baire set does so:

(20) $$\{g: g(T)=1, \quad \text{and} \quad g({}_jT)=0 \quad \text{for all } j\}.$$

Thus the set determined by (*b*) splits into two disjoint pieces, one being an element of $\mathscr{B}_0^+(W)$ because it is a subset of the null set W^*, and the other being an element of $\mathscr{B}_0^+(W)$ because it is the intersection of W and a Baire set. Thus the first condition in Theorem 15 is satisfied in any case, because we have assumed at the outset that the avoidance function was completely monotonic, while the third condition is satisfied because of the properties of **S**.

We now turn to the second of the conditions in Theorem 15. For fixed T we know that $(\mathbf{S}g)(T) \neq g(T)$ if and only if $(\mathbf{R}g)(T)=1$, and (as we have already observed) the points g in $2^{\mathcal{T}}$ which behave in this way are the elements of a $\mathscr{B}_0^+(W)$-set whose pr_W-measure is equal to that of the Baire set (20). Thus the desired nsc is precisely, that the Baire set (20) should have zero measure (for each $T \in \mathcal{T}$) in relation to the canonical (Daniell–Kolmogorov) model

$$(2^{\mathcal{T}}, \mathscr{B}_0, \mathrm{pr}; 2, \mathcal{T}; g_T \ (T \in \mathcal{T})).$$

This can of course be translated into a nsc bearing on the avoidance function, and we shall carry out the translation in a moment.

Let us first notice that when the nsc is satisfied then we can transfer our attention to the model

(21) $$(2^{\mathcal{T}}, \mathscr{B}_0^+(S), \mathrm{pr}_S; 2, \mathcal{T}; g_T \ (T \in \mathcal{T})),$$

which we shall call *the canonical model for a random strong incidence function* (*or random $\mathcal{T}$-closed set*). This is possible in virtue of the fact that S must now be $(\mathscr{B}_0, \mathrm{pr})$-thick, and because of our earlier general remarks about standard extensions and standard modifications. We still retain the option of discussing random events associated with weak incidence functions, because W covers S and so must be $\mathscr{B}_0^+(S)$-measurable and pr_S-full.

There is of course the alternative possibility of retracting the canonical model onto S as probability-space, but there is little to be said for this because S, unlike W, is not compact. In some circumstances it might be found desirable to retract as far as W, and to take the completed Baire subsets of this compact space W as the measurable sets, and we shall do this from time to time.

We call the model (21) *canonical* because there is a unique naturally defined such model within every name-class of stochastic processes representing a random $\mathscr{T}$-closed set. Indeed, if $(\Omega, \mathscr{F}, \ldots)$ is a version carrying zero–one random variables $Y_T (T \in \mathscr{T})$ such that $Y(\omega)$ is a strong incidence function whenever ω lies outside a measurable null set, and if $\mathscr{F}$ is complete, then the mapping

$$\Phi: \Omega \to 2^{\mathscr{T}}$$

determined by

$$\Phi(\omega) = (\ldots, Y_T(\omega), \ldots)$$

is a measurable and measure-preserving map of $(\Omega, \mathscr{F}, \ldots)$ into the model (21), with

$$g_T(\Phi(\omega)) = Y_T(\omega),$$

so that the model (21) is a canonical member of the same name-class which in every way is fit to replace our *version trouvée*. Moreover, every family of random variables over $(2^{\mathscr{T}}, \mathscr{B}_0^+(S), \mathrm{pr}_S)$ can be carried back to the Ω-version without change of joint distribution. We have therefore nothing to lose in using the model (21), and anything we gain in using it can be transferred to a non-canonical version if for some reason that should prove to be desirable.

We now have to throw the nsc on the avoidance function into a convenient form, and in so doing it will be highly desirable to remove all reference to the traps ${}_jT$, because we have always stressed the fact that these are merely accidental features of the argument which should play no part whatsoever in the final result.

We start by recalling that the natural domain of the avoidance function $A(.)$ is $\mathscr{T}_+$, the class whose elements are unions θ of finite collections φ of traps. If we take the obvious (directed) partial ordering by set-inclusion, then

$$\{A(\theta): \theta \subset T, \theta \neq T\}$$

is a monotonically decreasing net of real numbers which is bounded below, as θ 'swells' towards T, by $A(T)$. As θ swells towards T, $A(\theta)$ will converge to a limit which is also the infimum of the net, and we shall call this $A(T-)$.

Now the probability of the Baire set at (20) has the value

$$\lim_{j \to \infty} A({}_1T \cup {}_2T \cup \ldots \cup {}_jT) - A(T);$$

this is easily proved on remembering that almost every g in $2^{\mathscr{T}}$ is a wif. The value of this probability must clearly be greater than or equal to $A(T-) - A(T)$, which in turn is always non-negative. Thus $A(T-) = A(T)$ is a *necessary* condition for the mapping $\mathbf{S}$ to be a standard modification, but it is not sufficient (see Example 9 below). We therefore adopt a slightly less naïve approach.

Let Θ denote (for *fixed* T) the class of θs which can be represented as the union of finitely many c-traps each of whose $\mathscr{T}$-closures are covered by T. The class Θ is not empty, for it contains every finite union of traps ${}_jT$, where these are the traps specially associated with T in Theorem 13.

If we write $\theta_1 < \theta_2$ for pairs of members of Θ which are such that θ_1 can be expressed as the union of finitely many c-traps each of whose $\mathscr{T}$-closures are covered by θ_2, then it is clear that we shall have

$$\theta_1 < \theta_2 \quad \text{only if} \quad \theta_1 \subset \theta_2,$$

although the converse implication will not hold, in general. Accordingly, we see that Θ is partly ordered by $<$, and indeed it is *directed* by $<$ in an 'upwards' direction.

For let θ_1 and θ_2 be members of Θ, and let ${}_jT$ denote a generic member of the special sequence of c-traps associated with T by Theorem 13. Now (for $i = 1, 2$) θ_i can be expressed as a union of finitely many c-traps each of whose $\mathscr{T}$-closures are covered by T, and so are covered by the whole collection of ${}_jT$s, whence also by finitely many of the ${}_jT$s. If we assemble enough, but still finitely many, of the ${}_jT$s to achieve this covering both for $i = 1$ and $i = 2$, then on forming their union we shall have a member θ_{12} of Θ such that

$$\theta_1 < \theta_{12} \quad \text{and} \quad \theta_2 < \theta_{12},$$

as required.

We now consider the net

$$\{A(\theta) \colon \theta \in \Theta\};$$

this is monotonic decreasing, following the direction of Θ, and it is bounded below by $A(T)$. Thus, as θ swells towards T, the net converges to its infimum, which we shall call $A(T\dot{-})$, and we shall have $A(T\dot{-}) \geqslant A(T)$. We have already noticed that every finite union of traps of the special sequence $\{{}_jT \colon j = 1, 2, \ldots\}$ is a member of Θ, and so the probability of the Baire set at (20) is not less than

$$A(T\dot{-}) - A(T).$$

We shall now show that it has exactly this value.

Given $\varepsilon > 0$, we can find a positive integer k and some c-traps $T_1, T_2, \ldots, T_k$ (*not necessarily from the special sequence associated with* T) such that

$$\text{Cl}\,(T_h; \mathscr{T}) \subset T \quad \text{for } h = 1, 2, \ldots, k$$

and

$$A(T\dot{-}) + \varepsilon > A(T_1 \cup T_2 \cup \ldots \cup T_k) \geqslant A(T\dot{-}).$$

But, for every h,

$$\mathrm{Cl}(T_h;\mathscr{T})\subset T=\bigcup_j {}_jT,$$

and each T_h is a c-trap, so that the coverings claimed here for each h can all be refined to finite coverings. On performing this refinement for $h=1,2,\ldots,k$, and uniting the results, we obtain an element

$$\theta={}_1T\cup{}_2T\cup\ldots\cup{}_nT$$

of Θ which covers

$$T_1\cup T_2\cup\ldots\cup T_k,$$

and so satisfies the inequalities,

$$A(T\dot{-})+\varepsilon>A(T_1\cup T_2\cup\ldots\cup T_k)\geqslant A({}_1T\cup{}_2T\cup\ldots\cup{}_nT)\geqslant A(T\dot{-}),$$

and now the desired result follows at once.

This analysis suggests that we should call an avoidance function $A(.)$ *continuous from below at T* when $A(T\dot{-})=A(T)$. In this terminology we can assert

Theorem 16 (Second Fundamental Theorem). *Let $A(.)$ be a normalized completely monotonic avoidance function associated with a trapping-space $(C,\mathscr{T})$ and so characterizing a canonical model for a random weak incidence function. Then the following statements are equivalent.*

(*i*) *$A(.)$ is continuous from below at every trap $T\in\mathscr{T}$.*

(*ii*) **S** *is a standard modification.*

(*iii*) *The set S of strong incidence functions is* $(\mathscr{B}_0,\mathrm{pr})$*-thick, and so $A(.)$ determines a canonical model* (21) *for a random $\mathscr{T}$-closed subset of* **C**.

Proof. We already know that (*i*) and (*ii*) are equivalent, and that (*ii*) implies (*iii*). If (*iii*) holds, then the Baire set (20) must have measure zero (because it is disjoint with S), and so (*i*) and (*ii*) hold.

In using Theorem 16 it should be noted that a generic element of $\mathscr{B}_0^+(S)$ can be expressed in the form

$$(B_0\cap S)\cup N_1\cup N_2,$$

where B_0 is a Baire set, N_1 is a subset of S^* and N_2 is a subset of a Baire set which has pr-measure zero. The pr_S-measure of this element of $\mathscr{B}_0^+(S)$ is equal to $\mathrm{pr}(B_0)$.

We have now completed our main programme. We shall conclude with one or two further examples, and some very brief indications of the way in which this theory can be used. Further illustrations, especially concerning stationary random sets, infinitely divisible random sets and random convex sets, will be presented elsewhere [15, 12, 13].

6.2.8 SOME EXAMPLES AND APPLICATIONS

The following example is the most important single illustration of the general theory, but we shall not do more than sketch some of the developments associated with it.

Example 9. (Random closed subsets of R.) *Let $C = R$ and let $\mathcal{T}$ consist of all bounded non-empty open intervals. This is a special case of Example* 8, *which we showed could be taken as a basis for a theory of random closed sets in any second-countable locally compact Hausdorff space (in the present case the real line R with its customary topology). Because we have taken the traps to be intervals, the finite unions θ of traps will be the finite disjoint unions of bounded non-empty open intervals, and thus the avoidance function $A(.)$ can be concretely represented as a sequence $A_1, A_2, \ldots$ of real-valued 'component' functions having respectively* 2, 4, ... *real arguments:*

$$A_1(\alpha_1, \beta_1) \quad (-\infty < \alpha_1 < \beta_1 < \infty),$$

$$A_2(\alpha_1, \beta_1; \alpha_2, \beta_2) \quad (-\infty < \alpha_1 < \beta_1 \leqslant \alpha_2 < \beta_2 < \infty),$$

and so on. For completeness it is desirable to add the component A_0; this is defined by $A_0(\emptyset) = 1$.

The requirement of complete monotony imposes a large number of weak inequalities on certain finite linear combinations of the components $A_0, A_1, \ldots, A_m, \ldots$ of the avoidance function A, and in what follows we shall suppose that these are all satisfied. We then know, from all that has gone before, that A is associated in a unique way with a canonical model for a random weak incidence function and therefore in manipulating A we are entitled to make full use of this fact, and to interpret $A(\theta)$ as the probability that a random wif g vanishes on each of the traps composing θ. This greatly simplifies the proofs which follow.

Now we want a canonical model for a random closed subset of the line, i.e. for a random *strong* incidence function, and we can have one, by following the prescription given in the last section, *if and only if A is continuous from below at each trap T.* We must interpret this nsc in terms of the components $A_0, A_1, A_2, \ldots, A_m, \ldots$ of A.

Suppose then that (α, β) is the trap T being tested (and then $-\infty < \alpha < \beta < \infty$). In the present situation every trap is a c-trap, and the

$\mathscr{T}$-closure (now the same thing as 'ordinary' closure) of the trap (γ, δ) will be covered by T if and only if

$$\alpha < \gamma < \delta < \beta.$$

In view of our preceding remarks we can now argue in terms of the associated random wif, g, as follows. The system Θ depending upon T comprises any θ whose left-hand end-point α_1 lies to the right of α and whose right-hand end-point β_m lies to the left of β, and we shall have

$$A_m(\alpha_1, \beta_1; \alpha_2, \beta_2; \ldots; \alpha_m, \beta_m) \geqslant A_1(\alpha_1, \beta_m) \geqslant A_1(\alpha, \beta)$$

whenever $(\alpha_1, \beta_1) \cup \ldots \cup (\alpha_m, \beta_m)$ belongs to Θ. This being so, the condition of continuity from below at T reduces to

$$\text{(22)} \qquad \lim_{\alpha_1 \downarrow \alpha, \beta_1 \uparrow \beta} A_1(\alpha_1, \beta_1) = A_1(\alpha, \beta).$$

But now,

$$A_1(\alpha_1, \beta_1) - A_1(\alpha, \beta)$$

is the probability that the random wif g makes (α_1, β_1) a 0-trap and (α, β) a 1-trap. Let E denote the set of wif's having these two properties, and choose fixed numbers α', β' such that $\alpha < \alpha' < \beta' < \beta$. In what follows we may assume without loss of generality that $\alpha < \alpha_1 < \alpha'$ and that $\beta' < \beta_1 < \beta$. We now observe that *E is covered by* the union of the two sets of wif's defined as follows:

(*i*) g makes (α_1, β') a 0-trap and (α, β') a 1-trap;

(*ii*) g makes (α', β_1) a 0-trap and (α', β) a 1-trap.

On the other hand, *E covers* each of the two sets of wif's defined as follows:

(*iii*) g makes (α_1, β) a 0-trap and (α, β) a 1-trap;

(*iv*) g makes (α, β_1) a 0-trap and (α, β) a 1-trap.

These simple observations show that the nsc (22) is equivalent to the apparently weaker condition that there is to be *separate* right-hand continuity at α and left-hand continuity at β (whereas equation (22) demanded *joint* continuity). We thus have

Theorem 17. *A normalized completely monotonic avoidance function is associated with a canonical model for a random closed set on the real line if and only if its first-order component is right-continuous in its first argument, and left-continuous in its second argument.*

This surprising result is very gratifying, for it means that in order to show that a wif-model is in fact a sif-model we have only to apply a test to the component $A_1(\alpha, \beta)$ of the avoidance function, but a greater surprise is in store. Obviously much interest is to be attached to random closed sets on the line which in some sense have a stochastic structure which is unaltered by the shift transformation of the line; we shall say that an avoidance function A is *stationary to the mth order* if all of $A_0, A_1, A_2, \ldots, A_m$ are invariant under a uniform shift of their arguments. We shall now prove

Theorem 18. *Suppose that we have a canonical model for a random* weak *incidence function on the bounded non-empty open intervals of the real line, and suppose that the associated avoidance function A is at least* first-order *stationary. Then the nsc of Theorem* 17 *is automatically satisfied, so that without further investigation we can take our model to represent a random closed set.*

Proof. As we have first-order stationarity we can replace the component A_1 of A by the function $a(.)$ on $(0, \infty)$ defined by

$$a(t) = A_1(\alpha, \alpha+t). \tag{23}$$

We shall thus have a model for a random closed set if and only if this function $a(.)$ is *left*-continuous. We now give an argument which shows that $a(.)$ is in fact forced to have this property, and that will prove the theorem.

Notice first that we already know, from complete monotony, that $a(.)$ is bounded by 0 and 1, is monotonic decreasing, and is convex from below in the sense that

$$a(t) - a(t+u) - a(t+v) + a(t+u+v) \geqslant 0 \quad (u, v, t > 0). \tag{24}$$

We shall find it instructive to give an independent proof of inequality (24), in its equivalent form

$$a(t) - a(t+u) \geqslant a(t+v) - a(t+v+u), \tag{25}$$

by exploiting the fact that the random element g can be taken to be a wif. First we note that the left-hand side of inequality (25) is the probability that

(26) $(0, t)$ is a 0-trap while $(0, t+u)$ is a 1-trap,

and that the right-hand side of inequality (25) is the probability that

(27) $(-v, t)$ is a 0-trap while $(-v, t+u)$ is a 1-trap.

I claim that any wif g which makes (27) true, *must* make (26) true; it is obvious that the required result will follow if this claim is justified.

Suppose then that g is a wif which makes (27) true and (26) false; then we must have

$$(-v, t) \text{ is a 0-trap,}$$

$$(-v, t+u) \text{ is a 1-trap,}$$

and

$$(0, t+u) \text{ is a 0-trap.}$$

(The last fact follows from the inclusion-relation $(0,t) \subset (-v,t)$, which prevents $(0,t)$ from being a 1-trap.) But now, because $t > 0$, we have

$$(-v, t+u) = (-v, t) \cup (0, t+u),$$

and so there is an assignment of 0s and 1s to traps which contradicts the wif character of g.

This gives the desired proof of the inequality (24), and the theorem then follows at once, because a 0–1 bounded monotonically decreasing function on $(0, \infty)$ which is convex from below is necessarily continuous on this interval.

This powerful result resembles, but lies much deeper than, a non-probabilistic theorem which asserts the rather trivial fact that if a given *fixed* wif g is invariant under the shift, then it must be a sif. This last result is quite obvious as soon as we notice that the existence of any one trap (α, β) at which g takes the value 0 implies that *all* traps must be 0-traps (for an arbitrary trap (γ, δ) can be covered by finitely many shifted copies of (a, β)). Thus there are only two possibilities; either $g(T) = 1$ for all traps or $g(T) = 0$ for all traps, and in either case g is a sif. The corresponding fixed closed sets are the whole line, and the empty set.

An analysis of non-sif wif's is full of interest and has been carried out in a few cases, of which Example 9 is one; the results of this investigation will be presented elsewhere [14].

We now return to the study of the a-function associated with a first-order stationary random closed set. We have seen that $a(.)$ must be continuous on $(0, \infty)$, but we have not shown, and it is not true, that $a(.)$ is necessarily continuous on $[0, \infty)$; it can happen that $a(t)$ has a limit as $t \downarrow 0$ which is less than $1 = a(0)$. (We did not define $a(.)$ at $t = 0$, but of course 1 is the only natural value to give it there, since $A(\emptyset) = 1$.) To see that this is so, let the random closed set be surely the whole line; then $a(t) = 0$ for all $t > 0$. The values

$$1 - a(0+) \quad \text{and} \quad a(\infty),$$

are in fact very interesting, for we have

Theorem 19. *For a first-order stationary random closed set on the line, the jump* $1-a(0+)$ *at* $t=0$ *is the probability that an arbitrary point will belong to the random set, while the limit* $a(\infty)$ *is the probability that the random set is vacuous.*

Proof. We invoke the sif interpretation and then argue as follows. First,

$$\begin{aligned} 1-a(0+) &= \lim\,(1-a(2^{1-n})) = \lim \operatorname{pr}(\{g\colon (t-2^{-n}, t+2^{-n}) \text{ is a 1-trap}\}),\\ &= \operatorname{pr}(\{g\colon (t-2^{-n}, t+2^{-n}) \text{ is a 1-trap for all } n\})\\ &= \operatorname{pr}(\{g\colon t \text{ does not belong to any 0-trap}\})\\ &= \operatorname{pr}(\{g\colon t \text{ belongs to the closed set associated with } g\}). \end{aligned}$$

This proves the first half of the theorem, and the second half is established by a similar argument; in detail,

$$\begin{aligned} a(\infty) &= \lim a(2n) = \lim \operatorname{pr}(\{g\colon (-n,n) \text{ is a 0-trap}\})\\ &= \operatorname{pr}(\{g\colon (-n,n) \text{ is a 0-trap for all } n\})\\ &= \operatorname{pr}(\{g\colon \text{there are no 1-traps}\})\\ &= \operatorname{pr}(\{g\colon \text{the closed set associated with } g \text{ is vacuous}\}). \end{aligned}$$

There are of course similar but less tidy results in the non-stationary case.

Let us now look at a very special example of a *strictly* stationary random closed set. Let b and c be positive real numbers, and let V be a random variable uniformly distributed over the interval $[0, b+c)$, and let F be the random closed set

$$\bigcup_{-\infty<m<\infty} [V+m(b+c), V+b+m(b+c)]. \tag{28}$$

Obviously this is strictly stationary, and it is easy to compute its a-function; in fact

$$a(t) = (c-t)^+/(b+c) \quad (0<t<\infty). \tag{29}$$

Notice, to illustrate our last theorem, that $1-a(0+) = b/(b+c)$, and that $a(\infty) = 0$. It will be convenient to write equation (29) in the form

$$a(t) = p\left(1-\frac{t}{l}\right)^+ \quad (0<t<\infty) \tag{30}$$

so that l is the length of a component interval of the complement of the random closed set, and p is the probability that an arbitrary point t will lie outside the random closed set. The interesting case $p = 1$ ($b = 0$) should be noted; this is a strictly stationary random lattice of step I.

We can interpret the two degenerate cases $I = 0, \infty$, as follows. When $I = \infty$, then equation (30) gives the a-function for the strictly stationary random closed set which is vacuous with probability p and is otherwise the whole of R. When $I = 0$, then equation (30) gives the a-function for a strictly stationary random closed set which is almost surely the whole line; notice that in this last case p is redundant.

Let us now 'mix' probability-spaces carrying strictly stationary random closed sets of the above sort for every value of I; we obtain a strictly stationary random closed set defined over the mixture-space, and its a-function will be of the form

$$(31) \qquad a(t) = \int_{(0,\infty)} p(I)\left(1-\frac{t}{I}\right)^{+} dF(I) + p(\infty)(1-F(\infty)) \quad (0<t<\infty),$$

$p(.)$ being any Borel measurable function bounded by 0 and 1, and $F(.)$ a df on $[0,\infty]$. On inspecting equation (31) it will be seen that we can rewrite it as

$$(32) \qquad a(t) = \int_{(0,\infty]} \left(1-\frac{t}{I}\right)^{+} \rho(dI) \quad (0<t<\infty),$$

where ρ is now a (possibly deficient) probability measure over the half-line $(0,\infty]$.

However, we know that the most general a-function for a *first-order* stationary random closed set is (0–1)-bounded, monotonic decreasing and convex from below, and a Choquet analysis easily shows that every such function can be represented uniquely in the form (32). (Cf. Ali and Silvey [1].) Thus we have

Theorem 20. *The a-function for a first-order stationary random closed set on the real line can be uniquely represented in the form* (32), *where* ρ *is a (possibly deficient) probability measure on* $(0,\infty]$. *A strictly stationary random closed set with this a-function can always be constructed by 'mixing' the simple arithmetic-progression-type models described above.*

It is important to realize that a strictly stationary random closed set need by no means be of the 'arithmetic progression type'. Thus the first-order component of the A-function does not characterize a random closed set even in the strictly stationary case.

We turn now to compactness principles, and prove the important

Theorem 21. *In the topology of simple convergence for* $R^{\mathcal{T}+}$, *the following are compact:*

(*i*) *the set of all avoidance functions associated with random weak incidence functions (on a trapping-space* $C, \mathscr{T}$*);*

(*ii*) *the set of all avoidance functions associated with first-order stationary random closed sets on R;*

(*iii*) *the set of all avoidance functions associated with mth order stationary random closed sets on R;*

(*iv*) *the set of all avoidance functions associated with strictly stationary random closed sets on R.*

Proof. (*i*) Use the definitions at equations (12) and (13), and Theorem 12. (*ii*)–(*iv*) Use the definitions together with Theorem 18.

This theorem allows us to introduce some valuable if abusive terminology. If we have a net of avoidance functions for random wif's, and if the net converges at each θ, then by (*i*) we know that the limit is an avoidance function for a random wif. Thus, if the avoidance functions for a net of random wifs converges at each θ, we can and shall call the random wif associated with the limiting avoidance function, *the weak limit of the convergent net of random wif's.*

In view of (*ii*)–(*iv*) we have the useful

Corollary. *Let a net of first-order stationary random closed sets on R have avoidance functions which converge at every* θ*. Then the limiting avoidance function determines a first-order stationary random closed set, which will be called the weak limit of the net of first-order stationary random closed sets.*

This corollary permits a complete analysis of infinitely divisible first-order stationary random closed sets on the line. We call a random $\mathscr{T}$-closed set on any trapping-space $(C, \mathscr{T})$ *infinitely divisible* when it can be represented, for every k, as the $\mathscr{T}$-closure of the *union* of k independent identically distributed random $\mathscr{T}$-closed sets; i.e. when we can write

(33) $$A(\theta) = (A^{(k)}(\theta))^k \quad (k = 1, 2, \ldots),$$

where $A^{(k)}(.)$ is an avoidance function for every k. Infinite divisibility for random point-processes and certain other stochastic processes has been studied by Matthes and by Lee [19, 17]. An exploration of the concept from the present point of view will be presented in [12]. Here we merely point out the important fact that continuity from below at each trap T must automatically hold for $A^{(k)}$ if it holds for A; this enables a large part of the analysis to be carried out without repeated verification of the continuity condition. The discussion of infinite divisibility can of course be made quite general; it is not restricted to the real-line case, although in that case and when first-order stationarity is present the corollary is a useful additional aid.

We promised to use Example 9 to show that the rather strict formulation of 'continuity from below' (in terms of $A(T\dot{-})$, instead of $A(T-)$) cannot be dispensed with; let us now do this.

It is clear that within the context of Example 9, the following function g_c is a wif, but not a sif:

> $g_c(T) = 0$ unless either $c \in T$, or c is the right-hand end-point of T; in these two exceptional cases $g_c(T) = 1$.

Here c is some fixed point on the real line R. Now the singleton $\{g_c\}$ is a Borel set (though not a Baire set) in $2^{\mathscr{T}}$, and so we can concentrate all our probability there, in the Kakutani model, and then retract to a canonical model for a random wif. This just means that we assign probability zero, or one, respectively, to each Baire set in $2^{\mathscr{T}}$ if among its elements there is no, or at least one, element g which agrees with g_c on the countable set of traps determining the Baire set.

Now put $\theta_n = (c-2+1/n, c-1/n)$; then $A(\theta_n) = 1$ for each $n \geqslant 1$. Also θ_n swells as n increases, and has the limit $T = (c-2, c)$, and $A(T) = 0$. There is no contradiction here, because θ_n is a c-trap whose closure is covered by T, and we are merely observing a consequence of the fact that

$$A(T\dot{-}) = 1 > 0 = A(T),$$

associated with the almost sure wif–non–sif character of the model. If however we had taken $\theta_n = (c-2+1/n, c)$, then once again θ_n swells to T as $n \to \infty$, but now

$$\lim A(\theta_n) = A(T-) = 0 = A(T);$$

thus $A(T-)$ is in general smaller than $A(T\dot{-})$, and that is why the naïve version of the continuity condition is inadequate.

Finally, here are a few very general comments on the axioms for a trapping-space used in this paper. It will be clear that TS(*iii*) and TS(*iv*) have only been used to deduce the crucial Theorem 13, and that an alternative (though perhaps not equivalent) theory could have been constructed by taking the conclusion of Theorem 13 as an axiom to replace *both* TS(*iii*) *and* TS(*iv*). This procedure might in fact result in an extension of the present theory, if the conclusion of Theorem 13 can hold even when TS(*iii*) is false; the conclusion of Theorem 13 *does* imply TS(*iv*), so it seems that it is TS(*iii*) which may be unnecessarily strong. However

TS(*iii*)*: *each trap T is a countable union of traps whose $\mathscr{T}$-closures are all covered by T,*

is not strong enough, for when combined with TS(*i*), (*ii*) and (*iv*) it will not yield Theorem 13. To see that this is so, look at

Example 10. *Let $C = R$, and let $\mathscr{T}$ consist of* all *non-empty subsets. Then every trap is its own $\mathscr{T}$-closure. It is obvious that* TS(*i*) *and* TS(*ii*) *hold, and it is clear that each singleton is a $\mathscr{T}$-closed c-trap, so that* TS(*iv*) *holds trivially. Also* TS(*iii*)* *holds because of the $\mathscr{T}$-closure of T. However, the c-traps are here exactly the finite subsets of R, and so a countable union of c-traps cannot make up an uncountable set. Thus Theorem* 13 *fails for any uncountable T* (*e.g. for $T = R$*).

These examples suggest that the best extension of the theory along the present lines may be that afforded by taking TS(*i*), TS(*ii*) and the conclusion of Theorem 13 as the new axioms. I do not personally care for this solution, because I like axioms to be visibly satisfied or not, in any particular case; also I value the separation of the countability and pseudo-local-compactness axioms into TS(*iii*) and TS(*iv*), but I accept that this is merely a matter of taste.

What is *not* a matter of taste is that no variation on the present scheme seems able to deal with the important case $C = R^2$, $\mathscr{T}$ = all open half-planes; this is an important example, because the $\mathscr{T}$-closed sets are then the closed convex sets, and there would be applications to the Geometry of Numbers. As Kingman has pointed out, we can make $\mathscr{T}$ countable (all 'rationally located and oriented' open half-planes) if we are only interested in *compact* convex sets, as is very often the case. Even then there are no *c*-traps, so Theorem 13 fails. A more penetrating analysis of this example [13] shows that a solution can be worked out by classifying the non-sif wif's, and this suggests that the ultimate replacement for TS(*iii*)+(*iv*) will be an axiom of the form:

Each trap T can be associated with a sequence of traps $\{{}_1T, {}_2T, \ldots\}$ in such a way that, for each weak incidence function g, the following statements are equivalent:

(*a*) $g(T) = 1$, *and T can be covered by g-null traps;*

(*b*) $g(T) = 1$, *and* $g({}_jT) = 0$ *for each j.*

This would obviously be a perfectly good replacement for Theorem 13 except that in general the traps ${}_jT$ would then occur *explicitly* in the nsc making a random wif into a random $\mathscr{T}$-closed set. Our pseudo-topological formulation of the theory has the merit of avoiding that penalty. Devices for avoiding it in general are perhaps most likely to be found by pursuing the more exotic varieties of big game. *Solvitur venando!*

It is a pleasure to express my indebtedness to John Kingman and Klaus Krickeberg, from whose criticisms and suggestions this work has greatly benefited.

REFERENCES

1. S. M. Ali and S. D. Silvey, "Association between random variables and the dispersion of a Radon–Nikodym derivative", *J. R. statist. Soc. B* **27** (1965), 100–107 and 533.
2. D. S. Carter, "The exponential of a measure space", *Technical Rep. No.* 29 (1966), Dept. of Math., Oregon State University, Corvallis.
3. G. Choquet, "Theory of capacities", *Ann. Inst. Fourier* **5** (1955), 131–295.
4. R. Davidson, *Some Geometry and Analysis in Probability Theory*, Smith's Prize Essay, Cambridge (1967).
5. J. L. Doob, "Probability in function space", *Bull. Am. math. Soc.* **53** (1947), 15–30.
6. ——, *Stochastic Processes*, Wiley, New York (1953).
7. R. M. Dudley, "On measurability over product spaces", *Bull. Am. math. Soc.* **77** (1971), 271–274.
8. E. B. Dynkin, *Markov Processes, II*, Springer, Berlin (1965).
9. D. H. Fremlin and D. G. Kendall, "Measures on complete σ-algebras and outer measures derived from sub-σ-algebras", in preparation.
10. D. G. Kendall, "An introduction to stochastic geometry", 1.1 of this book, and "An introduction to stochastic analysis", 1.1 of *Stochastic Analysis*, Wiley (1973).
11. ——, "Separability and measurability for stochastic processes: a survey", 5.4 of *Stochastic Analysis*, Wiley (1973).
12. ——, "On infinitely divisible random sets", in preparation.
13. ——, "On random closed convex sets", in preparation.
14. ——, "On the structure of weak incidence functions", in preparation.
15. ——, "On stationary random sets", in preparation.
16. J. F. C. Kingman, "Markov population processes", *J. Appl. Prob.* **6** (1969), 1–18.
17. P. M. Lee, "Infinitely divisible stochastic processes", *Ztschr. Wahrsch'theorie & verw. Geb.* **7** (1967), 147–160.
18. Z. P. Mamuzić, *Introduction to General Topology*, Noordhoff, Groningen (1963).
19. K. Matthes, "Unbeschränkt teilbare Verteilungsgesetze stationärer zufälliger Punktfolgen", *Wiss. Z. Hochschule Elektrotechn. Ilmenau* **9** (1963), 235.
20. P.-A. Meyer, *Probability and Potentials*, Blaisdell, Waltham (1966).

21. E. Nelson, "Regular probability measures on function space", *Ann. of. Math.* **69** (1959), 630–643.
22. C. Ryll-Nardzewski, "Remarks on processes of calls", *Proc. 4th Berkeley Symposium* **2** (1961), 455–465.
23. S. J. Taylor, "The α-dimensional measure of the graph and the set of zeros of a Brownian path", *Proc. Cambridge Phil. Soc.* **51** (1955), 265–274.

Note. Other approaches to the treatment of random sets are mentioned in [10], and listed in the bibliographies there; to these should have been added that of D. S. Carter [2].

Note added in proof. The attention of the reader is drawn to an article "Ensembles fermés aléatoires, ensembles semi markoviens, et polyèdres poissoniens" by G. Mathéron which has just appeared in *Advances in Applied Probability*, **4**, 3.

7

CONCLUSION

7.1

Letter to F. Papangelou, Easter Day, 1970

R. DAVIDSON

Thank you very much for your farewell note. I'm sorry not to have written; of course things have been busy—I have had to write several lecture courses and am even now hammering out the syllabus for another—but it's no excuse. So how are you all and how is Ohio? My knowledge of the States is mostly derived from reading of the campaigns in the Civil War—and then the Columbus in Miss. was of greater strategic importance than that in Ohio.

Since you left I've been occupying my time as usual—semigroups and geometric stochastic processes. I tried to make progress on the problems left by my stay in Heidelberg—which are

(1) the main existence problem;

(2) a characterization of modulus-squared characteristic functions (*à la* Bochner's theorem).

(3) If you start off with a process of cars-with-velocities on a road, the joint distribution of positions and velocities being spatially homogeneous, and you then let the process run, can you say the number of cars passing you in consecutive seconds does NOT converge to 0 in probability? This has bearings—vague—on (1), but is clearly of great interest—being so elementary—in its own right. I can do it if the velocities of the cars are independent.

Then I turned to semigroups with the idea of, if possible, extracting a Delphic kernel from a more or less arbitrary (cancellative commutative topological) semigroup. But the stumbling block was the production of enough homomorphisms (continuous, that is) into the unit disc. I only realized rather late how beastly semigroups are (compared with groups)—there exists a compact commutative $\frac{1}{2}$ group with no non-trivial continuous homomorphisms, etc.

So then I moved off to Delphicize l.c.a. groups in the way I had used for laws on the line. This, as expected, works, i.e. we can, by quotienting out units and discarding elements with idempotent factors, reduce $M(G)$, where G is a 2nd countable l.c.a. group with a generating compact neighbourhood of e, to strongly Delphic form. I suppose I ought to write this up, but feel it is rather flogging a dead horse. The same ideas will work, of course, if we confine ourselves to measures concentrated on a fixed closed sub-$\frac{1}{2}$ group of G.

In geometric stochastic processes I have two problems, one posed by a physicist, the other by a doctor. In the physical problem, particles of gold are rained down on a quartz slab, where they migrate (presumably in independent Brownian motions) until they either evaporate, or meet other particles—when evaporation is arrested and the particles clump together; or they hit, and stick to, previous clumps. The physicist can only see clumps of a certain size, and wants to know the distribution of the number of these per unit area, after given time and with, I suppose, known rates of diffusion, evaporation and raining. There appear to be charming problems on the shape of clumps. I gave this one to my (unique) research student, but it is difficult to tear him away from the computer, and anyway he seems to like p-functions, etc., more.

The doctor's problem is a geometric continuous-state branching process (of arteries in the lung). He slices lungs and observes the distribution of small artery sizes, and wants to know as much as he can about the rules for forking of the arteries. So far I have thought of 2 models for him, and investigated the mean effects of them, but it is difficult to see how the models can be distinguished on the basis of his experiments.

I've just had an excellent week climbing in Snowdonia, and return refreshed to the construction of a new course of lectures on stochastic (Markov) processes for Part IB, partly on the lines of the Dynkin–Yushkevitch book which I read in Heidelberg—mostly at mealtimes. I'm very glad Klaus hurried through the translation, because the American edition costs \$15 against DM. 14.80 .... It is now extremely likely that I shall be visiting Russia (Ambartzumian, and Dynkin, anyhow, I think) this autumn, and I shall get D. and Y. in Russian, and also the excellent Armenian polylingual mathematical dictionary, if there are any left.

[*This letter was sent from Trinity College, Cambridge, to Papangelou* (*then in Ohio*)*; the editors are very grateful to Professor Papangelou for permission to publish this survey of Davidson's research plans at that time.*]

7.2

Rollo Davidson: 1944-1970

E. F. HARDING and D. G. KENDALL

Rollo Davidson was born on 8 October 1944 in Bristol, and spent his childhood at Thornbury, Gloucestershire. Like his father, his father's father, both his mother's brothers and his own younger brother, he took a Scholarship to Winchester. Here his career was characterized by breadth (Ancient History as well as Mathematics at A-level) and speed (he won the Senior Mathematical Prize at the age of 16). One of his school papers which has been preserved is a Stewart McDowall Essay (1962) on 'The Appeal of Science to the Victorian Intelligentsia'.

He matriculated as a Scholar of Trinity College at the University of Cambridge in October 1962. He was awarded the Percy Pemberton prize as the Trinity undergraduate most distinguished in his studies in his first year, and at the end of his second year he was already a Wrangler. It would have been normal at that time to spend the third year working for Part III of the Mathematical Tripos, but with characteristic independence Davidson chose to take the course for the Diploma in Mathematical Statistics instead (which he duly received, with distinction). It was this choice which determined the way he was to spend the next 5 years, in which brief period he made profound contributions to Probability Theory, a circumstance perhaps without parallel. It is idle to speculate on what he might have achieved, had he lived to attain full maturity in years and in his profession, but to his friends and colleagues there can be no doubt of the immense loss to learning, and to the society of scholars, occasioned by his tragically early death in 1970. By this time he had become a Smith's Prizeman (1967), a Research Fellow of Trinity College (1967), Ph.D. (1968), Assistant Lecturer in the Department of Pure Mathematics and Mathematical Statistics (1968), Lecturer in the Statistical Laboratory in the same Department (1969), and Fellow-elect of Churchill College (1970).

It is so usual for prodigies to be sophisticated, and perhaps even intolerable, that one must stress how far Davidson was from following this familiar pattern. Extremely diffident, overcoming a natural shyness by power of will alone, and far from self-confident in his mathematical powers, he did not at all realize until the last year or two that he had the capacity not merely to solve hard problems but to create new fields of enquiry, and to take the undisputed lead in them. In his work on Delphic semigroups he was the problem-solver, cracking one hard nut after another, but in his work on Stochastic Geometry, with which his name will always be linked, he quickly became, as Klaus Krickeberg has written, 'the reader one wrote for':

> 'I have systematized, made precise and generalized many things, but most of the basic intuitive ideas are his. I always assumed that we would go on like this for some time to come, pushing each other ahead and complementing each other, and while writing the article I thought of him as *the* reader of it, and imagined how he would be surprised and pleased with certain things.'

Davidson had an unpredictable variety of special interests outside mathematics, and any visitor to his rooms in Trinity would be struck by these. He was interested in trains, in physical geography, in old books (a 'regular' at David's stall on Saturday mornings, he collected especially old Alpine books), and above all things he adored mountains. Mathematical excursions quickly became the passport to mountain wanderings. In June 1969 a 'Tagung' at Oberwolfach was devoted to 'Integral Geometry and Geometrical Probability'; Krickeberg and D. G. K. organised it, but it was the existence of Davidson's thesis which inspired the project. There were many happy rambles over the Black Forest hills during that week, which proved scientifically very fruitful, for it was followed by three weeks of collaboration in Heidelberg between Davidson and Krickeberg, which led to many advances.

We mention 'physical geography' because Davidson's papers contain a correspondence with the Hydrographer of the Navy on the existence (or not) of bores on certain Chinese rivers. This may have been linked with conversations with Trinity colleagues, though his own interest in bores undoubtedly came from growing up (by the Severn) near such a good one. But he may well have been equally interested in the observer of the Chinese bores, Commander W. Usborne Moore, R.N. (H.M.S. *Rambler*) in 1888–92, and he would certainly have leaped at the chance of an antiquarian contribution to physical science; he would have delighted in

D. E. Cartwright's 'Tides and waves in the vicinity of Saint Helena' (*Phil. Trans. Roy. Soc. A*, **270** (1971), 603–649) and in the utilization there of observations cited from 'Maskelyne, N. (1762*b*)'.

In August 1969 Davidson went to Canada to participate in the Twelfth Biennial Seminar of the Canadian Mathematical Society which was held in Vancouver. Here he went to the Cascade Pass with Daryl Daley, and on one of the weekends took a party to climb The Lions (getting within a few hundred feet of the summit). At about this time he was discussing the possible truth of the Riemann hypothesis with Littlewood, and lending a hand with Littlewood's psychophysical experiments and their statistical analysis.

From this point onwards Davidson's mathematical and mountaineering notes get rather mixed up. There is a 1969 entry of a visit to Snowdon (probably at Easter time) in characteristic style:

> '0th day ... Bangor: Cod and Chips, Tea, Buns, 6s. 9d. Buns OK.
> ...
> 5th day ... Pen yr Olewen, C. D., C. L., F. Grach, down into cwm below Craig Ysfa, up to Bwlch, Pen Helig ... very fine: sun on top of rocks.'

In June 1970 he went to Skye with E. F. H., climbing Sgurr Alasdair 'by something like Collie's route' (*frontispiece*). Later 'on the 15th we moved over to Sligachan and did the classic expedition: S. nan Gillean by the Pinnacle and West ridges. Very pleasant scrambling . . . . The new Guide recommends a descent on the right (W.) side of the 3rd pinnacle, but this appears to be dictated by a desire to stop the holds on the usual route getting too smooth.' By avid reading he had absorbed the spirit of early British mountaineering, and in the Coolins he almost ritually enacted for himself the legendary traditions. The climbing apart, it was finding in the Glen Brittle Post Office log-book a Cambridge entry of the 1920s (with such names as Adrian, R. H. Fowler, Littlewood and many others) that perhaps more than anything closed the link with the old days. Agile, swift and light as an elf on the mountains, and totally untroubled by 'exposure', he had advanced in skill and achievement in climbing as in everything else: his climbs at home were also stages in the symbolical ascent to the great peaks abroad, and he now had the Alps clearly in his sight.

By now, also, Davidson was preparing for a Royal Society sponsored visit to Erevan, in the Armenian Soviet Socialist Republic. Here he hoped especially to meet and work with R. V. Ambartzumian, a distinguished young mathematician sharing both his mathematical and his mountaineering interests, but of course he had intended as well to visit the probabilists in Moscow, Leningrad and Kharkov. His notes on the proposed visit include the postscript: 'I'd like to cross a pass or two in the Caucasus'. But this expedition was not to take place.

In July he went to the Alps with the Cambridge University Mountaineering Club, accompanied by a Russian grammar and Paul-André Meyer's *Probabilités et Potentiel*, the latter decorated by mathematical comments in the margins and climbing notes on the fly-leaf. After a very successful meet, the party broke up to go their several ways, Davidson remaining with Michael Latham (a gifted young mathematician from Gonville and Caius College) near Pontresina, to climb the Piz Bernina.

On 29 July 1970, while they were descending from the summit, an accident cost them both their lives.

David Williams expressed all our feelings at the time when he wrote:

> 'I feel so angry—if 'angry' is the word (but anger *that* he died, not at *how*)—at such a loss for parents, friends, Cambridge, but, above all, for himself.'

From such feelings this book emerged. In the years that have gone to the making of it one has learned to see things in a slightly different perspective. Rollo's was a *magnificent* life; a flawless blend of personal relations, mathematics and mountain adventure. The hazards of the latter, never wholly to be avoided, are familiar to all, and to rail at its folly is to invite a reply which he himself might have made, in the words of one of the more sympathetic characters in contemporary fiction:

> 'If you always look over your shoulder, how can you still remain a human being?'

Author Index

Subject Index